W9-CKH-817

MEMBRANES

VOLUME 1

Macroscopic Systems and Models

MEMBRANES

A SERIES OF ADVANCES

Edited by

GEORGE EISENMAN

Volume 1: Macroscopic Systems and Models

Volume 2: Lipid Bilayers and Antibiotics

Other Volumes in Preparation

MEMBRANES

A SERIES OF ADVANCES

Volume 1

Macroscopic Systems and Models

edited by GEORGE EISENMAN

Department of Physiology
School of Medicine
University of California
Los Angeles, California

MARCEL DEKKER, INC. New York 1972

MARCEL DEKKER, INC.
95 Madison Avenue, New York, New York 10016

LIBRARY OF CONGRESS CATALOG CARD NUMBER 75-163920

ISBN 0-8247-1171-8

PRINTED IN THE UNITED STATES OF AMERICA

PREFACE TO VOLUMES 1 AND 2

These volumes initiate a series whose purpose is to provide a forum for the publication of significant creative advances and definitive syntheses bearing on the physics, chemistry, and biology of membranes. Reviews in the conventional sense will not be included; and since such advances occur irregularly, so will the series appear irregularly. Topical volumes will appear when timely, each of which will deal with a given aspect of membranes in a self-contained manner.

The study of membranes has a long history in physical chemistry and biology, and the initial volumes of this Series of Advances will reflect this history. Thus, Volume 1 brings up-to-date the knowledge for classical ion-exchange membranes having sufficient thickness that the concept of macroscopic electroneutrality is applicable and deals with a few properties of cell membranes which can be most easily related to these. Volume 2 then proceeds to the newly developed ultrathin lipid bilayer membranes having biological dimensions, for which macroscopic electroneutrality no longer holds. The development of these membranes made from biological molecules has made the relevance of studies of model membranes so apparent to the understanding of cell membranes that the study of artificial membranes (particularly bilayers) has now entered a phase of explosive development. Subsequent volumes will carry on this development, dealing not only with such emerging properties of bilayers as the kinetics underlying ion permeation but also with those properties of cell membranes which can be most meaningfully related to the studies on aritifical membranes as, for example, in Volume 3 where the electrical properties of artificial bilayers will be compared with appropriate measurements for cell membranes and where the structure and function of bilayer membranes will be considered in further detail. To make the organization and scope of these volumes clear, it will be useful to describe in more detail the material to be covered in Volumes 1 and 2.

Volume 1, "Macroscopic Systems and Models," is meant to supplement the comprehensive works on conventional membranes* by providing

*See particularly: F. Helfferich, Ion Exchange, McGraw-Hill, New York, 1962; G. Eisenman, Glass Electrodes for Hydrogen and Other Cations: Principles and Practice, Dekker, New York, 1967; N. Lakshminarayanaiah, Transport Phenomena in Membranes, Academic, New York, 1969.

definitive treatments of the most recent developments for such membranes. It begins with a chapter by Charles Bean on the physics of porous membranes, in which the transport properties of membranes having uncharged pores of simple and well-defined geometry are defined. Bean develops a theory based upon considerations of the fluid dynamics of the motions of spheres in a fine space in which water is assumed to act as a continuum; and he presents data on diffusion and flow through fine, well-characterized pores, indicating that the structure and dynamics of the water in these pores are the same as in bulk water up to at least 10 Å from the solid surface. The second chapter, "Transport Across Ion-Exchange Resin Membranes: The Frictional Model of Transport," by Meares, Thain, and Dawson, examines critically theories of membrane transport of the Nernst-Planck and pseudothermodynamic type, and compares these with treatments based on nonequilibrium thermodynamics, particularly Spiegler's frictional model of membrane transport. Precise experimental data are given for a representative organic cation-exchange membrane and the cross coefficients are deduced and analyzed. This chapter also discusses interaction between isotopes and the evaluation and interpretation of friction coefficients. The third chapter, by Sandblom and Orme, describes and classifies the properties of the liquid ion-exchanger membranes, particularly from the point of view of their use as electrodes and as models for biological membranes. It contains a detailed formulation and theoretical analysis of the transport properties and membrane potentials for monovalent ions and monovalent-divalent mixtures. The fourth chapter, by Garfinkel, deals with the cation-exchange properties of dry glass membranes, with particular emphasis on the significance of these for the membrane potentials of such materials. The studies reported are not only of fundamental interest but provide a basis for deepening the understanding of the structure and function of glass membranes in their usual hydrated state. Since the dry silicate membranes described here are the first glass systems for which the actual profiles of concentration of diffusing species have been characterized directly for comparison with theoretical expectations, this chapter provides a clear verification of the essential cation-exchange nature of the glass electrode. In the fifth chapter, Agin considers certain general aspects of the physical basis for electrical excitability in nerve fibers, with particular emphasis on the conditions for producing negative conductance in nonbiological electrodiffusion regimes, and speculates on two postulates which could account for such conductances in bilayer membranes and in the Squid axon, in both cases comparing the theoretical expectations with experimental results. In the sixth and final chapter, Spyropoulos undertakes the formidable task of characterizing the steady-state resting potentials of the plasma membranes of the Squid axon and of Chara, over the complete range of 2-cation solutions using single cells perfused with well-defined ionic concentrations. Such data should provide a starting point for meaningful comparison of the steady-state properties of living cell membranes with the thick physicochemical membranes described in chapters 1 through 4 of Volume 1 and the ultrathin bilayer membranes of the second volume.

Volume 2, subtitled "Lipid Bilayers and Antibiotics," describes the simplest, presently well-understood properties of such ultrathin membranes; while problems of greater complexity will be dealt with in Volume 3. The first chapter, by Läuger and Neumcke, gives a detailed theoretical analysis of the electric properties of lipid bilayer membranes in the presence of lipid-soluble ions but in the absence of added carriers. The second chapter, by Ciani, Eisenman, Laprade, and Szabo, broadens this to include the theoretically expected effects of carriers on the electrical properties of bilayer membranes and generalizes previous treatments based on chemical equilibria to include kinetics as well. This chapter demonstrates that two "domains" of experimental importance are expected to exist depending upon whether chemical equilibria of chemical kinetics dominate. For generality, both chapters 1 and 2 present in parallel the expectations for a continuous (Nernst-Planck) and a discontinuous (Eyring) model of the membrane, noting explicitly those situations where these treatments become strictly equivalent. Chapter 3, by Szabo, Eisenman, Krasne, Ciani, and Laprade, summarizes the experimentally observed effects of carriers on the electrical properties of bilayer membranes. It confines itself to the domain of experimental conditions in which an understanding of chemical equilibria suffices to account for the observed actions of carrier molecules (the domain in which kinetic effects become important has not yet been fully characterized, and is therefore left to the third volume). This chapter also discusses the equilibrium basis for the molecular origin of ion selectivity. The fourth chapter, by Simon and Morf, deals in considerable detail with the cation specificity of carrier antibiotics and is remarkably successful at reconstructing theoretically the observed selectivity of a variety of antibiotics and model compounds, as well as ionic-hydration energies, from elementary considerations of charge-dipole (and higher multipole) interactions in the first ligand layer as well as the work of cavity formation in the molecules. It also describes some practical electrodes based on the electrical effects of these molecules in thick membranes. The fifth chapter, by Finkelstein and Holz, deals with the "pore" rather than "carrier" mechanism for ion permeation; and examines the permeability properties (and likely molecular structure) of the aqueous pores created in bilayer membranes by Nystatin and Amphotericin B. The sixth chapter, by Ross Bean, examines the variable ion conductances characteristic of the protein-based channels of the molecule, EIM, particularly as these bear on vectorial transport mechanisms and kinetic problems of "excitability." The concluding chapter, by Lev, Malev, and Osipov, examines the electrochemical properties of thick membranes exposed to macrocyclic antibiotics and reconciles a number of important differences observed between the effects of these molecules on conventional thick membranes versus ultrathin bilayers. In this way it provides a bridge between bilayers and the more classical membranes described in Volume 1.

George Eisenman

Los Angeles, California
October, 1971

FOREWORD FROM TORSTEN TEORELL

To write a foreward to this series of advances on membrane processes might appear to be simple for a person who has followed this field over many decades; he ought to know the business. From a view of the field over the long time span of about fifty years, it seems as if the science of membrane permeability has undergone a remarkable development. Yet, the main problems and the main motivations for the exploration of membrane phenomena remain very much the same as in early days. By and large, the status in present days is characterized more by an enormous increase of quantitative, theoretical, and mathematical knowledge rather than by an increase of basic qualitative "first principles." Of course, a few new concepts have been conceived--for example, the relatively new science of irreversible thermodynamics, which came into the membrane field in the late nineteen-forties.

If one were writing an outline of the history of the development of membranology, a simple approach would be to arrange the main achievements in a chronological order, just like a yearly review article, but this would be a very dull performance both for the reader and for the writer. A more fascinating approach would be to try to discern the "driving forces" for the development of this particular area of scientific society. Why was a special contribution made? What kind of background did it have? What sort of goals were in mind? Such an analysis would be equally as difficult as to write the contemporary history of some aspect of political sciences or any branch of the creative arts. Too many external and purely personal factors cannot be accounted for. The story would probably be different depending on which specialist attempted to write it. Viewpoints will necessarily become subjective. So, from that point of view, a foreword-writer may be excused if he leans towards a subjective, self-confession type of recitation. The blame is on me, if I prefer to refer to my own experiences only, with tacit omissions of many fine achievements and famous names.

My own entry in the membrane science took place in the late nineteen-twenties, which rightly could be characterized as a time of breakthrough of physical chemistry into biology and medicine. In fact, cell and tissue membranes were as much in the limelight then as now. It was an optimistic time; one believed that many riddles of life would soon be solved in terms of physical chemistry. In the immediate background there were the solution theories of Nernst-Planck; the protein work by Loeb, with the application of the new Donnan equilibrium theory; the pH concept by Soerensen; the definitions of the isoelectric point by Michaelis; and many other stimulating concepts. The rapidly developing colloid science, and much more, had given material for some "bibles," which more than anything else influenced the young generation of general physiologists of that epoch. I refer to Hoeber's remarkable, comprehensive book, "Physikalische Chemie der Zelle und der Gewebe," Freundlich's large volume, "Kapillarchemie," and Michaelis', "Die Wasserstoffionenkonzentration." These books acquainted us with the classical membrane works and theories, for example, those by Haber (one of the early workers on glass electrodes), by Bernstein (applications of the Nernst theories on the electrophysiological potentials), and by Beutner (liquid membrane models). The studies of models and living-membrane permeability were mostly centered on the electrochemical parameters, probably because it was relatively easy to measure potentials and electrical resistances. The radiotracers, or convenient microanalytical methods, which allow direct transport studies, were not yet available; they came much later.

Two great names of the decades before World War II were Osterhout and Michaelis, both members of the Rockefeller Institute (now Rockefeller University). Osterhout and his collaborators had dealt with ingenious models of liquid membranes (guaiacol, etc.), which showed a selective permeability for potassium similar to that exhibited by plant cells. It was a case of "ion accumulation," as a favored term went in those days. Michaelis, on the other hand, had constructed his model of solid, extremely narrow-pore dried collodion and had demonstrated that these were selectively permeable for either cations ("negative membranes") or anions ("positive membranes"). He called his materials "Geladene Membrane." The existence of fixed charges had already been invoked in 1915 by Bethe and Toropoff, who clearly ascribed the charge to the electrokinetic potential, i.e., the existence of ionic double layers arising from ion adsorption. Both Osterhout and Michaelis employed in their theoretical discussions the simple Nernst diffusion-potential formula and interpreted various experimental results in terms of <u>changed</u> mobilities of the ions within the membranes, i.e., changes of transference numbers. These numbers were, however, concentration dependent. Already Beutner had observed this so-called "concentration effect" on apple peels, and this effect was quite pronounced on Michaelis' dried collodion membranes.

While I was working with Osterhout, I became puzzled by the concentration dependence and made an attempt to reconcile the earlier adsorption ideas, particularly those of Wilbrandt, with a framework of a "mixed potential," but without assuming any changes of the ion mobilities. This led to the first quantitative formula for fixed-charge membranes in 1935. In the following year a similar, more extensive treatment was independently given by Meyer and Sievers. These theories implied, in essence, that there existed two Donnan equilibrium potentials at the membrane boundaries and an intramembrane diffusion potential in between, which summed up to the total potential difference. In the light of the knowledge of today, as is exposed in this volume, the theory was primitive. Yet, it served the usual purpose, because it seemed to have stimulated researchers to extensions, refinements, or other methods of attack, particularly in the years after World War II.

From then on, the history of membrane material and membrane theories has become quite complicated. Whatever crossbreeding has occurred is meaningless to discuss. A good, complete history until the end of the nineteen-fifties is given in Helfferich's excellent book the "Ionenaustaucher." It seems fair to state that the leading names in this period period were Schlögl and Schmid, who also greatly extended the more comprehensive theories I had published in 1951 and 1953. They included water movements across fixed charge membranes; and Schlögl, in particular, has also made excellent contributions to the understanding of anomalous osmosis and other cases of "abnormal" water movements.

At this stage of our account we should turn back again to a very important pioneer work on membranes, namely the intensive studies by Sollner. Sollner, who was a pupil of Freundlich, already in the early nineteen-thirties had started investigations on anomalous osmosis and general membrane behavior. He was able to produce a more convenient type of membrane than the high-resistence, somewhat "temperamental" membranes of Michaelis. He called those membranes "permselective." In my opinion, it was the use of the Sollner membranes which immensely sparked the beginning of a new period of extensive work on the behavior of membranes in general. Sollner has also realized the potentialities of liquid ion-exchange membranes as biological models. Sollner's, and his pupils' lucid work dominated the field until the advent of the industrially-made ionic exchangers, which became available about 1950. The importance of Sollner's work cannot be overestimated in the history of membranology. (Incidentally, the term "membranology" was coined, probably for the first time, at a cordial meeting between Sollner and myself.)

Since ion exchangers became universally accessible, both as solid membranes and as liquids, an almost explosive expansion has taken place; and it is easy to loose track of all the "spin offs." There appeared new theoretical contributions (by Meares and many others); there came practical applications, for instance, for desalination; analytical procedures using selective membranes electrodes were invented (I think it was

Marshall, who in 1939 first tried to measure bivalent cations potentiometrically.)

This is the right point to introduce the Editor of this series, G. Eisenman, who entered the scene in the early 1960's. He had become interested in the use of certain glass membranes as selective potassium and sodium electrodes. With great intensity and profound imagination he and his collaborators have given a new turn to the knowledge of membranes. The selectivity for different ions has now been elucidated down to the atomic level. The implications of this advancement for biological selectivity problems is obvious. This is a really startling example of the cross-reactions in science: glass technology may perhaps help electrophysiologists to explain neurological phenomena!

In the beginning of this account we made an allusion to the motivation of workers in the membrane science. I believe that a great deal of the physicochemical work, like the glass membrane studies, have been strongly influenced by the well known Hodgkin-Huxley theory of 1952 for the mechanism of nerve actions. Their "ion theory" is the most widely accepted; it has a rigid formalism and general applicability. Still, there remain unknown factors to be explained, such as the nature of the voltage-dependent behavior of potassium and sodium, which are different. Also, other nerve-action theories have been suggested, for instance Tasaki's "macromolecular" hypothesis on conformation changes in the excitable membranes. The restoration processes after nerve impulse firing require also the postulation of specific "ion pumps."

These concepts introduced by physiologists have in turn stimulated workers in other fields to draw "blueprints" of "gating mechanisms" and pump machineries. Many of these new ideas have been furthered by the intimate knowledge of the membrane architecture obtained by electron microscopy and X-ray analysis. The biochemists have constructed models for "facilitated diffusion" or "carrier transport." Oscillatory phenomena in artificial and living membranes have become appreciated, the Israeli school of irreversible thermodynamics, headed by A. Katchalsky and O. Kedem, has in recent years made promising attacks on the problem of the effects of chemical reactions coupled to old fashioned passive transport. Experiments on "mobile sites," instead of fixed ones, have been advanced and new theories elaborated by Eisenman and his able co-workers. The fixed charge model of the old days is now merely a special specimen in the rapidly growing population of membranes.

It is quite noticeable when one looks back in perspective over the past years that there is a "fashion" in the choice of problems favored by the then-active generation of research workers. Some problems come back cyclically time after time, often under new names. I have mentioned the early work in the nineteen-thirties by Osterhout on selective ion accumulation with liquid membranes. In a way, the circle has become closed in our own time. Now the most fashionable and also promising, activities deal with the "bimolecular membranes" developed by Mueller and Rudin, which are, in essence, highly specialized liquid structures that can be "doped" with various substances (e.g., antibiotics) to give remarkable similarities with certain living membrane phenomena. But, at this point I seem to have entered into the front-line of present membranology and this is a topic which is to be covered in subsequent volumes in this Series.

The personal recitation above may not contain much information about the driving forces for the transport of membranology as such towards the future. Evidentally, this development will be a highly coupled system of interactions with many cross coefficients, attributable to biological, technological, and many other factors. An objective prediction of the possible trends of achievements in the future is of course impossible. Yet, using my blunt subjective opinion, I venture the prediction that the really new contributions to the membrane science will be headed by the biochemists and by forthcoming leaders of the nonequilibrium thermodynamics. Whatever might happen, I am convinced that this volume will serve to stimulate the next generation of membranologists and edify the older ones.

Torsten Teorell

Uppsala, Sweden
July 1, 1970

CONTRIBUTORS TO VOLUME 1

D. Agin, Department of Physiology, University of Chicago, Chicago, Illinois

Charles P. Bean, General Electric Research and Development Center, Schenectady, New York

D. G. Dawson*, Chemistry Department, The University, Old Aberdeen, Scotland

Harmon Garfinkel, Corning Glass Works, Technical Staffs, Sullivan Park, Corning, New York

P. Meares, Chemistry Department, The University, Old Aberdeen, Scotland

Frank Orme, Department of Physiology, University of California, Berkeley, California

John Sandblom, Departments of Physiology and Medical Biophysics, Biomedical Centre, University of Uppsala, Box 572, S-751 23 Uppsala, Sweden

C. S. Spyropoulos, University of Chicago, Department of Physiology; and Democritus Nuclear Research Center, Athens, Greece

Torsten Teorell, Institute of Physiology and Medical Physics, University of Uppsala, Uppsala, Sweden

J. F. Thain, School of Biological Sciences, University of East Anglia, Norwich, England

*Present address: I.C.I. Paints Division, Slough

CONTENTS

CONTENTS OF VOLUME 2

MEMBRANES

VOLUME 1

Macroscopic Systems and Models

Chapter 1

THE PHYSICS OF POROUS MEMBRANES--NEUTRAL PORES*

Charles P. Bean

General Electric Research and Development Center
Schenectady, New York

How blind is he who cannot see through a sieve.
Cervantes (Don Quixote, Part II)

I. INTRODUCTION

A membrane may be defined as a phase that acts as a barrier to the flow of matter or heat. In most cases, with the exception of pure liquids and solids, it is heterogeneous in structure. It is these heterogeneities that give rise to the rich variety of transport phenomena that distinguish

---

*This article was prepared as the first part of a two-part discussion of porous membranes. The author is in the process of preparing the second part, on charged pores, to be published at a later date.

membrane systems and make them so important in the process of life and increasingly important in science and technology. The task of membrane science is to discover the relationships between structure--to a molecular scale--and the properties of membranes.

Perhaps the simplest conceptual heterogeneity is that of a pore, i.e., a discontinuity in a solid impervious phase that can allow selective transport of some chemical species. While this sieving property has been employed for thousands of years (friezes on Egyptian tombs of 1500 to 1900 B.C. show sieving in the preparation of beer and wine (1), it is only relatively recently that a scientific understanding has been brought to this process. In addition recent years have brought a wide variety of porous membranes into technology and prototechnology. It is the purpose of this review to summarize the present state of the theory of porous membranes and give an indication of their types and uses. Two important restrictions in the coverage are made. First, in the main, only systems in contact with aqueous phases are discussed and secondly, transport that derives from temperature differences is not discussed. The first of the omissions causes the exclusion of what is probably the largest industrial use of porous membranes--that of the diffusion separation of ^{235}U from ^{238}U (2, 3).

This chapter is organized so that simple theoretical cases are treated and then the structure and properties of membranes are viewed in the light of these considerations. The object is to delineate those areas of firm knowledge and to point out areas of potential theoretical and experimental advance.

II. BACKGROUND

If we consider a membrane interposed between two uniform phases, net transport between these phases will occur only if there is a driving force. In a completely general way one must hypothesize a generalized chemical potential difference in one or more components between the two phases to accomplish this transport. [Indeed, one of the first applications Gibbs made of his concept of thermodynamic potential was to a semipermeable membrane (4).] For the isothermal case, we consider that these driving forces can come from differences in concentration, electrical potential, or pressure. The flows can be those of the chemical constituents, electrical current, or total volume transport. Although we recognize that these flows are interdependent in that the electrical current, for instance, is composed of ion flow, in addition to the electronic contribution which is potentially present, and that this ion flow is a subclass of the flow of chemical constituents, this division is useful. We may picture the relationships between these potential differences and the flows as shown in Fig. 1. The causative

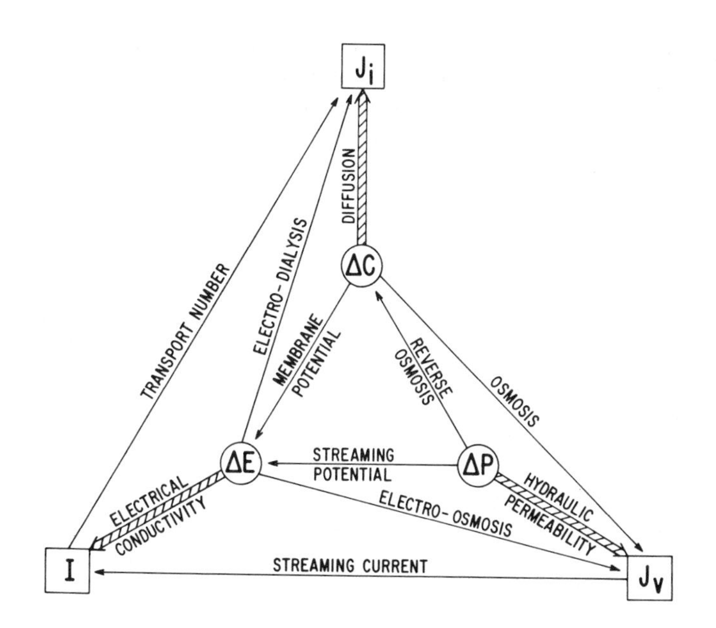

FIG. 1. A schematic representation of the isothermal transport properties of membranes. The driving forces of concentration, pressure, and voltage are shown encircled while the flows of constituents, volume, and current are represented by squares. The heavy arrows indicate the direct relations of diffusion, hydraulic permeability, and electrical conductivity. The finer arrows show eight indirect relationships. [Adapted from Fig. 1.5, Lakshinanarayanaiah (8).]

agents are encircled while the fluxes are in square boxes. Each arrow indicates a named causal relationship--some 11 in all. The heavy arrows represent the principal connections between force and flow while the light arrows, in the terms of irreversible thermodynamics, are off-diagonal components of an interaction matrix. The task of theory is to predict and correlate all of these effects in terms of the physical structure of the membrane and the molecular structure of the permeating species.

Before embarking on this program it may be useful to list some general references and earlier reviews that give various points of view and represent differing stages in the course of understanding porous membranes. The early review by Ferry (5) presents a complete and influential picture of the field of ultrafiltration as it appeared in the middle 1930s. The current state, with emphasis on biological membranes, is described by

Solomon (6). An extremely valuable monograph by Schlögl (7) describes the general properties of matter transport while a recent book by Lakshminarayanaiah (8) expands on his earlier article in Chemical Reviews (9) to give an almost encyclopedic account of experiment and theory in the field of membrane transport.

III. NEUTRAL PORES

A. Irreversible Thermodynamics of Transport Through Neutral Membranes

In recent years, a theory of the irreversible thermodynamics of membrane transport has been developed that provides a background for all mechanistic discussions. Based on the fundamental contributions of Onsager (10), with important contributions by Staverman (11) and Kirkwood (12), it has become codified by Katchalsky and his collaborators (13, 14). An excellent general review is given in the text by Katchalsky and Curran (15). We present some of the key results below and in the following section give an account of the extension of these concepts to a more physical, though generalized, frictional model.

As the first example we consider a system containing a solvent (water) and a dilute solute of nonelectrolyte that bathes--in possibly unequal concentrations--either side of a membrane that may also sustain a pressure difference. A schematic diagram is shown in Fig. 2.

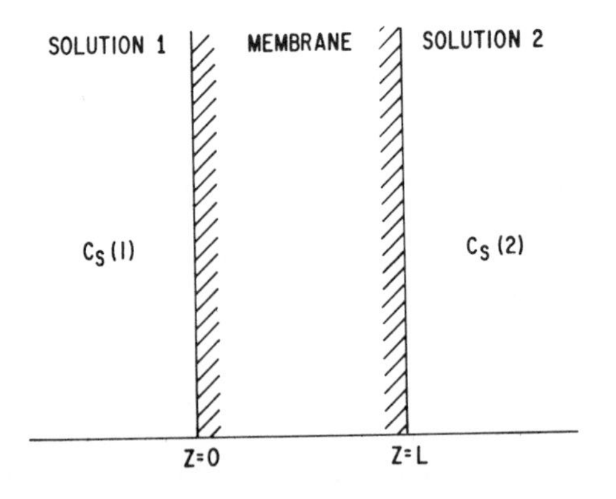

FIG. 2. Schematic representation of a membrane separating two regions of uniform concentration of nonelectrolyte.

We may define the fluxes and forces in several ways. For a square centimeter of membrane surface, a conventional method considers the rate of total volume transport J_V and the differential rate of transport of solute and solvent J_D, to be defined by

$$J_V = J_w \bar{V}_w + J_s \bar{V}_s$$
$$J_D = J_s / \bar{C}_s - J_w \bar{V}_w \tag{1}$$

Both J_V and J_D are expressed as centimeters per second. J_w and J_s are the flows from sides 1 to 2 in moles per square centimeter per second of solvent and solute, respectively, while $\bar{V}_w$ and $\bar{V}_s$ are the molar volumes of solvent and solute. The average concentration of solute, i.e. $[C_s(1) + C_s(2)]/2$, is denoted by $\bar{C}_s$. The unit of concentration is moles per cubic centimeter. We consider a pressure difference ΔP and a concentration difference represented by an osmotic pressure difference $\Delta\pi$ as driving forces. These pressures are defined by

$$\Delta P = P_1 - P_2$$
$$\Delta\pi = RT(C_s(1) - C_s(2)) \tag{2}$$

and may be expressed in dynes per square centimeter or atmospheres. R is the molar gas constant. If the driving forces are small enough the fluxes are linearly related to the forces and we assume phenomenological equations of the form

$$J_V = L_P \Delta P + L_{pD} \Delta\pi$$
$$J_D = L_{Dp} \Delta P + L_D \Delta\pi \tag{3}$$

where the Ls are constant coefficients for a given membrane, solvent, and solute. The results of irreversible thermodynamics are

$$L_{pD} = L_{Dp}$$

and (4)

$$L_{pD}^2 \leq L_p L_D$$

The principal consequence is that three measurements are necessary to characterize a membrane-solute-solvent system at a given temperature and concentration. The important reservations are that the driving forces are sufficiently small to make the linear relations appropriate and secondly that the baths are adequately stirred to make the assumptions of constant concentration in each compartment sufficiently correct. Neglect of this latter condition, the so called unstirred-layer problem, has vitiated many experiments (16). The principal experiments that may be performed to measure the coefficients are those of measuring the hydraulic permeability

$$(J_v)_{\Delta\pi = 0} = L_p\,\Delta P \tag{5}$$

and the steady-state pressure difference for zero flow occasioned by a difference in concentration, i.e.,

$$(\Delta P)_{J_v = 0} = -\,(L_{pD}/L_p)\Delta\pi \tag{6}$$

Staverman (11) has identified the term in parentheses as a reflection coefficient σ

$$\sigma = -(L_{pD}/L_p) \tag{7}$$

A measurement of solute transport with no fluid transport gives a solute permeability coefficient ω defined by

$$\omega = (J_s/\Delta\pi)_{J_v = 0} = \bar{C}_s\,(L_pL_D - L_{pD}^2)/L_p \tag{8}$$

As an example of this type of treatment, Kaufmann and Leonard (17) have made a careful study of properties of a 0.00754-cm-thick cellophane with various solute nonelectrolytes. In contrast to most previous work they have been very careful to eliminate the effects of unstirred layers. The determinations were made for concentrations of 0.05 M. Their results are shown in Table 1. The hydraulic permeability is seen to be primarily a function of temperature--it scales with the inverse of viscosity--and only weakly dependent on solute. The off-diagonal terms are small compared to the diagonal terms and both the solute permeability and reflection coefficient vary in the order expected for increasing molecular weight of the solute. The weakest temperature dependence is that of the reflection coefficient. It should be noted that the volume fraction of water in the membrane was determined to be 0.69 and nearly independent of both temperature and solute. The significance of these observations in a mechanistic sense will be clarified by the discussions of Sec. III. F.

TABLE 1

Phenomenological Coefficients For a Cellophane Membrane (17)

Solute	Temperature °C	L_D cm/atm-sec	L_{Dp} cm/atm-sec	L_p cm/atm-sec	L_{pD} cm/atm-sec	ω moles/atm-cm^2-sec	σ
Glucose	27	1.19×10^{-4}	1.77×10^{-6}	2.00×10^{-5}	1.77×10^{-6}	4.56×10^{-9}	0.0885
	37	1.53	1.93	2.44	1.93	6.00	0.0791
	47	1.73	2.17	3.05	2.17	6.80	0.0711
Sucrose	27	7.49×10^{-5}	2.00×10^{-6}	1.90×10^{-5}	2.00×10^{-6}	2.85×10^{-9}	0.105
	37	9.54	2.28	2.35	2.28	3.64	0.097
	47	10.96	2.72	3.03	2.72	4.24	0.090
Raffinose	37	6.56×10^{-5}	2.89×10^{-6}	2.30×10^{-5}	2.89×10^{-6}	2.50×10^{-9}	0.126

B. Frictional Interpretation of Phenomenological Coefficients

While the Nernst-Einstein relationship represents a relationship between diffusion and frictional constants, the first work of this type applied to membranes was performed by Spiegler (18) and extended by Kedem and Katchalsky (13). The system considered is a homogeneous phase of thickness L, wherein the local flows of solute and water are denoted by J_s and J_w, respectively, while the driving forces are the negative local gradients in chemical potential X_s and X_w. It is assumed that, by Newton's third law, these forces are counterbalanced by frictional forces expressed as a sum of pair-wise interactions between the entities of solute, solvent, and membrane. Further it is assumed that the frictional forces are proportional to the relative velocities of the elements, i.e.,

$$F_{ij} = -f_{ij} (v_i - v_j) \tag{9}$$

where the force is that exerted on a mole of component i by the relative motion of i and j. The frictional coefficients can be expressed as units of atmosphere - centimeter - second per mole. Choosing the membrane as reference, i.e., $v_m = 0$, we can write

$$\begin{aligned} X_s &= -F_{sw} - F_{sm} = f_{sw} (v_s - v_w) + f_{sm} v_s \\ X_w &= -F_{ws} - F_{wm} = f_{ws} (v_w - v_s) + f_{wm} v_w \end{aligned} \tag{10}$$

These forces can be integrated to give the chemical potential differences across the membrane with the assumption that the frictional coefficients are independent of concentration. Identification of the fluxes and forces with the phenomenological coefficients gives

$$\begin{aligned} \sigma &= 1 - \omega \bar{V}_s/L_p - K_D f_{sw}/\Phi_w (f_{sw} + f_{sm}) \\ \omega &= K_D/L (f_{sw} + f_{sm}) \\ L_p &= \Phi_w/L (f_{wm}/\bar{V}_w + \bar{C}_s (1-\sigma) f_{sm}) \end{aligned} \tag{11}$$

These equations contain two dimensionless quantities that have not been defined earlier, i.e., the distribution coefficient of solute between the water in the membrane and that in the solution K_D, and Φ_w the volume fraction of water in the membrane.

Table 2 shows the frictional coefficients for the membrane and solutes of Table 1. In this table, we note the same regularities as were noted

TABLE 2

Frictional Coefficients and Membrane Parameters For a Cellophane Membrane (17)

Solute	Temperature °C	K_D	Φ_w	f_{sw} atm-cm-sec/mole	f_{sm} atm-cm-sec/mole	f_{wm} atm-cm-sec/mole	f_{sw}/f^o_{sw}	f_{wm}/η cm^2/mole
Glucose	27	0.704	0.694	1.79×10^{10}	2.47×10^{9}	8.18×10^{7}	5.15	9.56×10^{15}
	37	0.695	0.689	1.36×10^{10}	1.65×10^{9}	6.69×10^{7}	4.76	9.61×10^{15}
	47	0.690	0.685	1.21×10^{10}	1.38×10^{9}	5.34×10^{7}	5.15	9.23×10^{15}
Sucrose	27	0.734	0.694	2.78×10^{10}	6.08×10^{9}	8.43×10^{7}	6.10	9.73×10^{15}
	37	0.721	0.689	2.18×10^{10}	4.40×10^{9}	6.79×10^{7}	5.75	9.64×10^{15}
	47	0.705	0.685	1.89×10^{10}	3.03×10^{9}	5.27×10^{7}	5.85	8.99×10^{15}
Raffinose	37	0.760	0.684	3.06×10^{10}	9.82×10^{9}	6.63×10^{7}	6.41	9.20×10^{15}

previously. The interaction between water and the membrane is primarily a function of temperature and follows the viscosity of the appropriate solutions as the constancy of the last column shows. The fact that the friction between the water and membrane is one to two orders of magnitude less than that between solute and water or solute and membrane emphasizes the cooperative nature of water movement. The next section on hydrodynamics treats this interaction. For a free solution, by the earlier mentioned Nernst-Einstein relationship, the frictional coefficient f^{o}_{sw} is given by

$$f^{o}_{sw} = RT/D_0 \tag{12}$$

where D_0 is the diffusion constant. The ratio of the observed frictional coefficient to that for a free solution is given in the second-to-last column and is seen to be a ratio of about five to six, increasing with increasing solute particle size and lastly, the interaction between solute and membrane is seen to be an order of magnitude less than that for solute and water. We discuss the significance of these observations later.

While there are many values in looking at the properties of membranes in this way, the key assumption of homogeneity is contrary to the structure of porous membranes and, hence, the results must be interpreted with structural insight. As a simple example, consider two porous membranes that are identical in structure except that one has some blind alleys. The phenomenological coefficients will be identical to the extent that normal diffusion and flow is unaltered by the blind alleys. The frictional coefficients will be changed, as inspection of Eq. (11) shows, by the consequent change in Φ_w.

C. Some Results from Fluid Dynamics

In this section we give some of the results of fluid mechanics, both theoretical and empirical, that are basic to the understanding of porous membranes. The fundamental relationships are the Navier-Stokes equations together with an equation of state. The Navier-Stokes formulation (19) results from the application of Newtonian mechanics to a deformable system characterized by an internal friction. In vectorial form these equations are

$$\rho \frac{\partial \underline{v}}{\partial t} + \rho (\underline{v}\ \mathrm{grad})\ \underline{v} - \eta \Delta \underline{v} + \mathrm{grad}\ P = \underline{f} \tag{13}$$

with the condition for an incompressible fluid that

$$\mathrm{div}\ \underline{v} = 0 \tag{14}$$

In these equations ρ is the density, $\underline{v}$ is the vectorial velocity, η is the viscosity, P is the pressure, and f is the externally applied force per unit volume. These equations have within them all incompressible fluid phenomena but unfortunately very few solutions exist. On the other hand, as Reynolds pointed out, it is possible to find scaling laws that permit the extrapolation of experiments in one regime of size, density, and velocity to another.

For most cases of interest to membrane physics the first two terms in Eq. (13) can be neglected owing to the small velocities involved. Neglect of these inertial terms can be justified if the Reynolds number by

$$Re = vD\rho/\eta \tag{15}$$

is less than, say one. In this equation, D is a characteristic of the problem. For instance, it could be a pore diameter. The resulting equations in Cartesian coordinates are

$$\frac{\partial P}{\partial x} = \eta\nabla^2 v_x + f_x, \quad \frac{\partial P}{\partial y} = \eta\nabla^2 v_y + f_y, \quad \frac{\partial P}{\partial z} = \eta\nabla^2 v_z + f_z \tag{16}$$

and

$$\frac{\partial v_x}{\partial x} + \frac{\partial v_y}{\partial y} + \frac{\partial v_z}{\partial z} = 0$$

Even these equations have few complete solutions. One of the simplest is that of the motion of a sphere in a still liquid. The result, Stokes Law, says that the force, to cause steady motion of a sphere of diameter d with a velocity v in a medium of viscosity η is given by

$$3\pi\eta d\underline{v} \tag{17}$$

Another simple case is that of flow down a long cylindrical tube of diameter D, wherein the velocity is axial and is given by

$$v_z = (-\partial P/\partial z)(D^2/4 - r^2)/4\eta \tag{18}$$

where the critical assumption is that the velocity is zero at the surface of the tube. The result of integrating this over the cross section of the cylinder gives the Hagen-Poiseuille law

$$Q = (-\partial P/\partial z)\pi D^4/128\eta \tag{19}$$

where Q is the volume flow. The average fluid velocity is just half of the maximum central velocity: $(\partial P/\partial z)D^2/16\eta$. These results can be generalized to various uniform cross sections other than the circular one assumed above. For instance, if the tube has an elliptical cross section with axes a and b, then D in Eq. (19) is replaced according to (20)

$$D = \sqrt{2}ab/(a^2 + b^2)^{\frac{1}{2}} \tag{20}$$

To examine the effects of particle motion in tubes, it is necessary to know the modifications of Stokes Law for motion of a sphere in a tube with or without flow. This problem has no analytic solution. Indeed, most calculations have considered only the special case of a sphere moving along the central axis. This problem is one of some venerability with contributions by Ladenburg (21) and Faxén (22). More recently Bohlin (23) has made a more exact calculation that has been reviewed by Faxén (24). The result is that the force that must be applied to move the sphere with a velocity v is given by

$$3\pi\eta d(v - v_{max}G(\alpha))/F(\alpha) \tag{21}$$

where the functions G and F are functions of the relative diameters of ball and tube, i.e., $\alpha = d/D$. v_{max} is the fluid velocity at the center of the tube far from the sphere. These functions can be expressed by series expansions

$$\begin{aligned} F(\alpha) &= 1 - 2.10443\alpha + 2.08877\alpha^3 - 0.94813\alpha^5 \\ &\quad - 1.372\alpha^6 + 3.87\alpha^8 - 4.19\alpha^{10} + \\ G(\alpha) &= 1 - 2\alpha^2/3 - 0.1628\alpha^5 - 0.4059\alpha^7 \\ &\quad + 0.5326\alpha^9 + 1.51\alpha^{10} + \end{aligned} \tag{22}$$

Figure 3 plots these functions. The function F(1) has the limiting value of zero while G(1) approaches one-half, inasmuch as the sphere acts as a piston and, as noted earlier, the average fluid velocity is one-half of the peak velocity. Fidleris and Whitmore (25) have experimentally verified $F(\alpha)$ to $d/D = 0.6$. Their results are plotted on Fig. 3. The question of spheres that are off-axis is much more complicated. Famularo (26) has calculated the coefficient of the leading term of $F(\alpha)$ as a function of radial position. This drag-correction coefficient as a function of position is shown in Fig. 4. It is rather remarkable that it has a broad minimum at about four-tenths of the way to the surface. The possibility of sphere rotation and migration in a flowing stream is somewhat complicated. Goldsmith and Mason (27) observed no radial displacement for rigid spheres at Reynolds numbers less than 10^{-6}. At Reynolds numbers of about 10^{-3}, however, Segré and

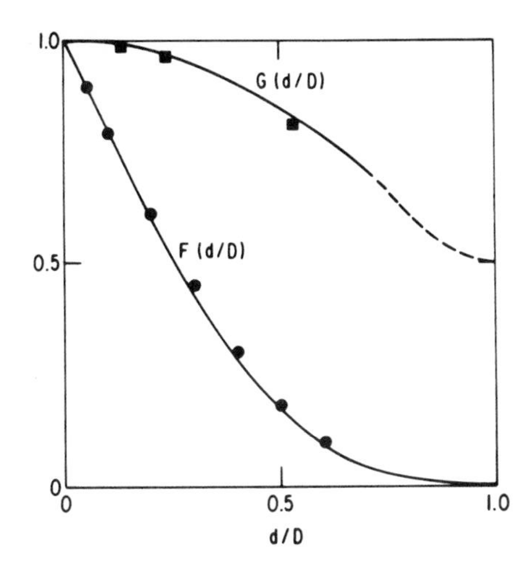

FIG. 3. Hydrodynamic correction factors for particle movement along the axis of a circular cylinder. The particles are spheres of diameter d in circular cylinders of diameter D. The F factor gives the reduction in particle mobility in cylinders with no net fluid transport. The G factor gives the ratio of particle velocity to peak fluid velocity for free movement of the particle. The dashed extension of this curve is an approximate extrapolation for $d/D \rightarrow 1$. [The solid squares are experimental points from Goldsmith and Mason (27) while the solid cicles represent the work of Fidleris and Whitmore (25).]

Silberberg (28) observed displacement toward a cylindrical shell half the size of the tube.

A last point from fluid dynamics is Einstein's remarkable calculation of the effective viscosity of a medium with a dilute suspension of spheres (29). If ϕ is the volume fraction of spheres in a homogeneous medium of viscosity η_0, the spheres interfere with the lines of flow to give an effective viscosity η given by

$$\eta = \eta_0 \, (1 + 2.5\Phi + ---) \tag{23}$$

Higginbotham, Oliver, and Ward (30) have verified this limiting behavior for macroscopic spheres but find that a modified law

$$\eta = \eta_0/(1 - 2.5\Phi) \tag{24}$$

holds much better for higher concentrations and indeed is valid to $\Phi = 0.3$. In addition, they show that the flux of fluid in a Hagen-Poiseuille

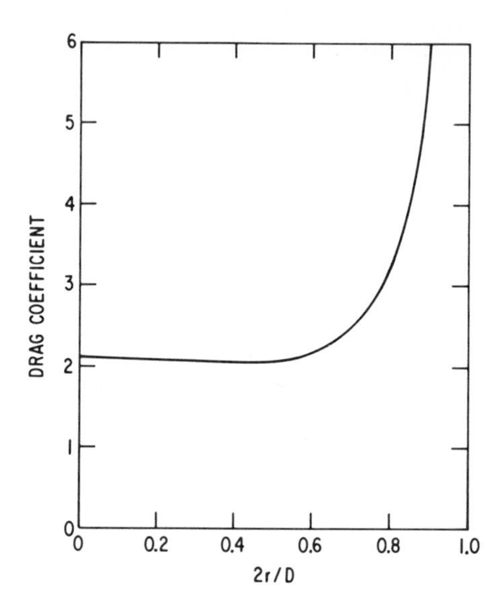

FIG. 4. Drag coefficient as a function of radial position in a pore. The leading term of the expansion of F(d/D) shows a slight minimum according to the calculations of Famularo (26).

experiment in a small-diameter tube gives a consistently larger flow or smaller effective viscosity than for a larger diameter. This effect is equivalent to that calculated from a postulated depleted layer of thickness equal to 0.35 of the sphere diameter at the tube wall, if that layer were considered to have only the intrinsic viscosity η_0 of the medium.

D. Solvent Transport--Diffusion and Flow

After a brief treatment of viscosity and diffusion of pure, unbounded water, this section treats the flow and diffusion of water through capillary systems. The emphasis is on the phenomena associated with uniform circular pores of a size large compared to the water molecule. The situations of porous media and very small pores is dealt with in later sections. An experimental series of the author's, previously unreported, is described in some detail.

It is a truism to observe that water has remarkable properties. Its melting point, freezing point, and liquid range distinguish it from homologous compounds. But particularly the fact that its solid form is less dense than the liquid form, and the presumably related density maximum at 4°C, cause it to be a curiosity. Owing to these peculiarities and the pervasive importance of water, many models have been proposed for the structure of water (31). For instance, the "flickering-cluster," the "water-hydrate," and "distorted-bond" models all have adherents, successes, and failures. [It is, perhaps, uncharitable to speculate whether the picturesqueness of the name compensates for the incompleteness of the theory (32).] Fortunately for our purposes, the viscosity of water is quite unremarkable (33). It proceeds monotonically from 1.7916×10^{-2} P at 0°C to 0.2783×10^{-2} P at 100°C. There are no discontinuities in value or slope over this range. It has a value of 0.8903 cP at 25°C and an activation energy of 3.9_4 kcal/g mole at that point. The self-diffusion of various isotopic forms of water has been measured by Wang and his collaborators (34). At 25°C the measurements may be expressed as $2.44 \pm 0.15 \times 10^{-5}$ cm^2/sec. In the study of self-diffusion in liquids, a correlation has been observed among molecular size, viscosity, and the self-diffusion coefficient. The correlation is suggested by the Stokes-Einstein relationship for the diffusion of spheres in a hydrodynamic continuum i.e.

$$D_0 = kT/3\pi\eta d \tag{25}$$

where D_0 is the diffusion coefficient at temperature T of spheres of diameter d in a medium of viscosity η. This relationship comes from the combination of Stokes Law, Eq. (17), and the Nernst-Einstein relationship. This cannot be directly applied to self-diffusion inasmuch as the assumptions of Stokes Law are not valid. The Nernst-Einstein relationship is general, however. Tyrell (35) has tabulated the ratio $kT/\pi D_0 \eta d^*$, where d^* is defined from the molar volume V_M as

$$N_A \pi d^{*3}/6 = V_M \tag{26}$$

where N_A is Avogadro's number. When the liquid is close-packed, the particle diameter and d* are related by

$$d = 0.905d^* \tag{27}$$

For some 25 liquids, the ratio Tyrell obtains is 1.5, with an average spread of about 20%. Water is 1.56. Two conclusions may be made from this correlation. First, the coefficient in the Stokes Law for motion of spheres among their own kind is about half of the hydrodynamic value and, secondly, water is not exceptional in its local organization. The effective diameter of the water molecule from these considerations is 3.3 Å while the distance between molecules derived from X-ray diffraction is 2.90-3.05 Å (36).

Granted that bulk water has quite normal viscous properties, there has been a question outstanding for many years concerning its properties at interfaces--in particular at polar interfaces. It has been hypothesized by many investigators that a long range structure is caused by a charged, say silicate, surface. The evidence for this point of view cannot be given in detail here; Kavanau gives a brief survey (37). As one example, from a study of the coupling between water and clay particles, Forslind (38) infers that rigid structure extends into the water for at least 300 Å. A review of various observations on silicate-water surface interactions is given by Low (39). While many of these studies show evidence for a region of high viscosity near a solid surface, in almost every case neither the geometry nor indirect electrical effects are known in sufficient detail to carry strong conviction. The Soviet physical chemist, Derjaguin, has made many studies (40) over the years that indicate water to have a rigid lattice extending about 1000 Å from a mica, glass, or quartz surface. These observations have resulted from use of a wide number of experimental techniques. In the latest series of experiments (41), the rigidity of thin films of water was measured by putting a quartz crystal (that formed one of the thin film boundaries) into resonant shear deformation. The natural frequency of the quartz was observed to be relatively unaffected by the film spacing until the separation was reduced to about 2000 Å, at which point an extremely strong coupling became apparent. This effect was absent with nonpolar liquids.

The importance of this point of view for the understanding of water transport is immense. All of the simple theories of water transport through fine pores would be incorrect by orders of magnitude if this effect were general. Because of the importance of this question, some space is devoted to a description of our experiments--experiments that do not corroborate this effect. Before describing these direct experiments on diffusion and flow in pores, it may be noted that the presence of this ice-like layer would imply an activation energy for transport that is greatly altered from that of bulk viscosity. Table 2 indicates that the water flow through the fine pores (~ 50 Å) of cellophane follows the viscosity within a few per cent from 27°C to 47°C, in which range the viscosity falls by 31%. Another experiment (42) is shown in Fig. 5. This plot shows the temperature dependence of ionic transport through pores of about 70Å diameter in mica as compared with ionic transport in bulk solution. These pores were created by the Price-Walker (43) technique of etching charged-particle tracks to form uniform pores of great length-to-width ratio. [A description of the techniques and kinetics of this method of pore formation is given by Bean, Doyle, and Entine (44).] The fact that the resistance of the membrane so closely follows the solution resistance from 25° - 2°C implies that the activation energy for transport with the pore is within 0.4% of that of the bulk solution. In particular, the formation of an ice-like structure at low temperatures would cause the low temperature points to go off the graph by orders of magnitude.

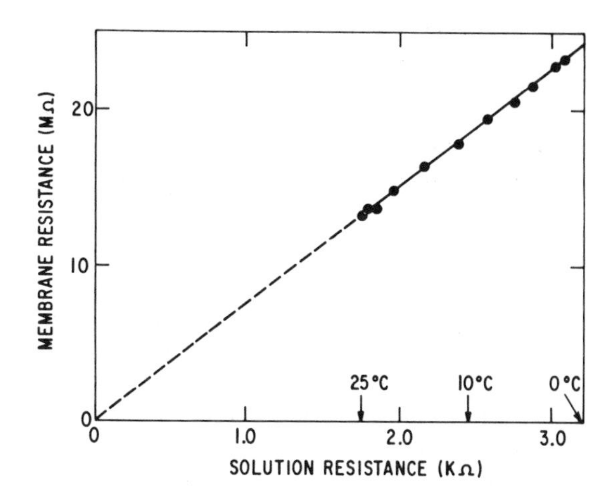

FIG. 5. Temperature course of electrical resistance of a fine pored mica membrane (42). The membrane resistance is plotted against the resistance between two electrodes in the 0.01 N KCl bathing solution. The membrane was $3.5 \cdot 10^{-4}$ cm thick and had $8.6_5 \cdot 10^4$ pores created by etching the sample in 35.7% HF for 11.5 sec at 25° C following irradiation with fission fragments from ^{252}Cf. The inferred pore radius is approximately 35Å.

It may be noted that these two observations are not necessarily responsive to the question, inasmuch as one case treats a primarily hydrogen-bonding surface while the other has a high ionic concentration. To remove these objections, we have measured the diffusion and flow of pure water through uniform and well characterized pores in the range of 300 to 3000 Å.

A system of pores will allow simple diffusion in the aqueous phase if a concentration difference exists along the pores. Application of a pressure will either enhance or retard this diffusion depending on the sign of the pressure difference relative to the concentration difference as implied by the phenomenological equations [see Eq. (3)]. This observation may be made quantitative and can serve as a check on hydrodynamic and diffusive properties of water in fine pores.

The diffusion equation for a moving system is well known (45) and consists of the normal diffusion equation with an added term proportional to the product of local velocity and concentration gradient, i.e.,

$$D_0 \Delta C - \underline{v} \cdot \mathrm{grad} C = \partial C / \partial t \qquad (28)$$

In one dimension, this becomes

$$D_0 \partial^2 C/\partial z^2 - v\partial C/\partial z = \partial C/\partial t \tag{29}$$

where D_0 is a diffusion constant, C is a local concentration as a function of transverse distance z and time t. For a condition of steady state, $C/\ t =$ 0, hence

$$D_0 \partial^2 C/\partial z^2 = v\partial C/\partial z \tag{30}$$

While this equation was set down by Hertz (46) and solved in limiting cases, the full solution, given below, was apparently first given by Manegold and Solf (47). It has also had many independent derivations (48). If we consider a membrane with a uniform velocity v and thickness L with concentrations C(1) and C(2) on the two sides of the membrane, the concentration profile from Eq. (30) is

$$C(z) = \frac{C(1)(\exp v/v^* - \exp(vz/v^*L)) + C(2)(\exp(vz/v^*L) - 1)}{\exp(v/v^*) - 1} \tag{31}$$

where

$$v^* = D_0/L \tag{32}$$

The functional relationship of Eq. (31) is plotted as Fig. 6 for various values of v/v^* that span the range from simple diffusion to almost transport by flow. If we define j as the local flow of solute matter per unit pore area and time, then by the reasoning that led to Eq. (28)

$$j = -D_0 \partial C/\partial z + vc \tag{33}$$

and from Eq. (31)

$$j = \frac{(D_0 v/Lv^*)(C(1)\exp(v/v^*) - C(2))}{\exp(v/v^*) - 1} \tag{34}$$

The two limiting cases of this flow

$$j \to (D_0/L)(C(1) - C(2)) + (C(1) + C(2))v/2, \quad v \ll v^* \tag{35}$$

$$j \to vC(1) \qquad\qquad , v \gg v^*$$

correspond to the simple cases of diffusion with a slight augmentation due to flow and transport by flow, respectively. The phenomenological equations [Eq. (3)] will apply only for the first limit. This augmentation or

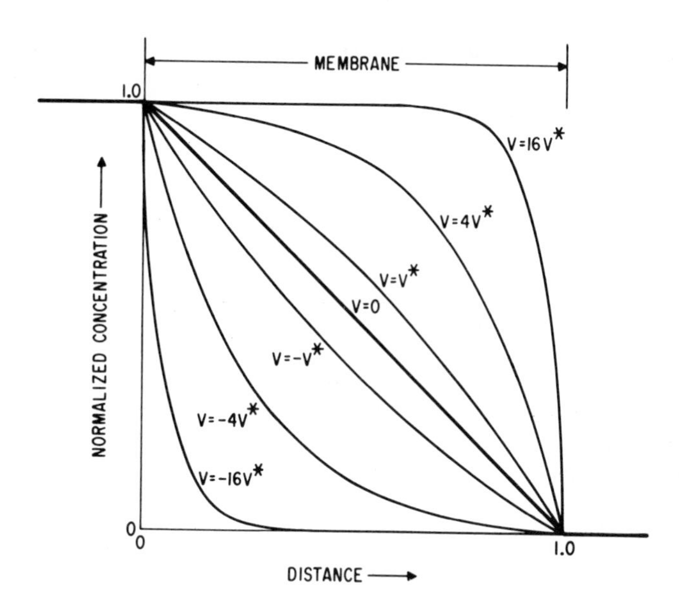

FIG. 6. Concentration of a diffusing species as function of distance across a membrane that has a uniform velocity V, within the solute filled phase of the membrane. Both concentration and distance are given in reduced units while the velocity is expressed in units of $V^* = D_0/L$, where D_0 is the appropriate diffusion constant for the diffusing species and L is the thickness of the membrane.

diminishment of diffusion by flow has been called solvent drag and has been studied in artificial (49) as well as natural (50) membranes.

If we consider the velocity v to be imparted by a pressure difference ΔP along the length L of a circular pore of diameter D, then the average velocity is, by Eq. (19)

$$\bar{v} = -(D^2/32\eta)(\Delta P/L) \tag{36}$$

Strictly speaking the development that led to Eq. (34) is invalid, inasmuch as the velocity difference across the pore will cause a concentration gradient within the cross section of the pore and hence make the one-dimensional treatment inaccurate. In practice, however, this inaccuracy is minimal for long pores owing to the rapidity of mixing across the pore compared with transit times down the pore. This point of view has been taken by Overman and Miller (51) in their study of this effect for flow and diffusion of D_2O-H_2O in a single macroscopic capillary. They measured the average concentration with the capillary under combined diffusion and flow and found excellent agreement with an integrated form of Eq. (31). One may define a characteristic pressure difference ΔP^* by

$$\Delta P^* = 32\eta D_0/D^2 \tag{37}$$

and rewrite Eq. (34) as

$$j = \frac{(D_0/L)(\Delta P/\Delta P^*)(C(1)\exp(\Delta P/\Delta P^*) - C(2))}{(\exp(\Delta P/\Delta P^*)) - 1} \tag{38}$$

Equations (37) and (38) are the working equations for understanding the effects of pressure on the flow of matter through a concentration gradient. Again it is strictly true only for self-diffusion; there is a slight correction for solvent diffusion that we treat in the next section. Figure 7 shows

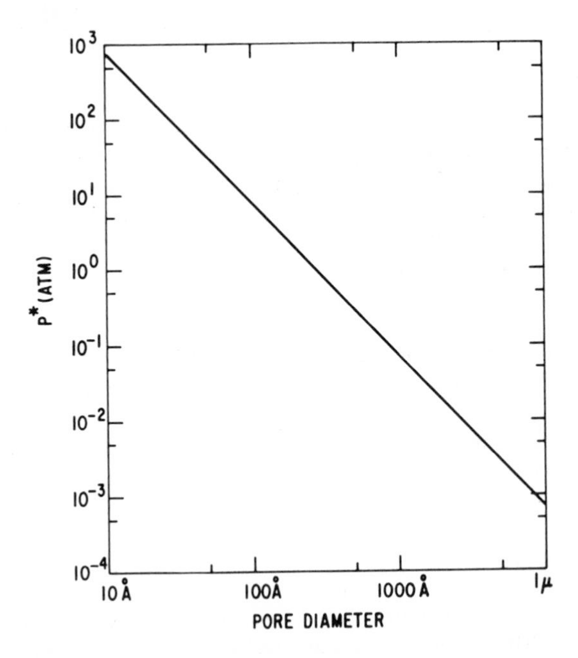

FIG. 7. Value of the parameter P^* as a function of pore diameter calculated for the transport of tritiated water molecules. P^* is a measure of the pressure difference required to significantly alter the diffusive transport of molecules across the membrane and is given by $P^* = 32\eta D_0/D^2$, where η is the viscosity, D_0 is the diffusion constant, and D is the pore diameter. The calculation is appropriate to 300°K.

ΔP^* as a function of pore diameter, calculated for the self-diffusion of water. Figure 8 shows the transport as a function of reduced pressure according to Eq. (38) with C(2) set equal to zero.

With a known number N of circular pores of diameter D, one can compare the results of a simple diffusion measurement ($\Delta P = 0$) with the results of fitting Eq. (38). Two effective diameters for diffusion and flow may be defined, and denoting them by D_D and D_F, respectively, we have

$$D_D^2 = 4L(J(o)/(C(1) - C(2)) \cdot D_o \cdot N\pi$$
$$D_F^2 = 32\eta D_o/\Delta P^* \qquad (39)$$

If the viscosity and diffusion constant differ from that of bulk solutions, the value of D_D derived from Eq. (39) will differ from the geometric radius. D_F is unaffected by these changes to the extent that the product ηD_o is independent of viscosity as Eq. (25) predicts. [This law is well obeyed by the temperature dependence of the self-diffusion of bulk water (34).]

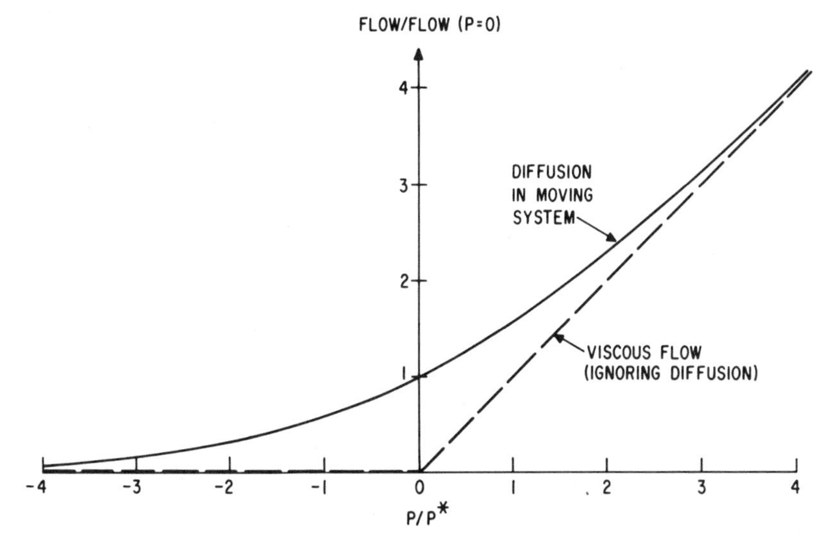

FIG. 8. Transport of a diffusing species from a region of high concentration to a region of zero concentration in a porous membrane as a function of the reduced pressure difference, P*. The transport is measured in units of the pressure-free diffusional transport. The solid curve gives the theoretical result expected for uniform pores while the dashed curve shows the result of simple viscous flow, ignoring diffusion.

Ideally then, one should find that the geometrical diameter D should equal D_D and D_F if the assumptions of normal hydrodynamic flow and normal viscosity are correct. Deviations from this equality signal the incorrectness of these assumptions.

Experiments on the diffusion and flow were performed by Bean and Doyle (52). The pores, as in the earlier experiment (42) were etched fission-fragment particle tracks in muscovite mica. The pores were collimated to less than 10° and while straight and uniform had diamond-shaped cross sections with equal sides and included angles of 65 and 115°. This diamond shape is not perfect in that there is some rounding of the corners in very fine pores. The experimental technique was to introduce a small amount of tritiated water into one compartment of a quartz cell containing quartz double-distilled water and to apply an appropriate pressure difference across the membrane separating the two stirred compartments. Periodic sampling of the initially nonradioactive compartment gave a flow rate of tritiated water. The results of two experiments are shown in Figs. 9 and 10 together with fits of the theoretical function given by Eq. (38). For

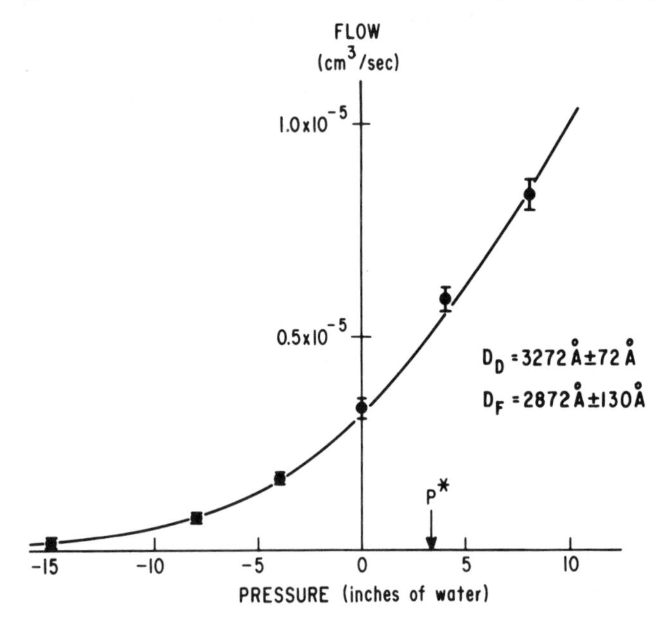

FIG. 9. Flow of water through a porous membrane as measured by radioactive tracer measurements. The total number of pores is $7.54 \cdot 10^4$ while the thickness of the mica sheet is $4.9\,\mu$. The pores were developed by etching a ^{252}Cf irradiated specimen for 600 sec in 35.7% hydrofluoric acid. The two measures of equivalent diameter for diffusion and flow are described by Eq. (39).

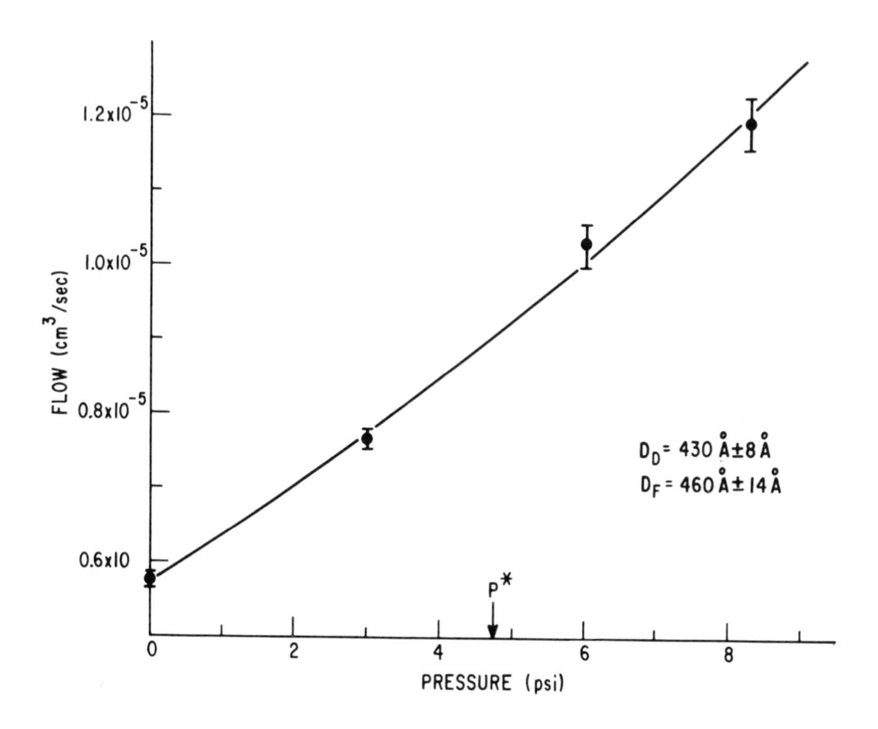

FIG. 10. Flow of water through a porous membrane as measured by radioactive tracer measurements. The total number of pores is $7.72 \cdot 10^6$ while the thickness of the mica sheet is $4.9\,\mu$. The pores were developed by etching a ^{252}Cf irradiated specimen for 100 sec in 35.7% hydrofluoric acid. The two measures of equivalent diameter for diffusion and flow are described by Eq. (39).

the larger pore of Fig. 9, both positive and negative pressures were applied, while for the smaller pores of Fig. 10 only positive or augmenting pressures were used. On the graphs are given values of D_D and D_F as deduced from Eq. (39). The errors include statistical ones as well as errors of curve fitting. A summary of the measures is shown in Table 3. Two additional measures are reported there. Replicas of the smallest pores were measured with an electron microscope to give the diameter of an equivalent area of 296 ±12Å, which agreed within experimental error to the effective diameters deduced from diffusion and flow. The comparison between diffusion diameter and flow diameter should not be exact owing to the noncircular nature of the pores. We do not have a solution for

TABLE 3

Various Measures of Pore Size, in Angstroms, for Diffusion and Flow Through Pores in Mica

Sample	1	2	3	4	5
Equivalent flow diameter, D_F (uncorrected for shape)	2870±130	1070±44	706±32	460±14	300±16
Equivalent diffusion diameter, D_D	3272±72	1280±30	734±32	430±8	294±6
Equivalent flow diameter, D_F (corrected for shape)	3156±142	1180±34	778±36	498±16	270±16
Equivalent diameter by direct observation	--	--	--	--	296±12
Equivalent diameter by resistivity	--	1072±40	--	--	--

hydrodynamic flow for the diamond shape but reasoning from the results for an elliptical cross section given by Eq. (20), the flow diameter should be increased about 10% for comparison with the diffusion diameter. A last measure was obtained for sample 2 by adding KCl to make a 0.01 N solution. The intent was to see if either the flow or diffusion was affected. They were not. However, the resistance was measured and converted to the equivalent pore diameter shown. All in all the results, summarized in Fig. 11, are consistent with normal hydrodynamics and, for the smallest pore, imply that any rigid layer must be 10 Å or less in size. Thus there is no evidence for the long range ordering that has been so often invoked. It is interesting that Mysels (53), in a study of the draining of soap films, came to the same conclusion that any rigid layer must be less than 10 Å in thickness.

The experimental results discussed above utilize a knowledge of the number of pores, the membrane thickness, and the fact that the pores were normal to the membrane. In principle, given normal hydrodynamics, one should be able to calculate a measure of the pore size by comparisons of transport or structural properties that depend in different ways on the pore

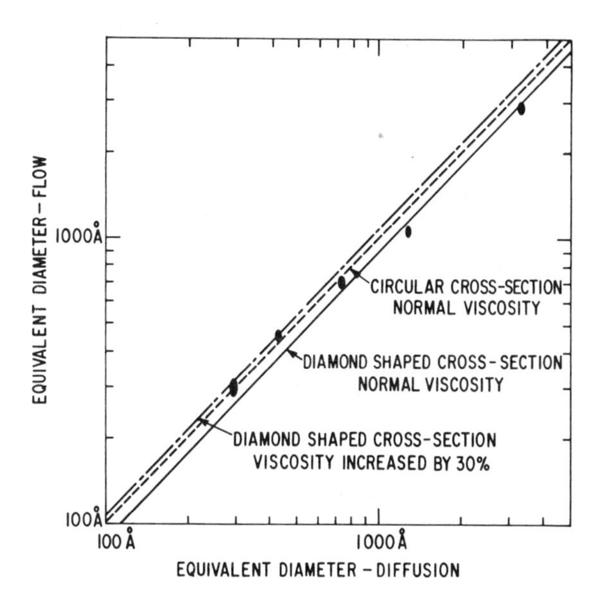

FIG. 11. A plot of equivalent diameters for flow and diffusion for five samples of porous mica. The size of the sample points shows estimated errors. The plotted lines are those theoretically expected under the assumptions noted on the figure.

diameter. In the earliest of these techniques, Ferry (5) compared the volume fraction of water in a membrane and the hydraulic permeability. For a model of straight through cylindrical pores, say N per square centimeter

$$\phi_w = N\pi D^2/4 \tag{40}$$

while the hydraulic permeability in cgs units is given by

$$L_p = N\pi D^4/128\eta L \tag{41}$$

and hence, for this case

$$D = (32\eta L L_p/\phi_w)^{\frac{1}{2}} \tag{42}$$

If the structure were one of randomly oriented pores with a small volume fraction, the coefficient would be 96 instead of 32 (54). Variations from uniformity and from cylindrical shape would produce variations in the meaning of D. The availability of isotopes of water in the postwar years allowed a more significant experiment to be performed in comparing the diffusive and osmotic or pressure coefficients. This approach was initiated by Pappenheimer, Renkin, and Borrero (55) and Koefoed-Johnsen and Ussing (56) with important contributions by many others (57-59). If we consider a system of straight-through pores, then by Eqs. (8) and (33), in cgs units

$$\omega_{HTO} = N\pi D^2 D_0/4LRT \tag{43}$$

While the hydraulic coefficient for large pores is given by Eq. (41), we must consider the fact that water may diffuse as well as flow down the chemical potential gradient caused by physical or osmotic pressure differences across the pores. Equation (35) shows the total flow for $v \ll v^*$ to be the sum of two terms. The driving force in the first term can be generalized to a chemical potential difference that includes the pressure and hence can exist even when, as in a simple flow experiment, C(1) is equal to C(2). The result of this inclusion is

$$L_p = N\pi D^4/128\eta L + N\pi D^2 D_0 \bar{V}_w/4RLT. \tag{44}$$

A dimensionless ratio of the flow and diffusive permeabilities may be formed.

$$g = L_p/\bar{V}_w \omega_{HTO} = 1 + D^2 RT/32\eta D_{HTO} \bar{V}_w \tag{45}$$

or by rearrangement

$$D = D^*(g-1)^{\frac{1}{2}} \tag{46}$$

Where D^* is a diameter related to that of the water molecule, i.e.,

$$D^* = (32\eta D_{HTO} \bar{V}_w/RT)^{\frac{1}{2}} \tag{47}$$

and substituting the 300°K values for the constants in Eq. (47)

$$D^* = 7.1\ \text{Å} \tag{48}$$

Thus, as Carl Wagner observed (60), the diffusive and viscous flows are about equal for a pore of twice the molecular diameter. Equation (46) has a somewhat wider range of applicability than its derivation suggests. First of all, the assumption of straight-through pores is not essential; pores at various angles with small amounts of intersection will not alter g. Secondly, any alteration of viscosity from its normal value will tend by Walden's law to keep the product $\eta D_{HTO}/T$ unaffected (34) and hence preserve the validity of Eq. (46). Lastly the limit of D = 0 or $L_p = \omega_{HTO}\overline{V}_w$ corresponds to what one would expect for diffusion of isolated molecules through a matrix. The main difficulty in interpretation concerns the effects of a nonuniform structure and of sizes comparable to the water molecule. We approach the former question in Part II on complex structures and merely note here that for a system of parallel pores of different diameters

$$D^2 = \overline{D^4}/\overline{D^2} \tag{49}$$

where the bars refer to number averages. The question of tight or single file pores is discussed in Section III. 6.

A few experimental results may be quoted to give a feeling for the magnitudes and consistency of these measurements. From the results of Kauffman and Leonard (17) on cellophane given in Tables 1 and 2, the diameters inferred from Eq. (42) or its modification are 24-42 Å. Cellophane has been studied by many workers; Renkin (57) and Durbin obtained values of 62 and 82 Å, respectively. Thau, Bloch and Kedem (61) obtained a g of 80 or an equivalent diameter of 63 Å. On the other extreme, bilayer lipid membranes have been shown by Cass and Finkelstein (62) to have a value of g = 1.00 ±0.05, or to correspond to a solution-diffusion model. [The fact that unstirred layers are important, particularly for the evaluation of ω_{HTO}, is shown by the experience of Huang and Thompson (63) who found values of g ranging from 4 to about 25. Cass and Finklestein attribute this to inadequate stirring in the diffusion experiment and could reproduce these effects with a thick supporting septum of the type employed by Huang and Thompson.] Very recently, Holtz and Finkelstein (64) have observed that certain polyene antibiotics (nystatin and amphotericin B) markedly increase the permeability of lipid films to ions, water, and nonelectrolytes. For a standard condition nystatin gives a g of 3.8 (12 Å by Eq. (46) and for amphotericin B, a value of g = 3.5 (11 Å). Diffusion measurements showed that glucose (~7 Å) was completely rejected. This discovery is exciting in that the simple interpretation of identical pores is quite convincing and the system will allow experimentation on known fine pores.

It must be admitted, however, that the experimental situation on nonporous membranes is not completely clear in that Thau, Bloch, and Kedem (61) found g = 1.8 for tributyl phosphate and Lakshminarayanaiah and White (65) found gs of 5.8 to 8.4 for stirred hydrocarbon layers, while general physical reasoning calls for g = 1 in dilute nonporous systems.

E. Particle and Molecular Transport

The results of earlier sections may be used to calculate, in the hydrodynamic approximation, the transport of spheres through cylindrical pores.

The membrane model employed is that of a solid phase pierced by N identical pores per unit area. The pores, of diameter D, are assumed to traverse straight through the thickness L of the solid phase. This model is almost never fulfilled in practice; one exception is the case of etched-particle tracks while another may be that of the bilayer membranes treated with nystatin or amphotericin B (64). This model, as has been recognized many times, has the virtues of being definite and of having the capacity for some generalization. In addition, the spheres (of diameter d) are assumed to move in a hydrodynamic continuum characterized by a viscosity η. The influence of the finite-size water molecules, in an extreme case, will be treated in Sec. III.6.

The program is: to calculate--as closely as possible--the transport coefficient L_p, ω, and σ for the ideal system described above; to indicate the generality of the results; and to make a comparison with experimental results. The most direct method of making the calculation is to evaluate the frictional coefficients of Eq. (10) and to use Eq. (11) to transcribe these to transport coefficients.

The hydraulic permeability as given in the preceding section is in cgs units

$$L_p = N\pi D^4/128\eta L \tag{50}$$

In this formula the sphere or solute concentration is assumed to be so low that the bulk viscosity is unaffected and, in addition, that adsorption on the pore walls does not reduce the effective diameter. The full expression of Eq. (44) is not employed inasmuch as this expression includes a measure of the solvent size.

Equation (21) gives an expression for the force on a single sphere on the pore axis. This can be generalized to a molar force as a function of radial position (again in cgs units) with N_A denoting Avogadro's number

$$X_s(r) = \left[\frac{3\pi N_A \eta d}{F(d/D,r)} v_s - \frac{3\pi N_A \eta d G(d/D,r)}{F(d/D,r)} v_w(r)\right] \tag{51}$$

The molecular force that appears in Eq. (10) is the average of this force averaged over all values of $v_w(r)$ and available positions for the particles. From the first term within the square brackets of Eq. (10) we may write

$$f_{sw} + f_{sm} = 3\pi N_A \eta d \int_0^{(D-d)/2} (1/F(d/D,r))2\pi r dr \bigg/ \int_0^{(D-d)/2} 2\pi r dr \tag{52}$$

The integrals imply that it is equally probable to find the center of a sphere at any radial position in the pore except with a distance of the sphere radius from the pore wall. To use Eq. (11) to determine the solute permeability coefficient ω, we note that the distribution coefficient K_D is less than the volume fraction of water, owing to the finite size of solute particles. Simple geometrical considerations give

$$K_D = (N\pi D^2/4)(1-d/D)^2 \tag{53}$$

where the first term is the volume fraction of water as given by Eq. (40). Putting Eqs. (11), (52), and (53) together gives

$$\omega = (ND^2/12N_A \eta dL)(1-d/D)^2 (D_{eff}/D_o) \tag{54}$$

where

$$D_{eff}/D_o = \int_0^{(D-d)/2} 2\pi r dr \Big/ \int_0^{(D-d)/2} (1/F(d/D, r)) 2\pi r dr \tag{55}$$

The terms in Eq. (54) have a straight forward interpretation. The first is simple Stokes-Einstein diffusion for very large pores. The second accounts for steric restrictions while the third contains all the hydrodynamic interaction between the sphere and pore. The steric correction was first made by Ferry (65) while the hydrodynamic corrections have been applied more recently (55, 57) but in a somewhat incomplete form. In these and all subsequent treatments, the expression for D_{eff}/D_o is replaced by various approximations to F(d/D, 0). In this approximation we may write

$$\omega/\omega_o \doteq (1-d/D)^2 F(d/D) \tag{56}$$

where ω_o is the solute permeability without interaction effects. This expression may be expanded to give

$$\omega/\omega_o \doteq 1 - 4.1044(d/D) + 5.2089(d/D)^2$$
$$- 0.1566(d/D)^3 - 4.1775(d/D)^4 + 1.406(d/D)^5 \tag{57}$$
$$+ 0.524(d/D)^6 +$$

Figure 12 plots this function as dashed curve b, while curve a represents the steric interaction alone. Curve c is an approximation to the correct curve by numerically performing the integration of Eq. (55) using the function of Fig. 4. The inhibition to diffusion is significantly larger than that of the usual formula [Eq. (57)].

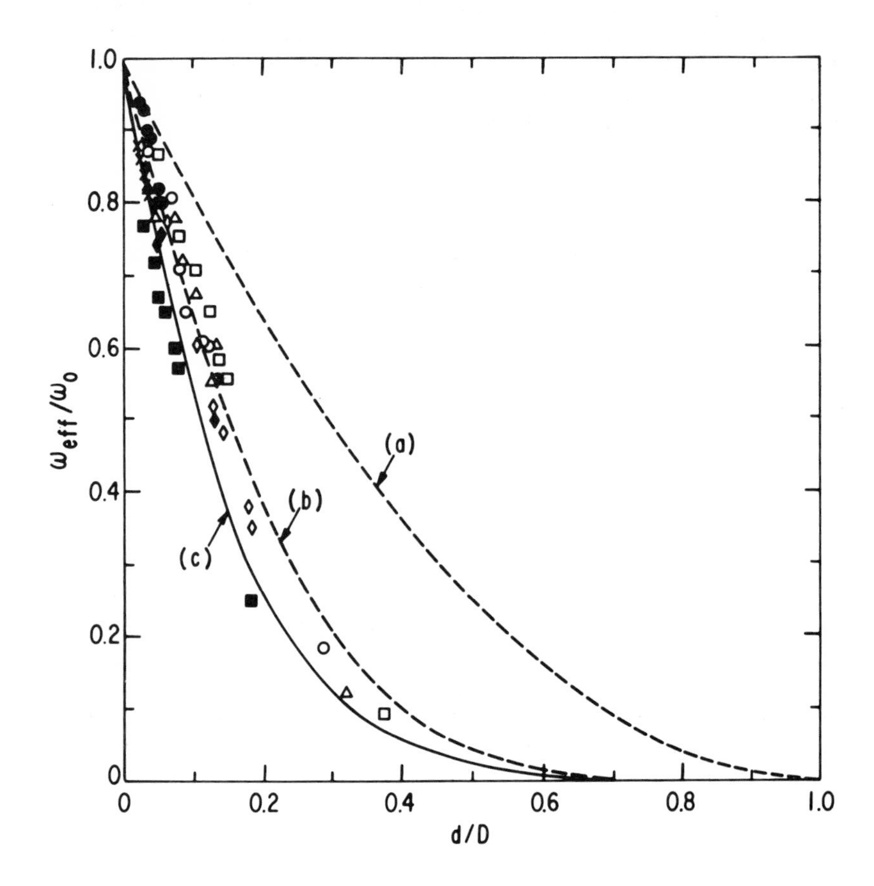

FIG. 12. Restricted diffusion as a function of the ratio of sphere to pore size. This plots the diffusion per unit pore area relative to that for a large pore under the assumptions: (a) steric hindrance alone, (b) steric hindrance and hydrodynamic hindrance with the hydrodynamic hindrance calculated from its axial value, and (c) steric hindrance and hydrodynamic hindrance with the radial variation of hydrodynamic hindrance included. The last curve is approximate. The experimental points are from Beck and Schultz [R. E. Beck and J. S. Schultz, Science, 170, 1302 (1970)]. The diffusing species are various small nonelectrolytes while the pores are etched particle tracks in mica.

To derive an expression for the reflection coefficient σ, in the same manner as that for ω, one must consider very carefully the averaging process by which transcription is made from the local forces of Eq. (51) to the average force of Eq. (10). It is more direct for this case to calculate the coefficient L_{Dp} of Eq. (3) and use Eq. (7) to derive σ. Consider a pore with equal concentrations of spheres on either side sustaining a pressure difference. For each sphere on the inside the net force is zero; hence by a generalization of Eq. (21)

$$v_s(r) = v_w(r)G(d/D, r) \tag{58}$$

where $v_s(r)$ and $v_w(r)$ are the solute velocity and fluid velocity far from the sphere, respectively. The total flux of solute is

$$J_s = NC_s\int_0^{(D-d)/2} v_s(r)2\pi r dr = NC_s\int_0^{(D-d)/2} v_w(r)G(d/D, r)2\pi r dr \tag{59}$$

hence

$$J_D\Big|_{\Delta\pi=0} = N\int_0^{(D-d)/2} v_w(r)G(d/D, r)2\pi r dr - N\int_0^{D/2} v_w(r)2\pi r dr \tag{60}$$

and

$$J_v\Big|_{\Delta\pi=0} = N\int_0^{D/2} v_w(r)2\pi r dr + NC_s\overline{V}_s\int_0^{(D-d)/2} v_w(r)G(d/D, r)2\pi r dr \tag{61}$$

By Eqs. (3) and (7)

$$\sigma = -J_D/J_v$$

$$\doteq 1 - \int_0^{(D-d)/2} v_w(r)G(d/D, r)2\pi r dr \Big/ \int_0^{D/2} v_w(r)2\pi r dr \tag{62}$$

in the limit that $C_s\overline{V}_s \ll 1$. The function $G(d/D, r)$ is not known off the axis but it is probably very uniform. [Unpublished experiments in collaboration with R. W. DeBlois and D. C. Golibersuch on the passage of single polystyrene spheres through single fine pores (66) lead to this conclusion.] In this limit we can write, using Eq. (18)

$$\sigma = 1 - G(d/D)(1-4(d/D)^2 + 4(d/D)^3 - (d/D)^4) \tag{63}$$

The last term in this equation was first derived by Ferry (67) on the basis of steric restriction with Poiseuille flow. Following Renkin (57), all workers have used a form essentially equivalent to Eq. (63) except that G(d/D) is replaced by F(d/D). This substitution was made owing to the lack of detailed understanding of the hydrodynamic situation and has a profound effect on the interpretation of data. Figure 13 shows the various approximations. Curve (a) is the steric factor alone while curve (b) is the usually employed expression and curve (c) is the result of Eq. (63). A prime difference between (b) and (c) is their linear and quadradic forms near the origin. Kedem and Katchalsky (13) and Dainty and Ginzburg (68) have pointed out that for a solution-diffusion model of independent motion of solute and solvent

$$\sigma_{s-d} = \omega \bar{V}_s / L_p \tag{64}$$

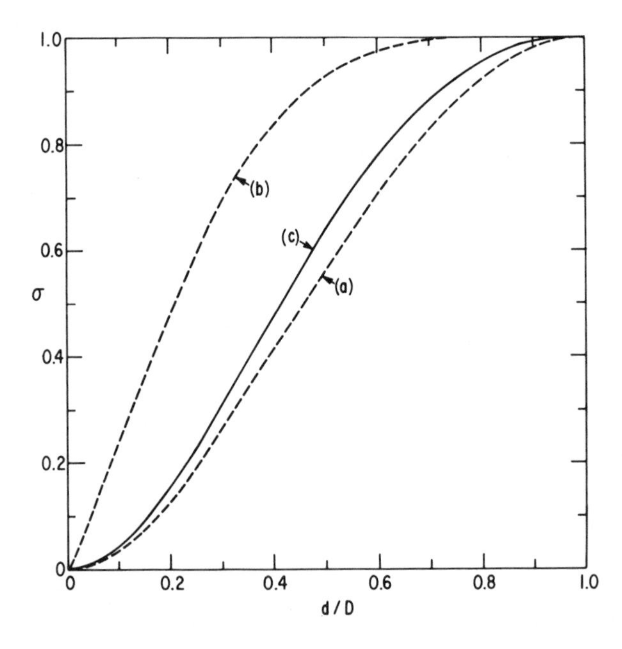

FIG. 13. Reflection coefficient in a hydrodynamic continuum as a function of reduced particle size. The theoretical calculations are for spherical particles of diameter d in a cylindrical pore of diameter D. The dashed curve (a) contains only the steric factor while the solid curve (c) includes the hydrodynamic correction of Eq. (63). The dashed curve (b) contains the incorrect hydrodynamic correction F(d/D).

This result may be obtained from the first part of Eq. (11) by setting f_{sw} equal to zero or by direct calculation. A further statement is that for any pore model, σ must be greater than σ_{s-d}. Since for the pore model here (with $\overline{V}_s = N_A \pi d^3/6$)

$$\omega\overline{V}_s/L_P = (16/9)(d/D)^2(1-d/D)^2 F(d/D) \tag{65}$$

in the approximation of ignoring the radial dependence of F(d/D). Table 4 plots values of $\omega\overline{V}_s/L_p$ and σ to show that this inequality obtains.

The results given above, particularly in the dimensionless quantities of σ, $\omega\overline{V}_s/L_p$, and K_D/Φ_w, have a generality beyond their derivation. The pores need not be straight, of the same length, or completely without intersections to have these relations hold. In that spirit we can look at the experimental results to see the areas of general agreement and conflict. One general result is that for a membrane of constant structure all the quantities above should be independent of temperature. Referring to Kaufmann and Leonard's work on cellophane reported in Tables 1 and 2 we see that

TABLE 4

Parameters for a Cylindrical Pore Model Calculated with Neglect of Radial Dependence of F(d/D) and G(d/D)

d/D	$\omega\overline{V}_s/L_p$	σ
0	0	0
0.1	0.0114	0.0425
0.2	0.0271	0.153
0.3	0.3331	0.305
0.4	0.0287	0.473
0.5	0.0190	0.638
0.6	0.0095	0.779
0.7	0.0023	0.884
0.8	--	0.948
0.9	--	0.982
1.0	--	1.000

ϕ_W is temperature independent-implying a constant structure. The reflection coefficients, however, show a distinct temperature dependence, as do the other quantities. The clearest qualitative disagreement is seen in the comparison of K_D and ϕ_W. By the assumptions that led to Eq. (53), K_D must be smaller than ϕ_W owing to steric exclusion. The fact that Table 2 shows it to be greater, with a difference that increases with lowered temperature and increased molecular weight, argues strongly in favor of significant adsorption of the carbohydrates on the cellophane surfaces. Such an effect, as noted by Dainty and Ginzburg (68), can cause significant changes in inferred parameters. To take a more quantitiative confrontation, let us examine Durbin's data on reflection of various nonelectrolytes by three porous membranes: dialysis, cellophane, and wet gel. To make a comparison it is necessary to have a scale of molecular sizes. Table 5 gives various estimates of molecular size: from density (69), assuming close packing of spheres; from diffusion data (69) and application of the Stokes-Einstein relationship [Eq. (25)]; and from measurement of models (70) and application (70) of Einstein's viscosity formula [Eq. (23)]. The last column is a list of rather arbitrarily chosen diameters intended to give a fair measure of molecular size. Figure 14 shows a best fit of Durbin's data to the theoretical expression of Eq. (63) with pore diameter as a fitted parameter. The value of 73 Å agrees fairly well with the 82 Å from water diffusion and flow quoted in Section III.D. (The cellophane of Kauffman and Leonard would be inferred to have a diameter of about 55 Å.) The diffusion and flow values for dialysis and wet gel were 33 and 118 Å respectively. Most importantly, this figure seems to show clearly that the limiting behavior for small particles is the square law of Eq. (63) rather than the linear relationship implicit in the older formulation. As a last mode of

TABLE 5

Molecular Diameters in Angstroms

	Density (69)	Stokes-Einstein (69)	Models (70)	Viscosity (70)	Chosen
Urea	4.7	3.6	4.6	3.6	3.9
Glucose	6.5	7.1	7.0	--	7.0
Sucrose	8.0	9.3	9.0	10.4	9.0
Raffinose	9.8	11.2	--	12.0	11.0
Albumin (bovine serum)	--	74	--	--	74

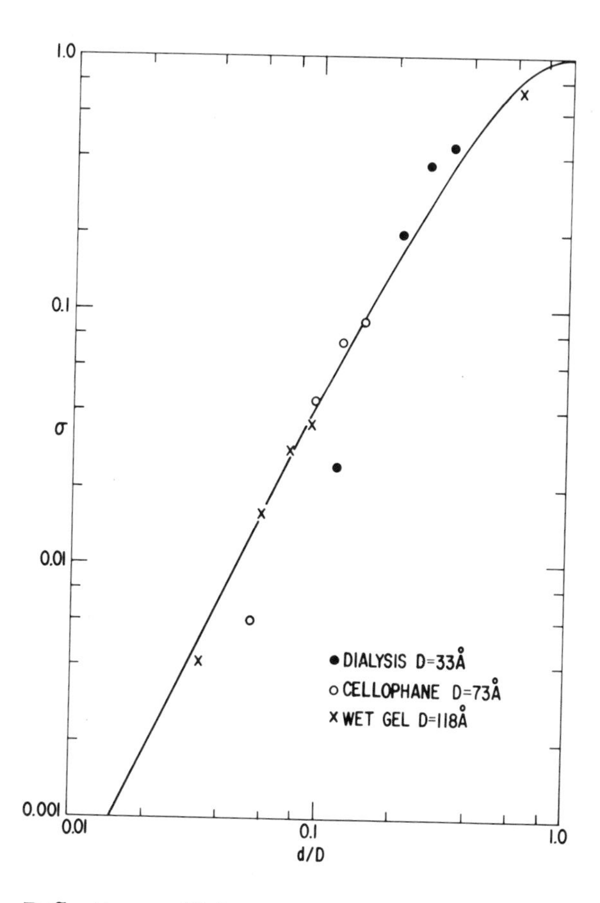

FIG. 14. Reflection coefficient as a function of reduced particle size. The solid curve is theoretical [Eq. (63)], while the points are taken from the work of Durbin [R. P. Durbin, J. Gen. Physiol. 44, 315 (1960)]. The membranes are dialysis tubing, cellophane, and wet gel while the molecules are those listed in Table 5 with the assumed diameters there indicated. The best fits are obtained by assuming diameters of 33, 73 and 118 Å for the three membranes.

comparison one can use Table 4 to construct a dimensionless graph of $\omega\overline{V}_s/L_p$ against σ and not require estimates of d. Figure 15 is such a plot. The circular-pore membrane theory forms one locus of points while the solution-diffusion model forms another. The experimental points are seen to form a band roughly paralleling the prediction of the cylindrical pore model but about a factor of two higher, and in all quoted cases, about a

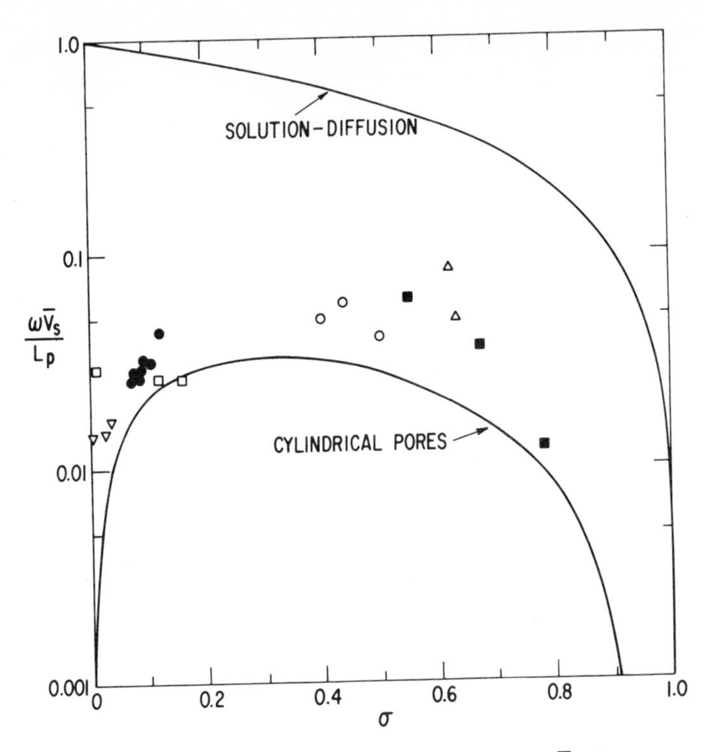

FIG. 15. A plot of the dimensionless parameters $\omega\bar{V}_S/L_p$ and σ. The solid curves are theoretical for a pure solution-diffusion model and for a uniform cylindrical pore model in the hydrodynamic approximation. The experimental points are: ● cellophane (17), o nitella (71), △ human red blood cells (72), □Visking dialysis tubing (73), ▽ DuPont wet gel (73), and ■ nystatin modified lipid membranes (64).

factor of ten below that predicted for the solution-diffusion model. It is too early to state the significance or guess at the generality of this result. Whether surface adsorption, variations of pore shape, size, and connectivity will explain these results must await further experiment and theory. It does seem clear, however, that a pore model will serve as a point of departure for future understanding.

F. Ionic Transport Through Neutral Pores (52)

If a pore of no surface charge in an electrically insulating material is filled and in equilibrium with an ionic solution, one would expect that its conductance could be calculated from the conductivity σ_0 and geometry of the pore; i.e., for a long pore of length L and diameter D (i.e., L>>D) this naive expectation by Ohm's Law gives for the conductance, R^{-1}

$$R^{-1} = \sigma_0 \pi D^2 / 4L \tag{66}$$

To the extent that the pore is not infinitely long, one can use an effectively pore length L* given by (74)

$$L^* = L + 0.805D \tag{67}$$

Similarly one would expect that the application of a pressure difference across such a pore or system of pores separating two identical solutions would cause a flow in which the effluent concentration would be equal to the bathing concentrations and hence that no straining of the ions would occur. This conclusion, of course, is modified by the knowledge that ions have finite sizes and that with very small pores the considerations of the previous section would become important. This section concentrates, however, on the consequences of the ions having a charge and the effects of the long range coulomb potential.

The basic equation for calculating the energy of an ion in various surroundings is that of the energy in the electrostatic field, namely

$$w = \int (\mathcal{E} . \mathcal{D} / 8\pi) dv \tag{68}$$

where w is the energy in ergs per square centimeter, $\mathcal{E}$ is the electrostatic field in statvolts per centimeter, $\mathcal{D}$ is the electrical displacement in electrostatic units, and dv is an element of volume. The integration is overall space external to the charges that give rise to the field. In a medium whose response to an electric field can be characterized by an isotropic dielectric constant ϵ

$$\mathcal{D} = \epsilon \, \mathcal{E} \tag{69}$$

For an ion of diameter d and charge q, in an infinite medium of dielectric constant ϵ_1,

$$w = q^2 / \epsilon_1 d \tag{70}$$

It is the lowering of this energy (the solvation energy) when an ion is transferred from vacuum to water, i.e., for $\epsilon_1 = 1$ to $\epsilon_1 = 78.5$ (25°C), which accounts for the high degree of solubility of most ions in water (75). There are questions, of course, concerning the exact meaning of the ionic diameter d and the extent to which the dielectric constant ϵ_1 describes the dielectric response near the ion. If we consider a heterogeneous medium that consists of a spherical cavity of diameter D filled with a medium of dielectric constant ϵ_1 and with an infinite medium of dielectric constant ϵ_2 outside, the energy of a charge q at the center is from Eq. (68)

$$w = q^2/\epsilon_1 d + (q^2/D)(1/\epsilon_2 - 1/\epsilon_1) \tag{71}$$

If ϵ_2 is less than ϵ_1 there is an energy increase on moving an ion from the infinite medium to the center of the cavity of

$$\Delta w = (q^2/D)(1/\epsilon_2 - 1/\epsilon_1) \tag{72}$$

$$= (q^2/\epsilon_2 D) \cdot P(\epsilon_2/\epsilon_1)$$

where $P(\epsilon_2/\epsilon_1) = 1 - \epsilon_2/\epsilon_1$. We shall see that the latter equation (72), is a general form for all geometries with an appropriate $P(\epsilon_2/\epsilon_1)$ for each geometry. We note that the uncertainties of the self-energy term [Eq. (70)] do not appear in this difference of energy. For dilute solutions the ratio of the concentration of ions C, at the center of such a cavity to the concentration in the medium C_0 is

$$C/C_0 = \exp(-\Delta w/kT) \tag{73}$$

by the well known Boltzmann factor. In this expression, k is Boltzmann's constant ($1.38 \cdot 10^{-16}$ erg deg^{-1}) while T is the absolute temperature. This expression is, of course, valid for any Δw. As noted by Wagner (76) in his pioneering work concerning ion concentration near air interfaces, for a charge a distance x from a plane interface separating regions of dielectric constant ϵ_1 and ϵ_2, the method of images gives

$$\Delta w = q^2[(\epsilon_1 - \epsilon_2)/(\epsilon_1 + \epsilon_2)]/4\epsilon_1 x \tag{74}$$

For this case the $P(\epsilon_2/\epsilon_1)$ equivalent to that above is $P(\epsilon_2/\epsilon_1) = (\epsilon_2/\epsilon_1)(1 - \epsilon_2/\epsilon_1)/2(1 + \epsilon_2/\epsilon_1)$. Recently Parsegian (77) has calculated the equivalent function for an ion on the axis of a long circular pore. This calculation, based on Eq. (68), is numerical with no closed form available. The results of these calculations are shown in Fig. 16. It can be noted that the pore is much more effective than the plane interface although, of course, it is less effective than the spherical cavity. Figure 17 shows a calculation for 25°C of C/C_0 for dilute solutions of various electrolytes as a function of pore size for pores in material of $\epsilon_2 = 3$ and $\epsilon_2 = 10$. These two values are taken as typical for plastics and ionic solids, respectively. The symbols 1-1, 2-1, etc., stand for the degrees of ionization of the components of the solution. As expected, the higher charges are excluded more than the smaller by a large factor, since the exponent in Eq. (73) is proportional to the square of the charge. (With gross electroneutrality this factor

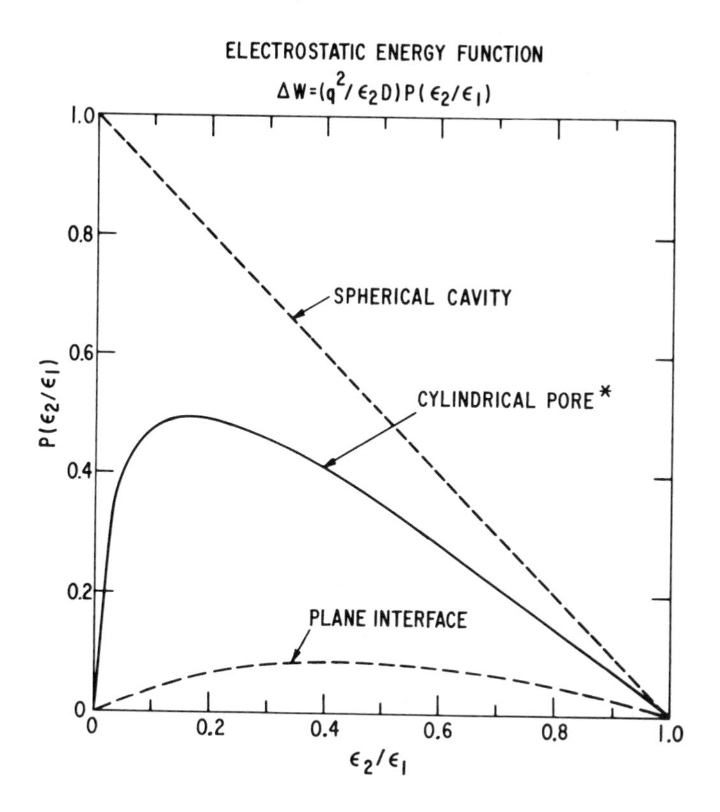

FIG. 16. Plot of electrostatic energy function for various geometries. The graph shows the energy required to move a charge q from an infinite medium of dielectric constant ϵ_1, to a specific distance D/2 from a matrix of lower dielectric constant ϵ_2, in the geometries noted. The first two cases are those of an ion at the center of a spherical cavity and long cylindrical pore of diameter D while the third is that of an ion at a distance D/2 from a plane interface. The calculation for the cylindrical pore is by Parsegian (77).

becomes q+ · q-, where q+ and q- are the charges of positive and negative ions.)

In the same spirit that we earlier used the Faxén correction as applying to the entire cross section of the pore although it was calculated for axial motion, we may use the concentration correction described above to derive an upper limit for the conductance of a neutral pore in a dilute solution. We may express this in terms of an effective pore diameter which, when used with Eq. (66), and of the conductivity of the bathing solution would give the

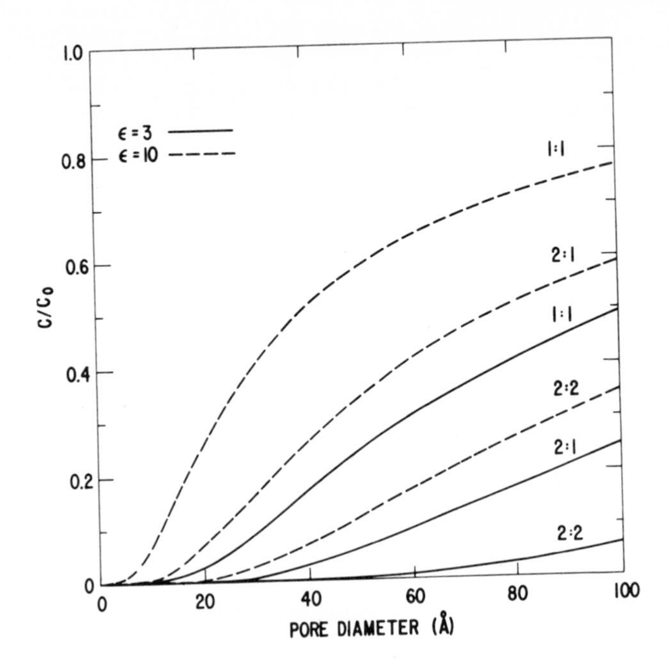

FIG. 17. Plot of the relative concentration within a neutral pore for dilute salt solutions at 25°C. The dashed curves are for a matrix dielectric constant of 10 while the solid ones assume a matrix dielectric constant of 3. The symbols beside each curve indicate the degree of ionization of the ionic components. These curves represent lower limits since the relevant energies are calculated on the axis of the pore and steric effects are neglected.

observed pore conductance. Figure 18 shows the effective diameters for neutral pores in mica [$\epsilon_2 = 7.4$, $P(\epsilon_2/\epsilon_1) = 0.46$] as a function of the "true" pore size. The dashed curve represents equivalence while the line marked 0-0 is the Ferry-Faxén correction for ions whose diameter is taken arbitrarily to be 3 Å. The other lines represent various ionic combinations. A principal conclusion from this figure is that the electrostatic correction is much greater than the effect of finite ion size.

The fact that the ionic concentration is less within the pores than in the bathing solution leads to the fact that they will show a nonzero reflection coefficient for salt solution filtration. This is one mechanism for reverse osmosis purification of salt solutions. Other mechanisms that have been proposed include preferential solution and diffusion of water in a homogeneous membrane phase and the effects of small charged pores (78).

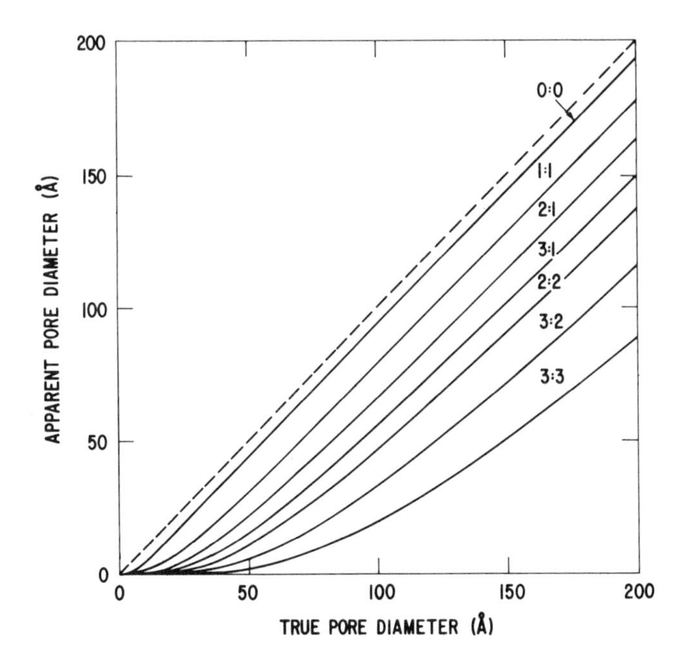

FIG. 18. Relationship between the apparent diameter of a neutral pore in mica as deduced from resistivity measurements and the true diameter. This curve includes both the Ferry-Faxén correction (0:0) and that for various charge states. Its validity is limited to dilute solutions.

The mechanism at hand was postulated, in recent times, by Sourirajan (79) to account for the properties of cellulose acetate. He put a critical diameter at about 8Å. Glueckauf (80) made the first serious attempt to calculate the rejection of neutral pores although his work predated Parsegian's calculation. In simplified form, if one ignores the effects of finite ion size by Eq. (62), the reflection coefficient σ is given by

$$\sigma = 1 - \bar{C}/C_0 \tag{75}$$

where $\bar{C}$ is the average concentration, with due account of flow profile, within the pore. If we equate $\bar{C}$ with the C calculated above

$$\sigma \doteq 1 - \exp(-D^*/D) \tag{76}$$

where

$$D^* = q^2 P(\epsilon_2/\epsilon_1)/\epsilon_2 kT \tag{77}$$

Sample calculations of this reflection coefficient are shown in Fig. 19. In a reverse osmosis experiment one has input and efflux concentrations that differ so that diffusion and pressure-induced flow compete in regulating the efflux concentration as in the treatment of Sec. III.D. In addition, there is an osmotic pressure difference. These items can be calculated on the principles outlined here and rejection as a function of pressure difference calculated for various pore sizes. The results of such a calculation (52) is shown in Fig. 20 together with experimental results of Breton (81) on a 40%-acetyl cellulose-acetate membrane. The comparison suggests that cellulose acetate may derive its properties from pores about 10 Å in diameter. While this is a somewhat unconventional view, Meares (82) has attributed most of the water flow to the flow through pores and the postulated size is consistent with a comparison of osmotic and diffusive flow of water carried out by Thau, Bloch, and Kedem (61) which gave $g = 2.5$, or by Eq. (46) an inferred diameter of 8.7 Å.

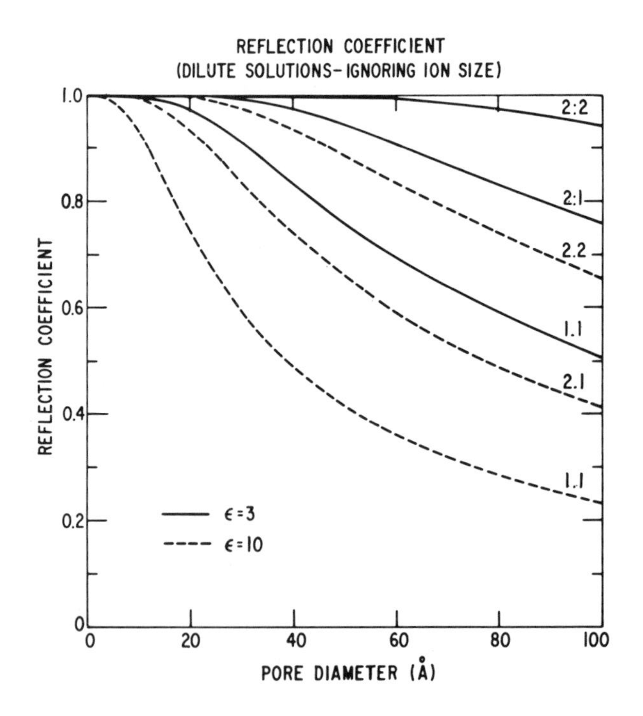

FIG. 19. Plot of the intrinsic rejection at room temperature of various types of salt solutions by neutral pores in matrices of dielectric constant 3 and 10.

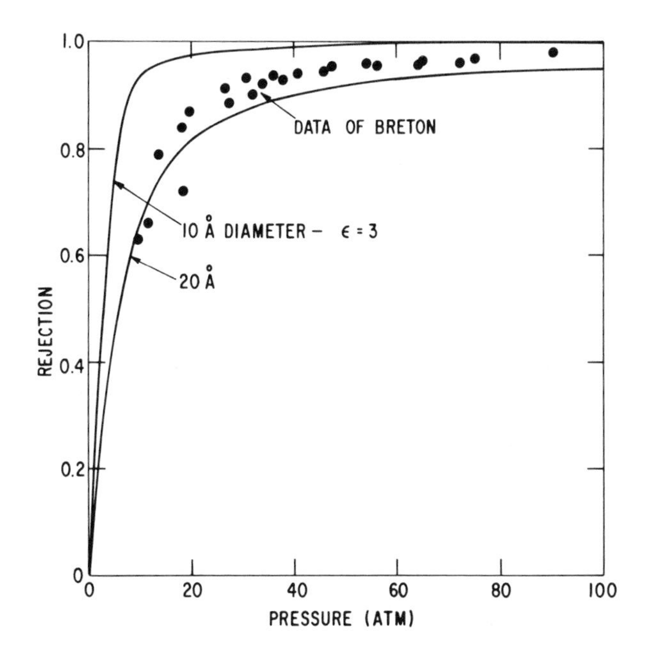

FIG. 20. Theoretical plot of the course of rejection as a function of pressure for 0.1 N NaCl for pore sizes of 10 and 20 Å in a matrix of dielectric constant 3. The data of Breton (81) on cellulose acetate is plotted for comparison.

There are two points of caution about the application of these ideas. First, the calculations apply to an isolated pore. High densities of pores would reduce the effect in that the effective dielectric constant of the matrix would be between that of the matrix material proper and that of the solution. Secondly, high salt concentrations will reduce the effects by electrostatic shielding.

G. Fine Pores

In this section we concentrate on the effects of pore sizes that are comparable to the size of the solvent molecule. The understanding in this area is not deep owing to the lack of a basic knowledge of the properties of strongly interacting systems with a finite number of elements and the lack of experiments on completely characterized systems in this size range. The importance of study in this regime is indicated by the fact that equivalent pore diameters in biological membranes are generally inferred to be two or four times (6) the diameter associated with the water molecule and, as pointed out in the previous section, important technical advances in reverse osmosis membranes may be associated with an understanding and creation of pores in this size range. Lastly, many ion exchange membranes have water channels in this general range (83).

We have introduced the topic briefly in Sec. III.D. where flow under an osmotic pressure gradient was assumed to have a diffusional component. This led to a transition from osmotic to diffusional flow for channels of about 7 Å in diameter, or about twice the solvent diameter. The argument employed was, in the main, one of plausibility. Longuet-Higgins and Austin (84) have treated the statistical mechanics of transport of solvent in a single file pore under an osmotic or pressure gradient and derive a flux corresponding to a hydraulic permeability coefficient of

$$L_p = NmD_o \bar{V}_w^2 / L^2 N_A RT \tag{78}$$

where m is the number of solvent molecules in the pore, D_o is a diffusion constant appropriate to their motion, and the other symbols have their usual meaning. If m is, for a filled pore, approximated by

$$m = \pi D^2 N_A / \bar{V}_w \tag{79}$$

the second term of Eq. (44) is recovered. Thus the naive diffusion model and this more carefully conceived model agree in the limit of filled pores. Longuet-Higgins and Austin argue that the diffusion constant D_o will be close to that of the free solution in that bonding forces must be similar. They apparently believe that diffusion is an alternative to viscous flow rather than having total flow with viscous and diffusive components whose relative magnitudes are a function of solvent to pore diameter.

Paganelli and Solomon (85), in their important study of tracer diffusion of water into the human red cell, considered its magnitude in comparison to osmotic flow and made modifications to the simple equations for diffusion and flow to account for the finite size of the water molecule. They employed the equations Renkin (57) set down for the restricted diffusion and flow of solute molecules and boldly applied them to the solvent itself. Specifically, they assumed that diffusion was reduced by a factor of

$$(1 - d_w/D)^2 \; F(d_w/D) \tag{80}$$

from that expected from simple geometrical considerations. (In this equation d_W is the diameter of the water molecule and D is the diameter of the pore.) In a similar manner they proposed that the flow was reduced by

$$(1 - 4(d_W/D)^2 + 4(d_W/D)^3 - (d_W/D)^4)\, F(d_W/D) \tag{81}$$

These equations pose some problems in interpretation. With regard to the first factor of Eq. (80), it merely states that molecular centers cannot approach more closely than their radius to the wall and is valid in the limit that $D \gg d_W$. The second is more obscure. Derived for dilute spheres in a continuum, its application to dense solutions of identical spheres can have at best only heuristic significance. It would seem reasonable, as a first approximation to dispense with this factor. The correction to hydrodynamic flow has even more difficulties. The first term corresponds to hard spheres moving with a parabolic velocity profile across the pore. The first layer of spheres, for instance, moves with a finite velocity. Again this is not unreasonable, although the details of the solvent-solvent and solvent-wall interaction will cause alterations in this expectation. The assumption of a F(d/D) in hydraulic flow as noted in Sec. III.D. is incorrect and G(d/D) should be used for consistency. Again, however, it is not clear that any factor should be employed. As one application of Eqs. (80) and (81), Paganelli and Solomon consider the alterations of Eq. (46) for finite size solvent molecules. Fortunately, here one takes the ratio of Eqs. (80) and (81) and the $F(d_W/D)$ factors cancel so that the same result would be found were they neglected. The result of this correction is to obtain a corrected value for pore size from Eq. (46) of

$$D = -d_W + (D^2_{eff} + 2d^2_W)^{\frac{1}{2}} \tag{82}$$

where D_{eff} is the value obtained from Eq. (46). When these factors do not cancel as in the interpretation of reflection coefficients where Rich et al. (86) used

$$\sigma = 1 - \frac{\omega_s \bar{V}_s}{L_p} - \frac{[(1-4(d_s/D)^2 + 4(d_s/D)^3 - (d_s/D)^4](F(d_s/D)}{[(1-4(d_W/D)^2 + 4(d_W/D)^3 - (d_W/D)^4](F(d_W/D)} \tag{83}$$

the situation is even more complex. This equation, in light of the considerations in this review, would be closer to the truth with $F(d_s/D)$ replaced by $G(d_s/D)$ and with $F(d_W/D)$ eliminated. In spite of the possible deficiencies of Eq. (83), the results for effective pore diameters (6), say 8.6 Å for human red cell and 12.4 Å for canine cells, are in remarkable coincidence with those from diffusion and osmosis of 9.0 and 11.8 Å, respectively. This may occur because the errors in Eq. (83) were substantially compensating.

In summary, the situation with regard to both theory and experiment for fine pores is somewhat underdeveloped with opportunity for significant contributions.

IV. SUMMARY AND CONCLUSIONS

In summary, this review of the transport properties of neutral pores shows that a firm base for continued experimental and theoretical advance has been laid. The concepts of irreversible thermodynamics provide an invaluable basis for mechanistic discussions although their domain of validity is limited by the necessary assumption of linear relationships between fluxes and flows. The frictional model for interpretation of the phenomenological coefficients has the virtue of making contact with microscopic parameters but has the concommitant disadvantage that this interpretation requires some structural insight. The more microscopic interpretation of transport derived from considerations of fluid mechanics has been carried somewhat further than previous treatments. Emphasis has been placed on the fact that earlier attempts ignored the essential difference between the restricted dynamics of a sphere in a pore for the conditions of no fluid motion and Poiseuille flow, respectively. While the theory of dynamics is incomplete in that off-axis motions have not been treated, first order expressions for transport coefficients are given that differ significantly from those that have been commonly employed. Central to the employment of fluid dynamics to the motions in fine space is the assumption that water acts as a continuum. New experiments on diffusion and flow through fine, characterized pores indicate that the structure and dynamics of water are those of the bulk to at least 10 Å from a solid surface. The dynamics of water on a finer scale is largely unknown both theoretically and experimentally. Comparison of theory and experiment for fluid and molecular transport through cellophane, gel, and dialysis membranes shows a first-order agreement with the expectations of the model derived from fluid mechanics but there are strong indications that this model is incomplete in its neglect of solute-surface adsorption and in its limitation to regular structures. The question of ions in neutral pores has been treated in some detail with the somewhat surprising conclusion that electrostatic effects can significantly limit the entry of ions (from dilute solution) into a fine pore. The resulting alteration in pore conductance has been calculated, together with a calculation of the reflection coefficient and consequent behavior in reverse osmosis experiments. The situation of transport in pores of only a few water molecules in diameter--mainly biological--has been briefly reviewed with the principal conclusion that further experimental and theoretical work will be needed to put this study on a firm footing.

ACKNOWLEDGMENTS

I wish to thank M. V. Doyle for her collaboration in the experiments described in Sec. III.D. as well as her invaluable assistance with the calculations and figures of this review. The Office of Saline Water (H. E. Podall, Contract Monitor) supported the diffusion and flow measurement of Sec. III.D. as well as the calculations of Sec. III.F. The stimulation of the Membrane Biophysics Course (H. Passow and R. Stämpfli, Directors) held in the fall of 1966 at The Saarland University, Homburg/Saar, was most influential in directing my attention to the important questions of transport through biological membranes. Lastly, I wish to thank the editor of this volume for the impetus to prepare this review.

LIST OF PRINCIPAL SYMBOLS

a	major semiaxis of ellipse (cm)
b	minor semiaxis of ellipse (cm)
C	concentration ($mole/cm^3$)
C_s	concentration of solute ($mole/cm^3$)
D	pore diameter (cm)
D_D	equivalent diffusion diameter (cm)
D_F	equivalent flow diameter (cm)
D*	characteristic length in liquids, $(32\eta D_{HTO}\bar{V}_w/RT)^{\frac{1}{2}}$ (cm)
D_o	diffusion constant (cm^2/sec)
D_{eff}	effective diffusion constant (cm^2/sec)
D_{HTO}	diffusion constant of tritiated water (cm^2/sec)
$\mathcal{D}$	electrical displacement (esu)
d	diameter of particle or ion (cm)
d*	characteristic length associated with particles from Eq. (26) (cm)

d_s	diameter of solute molecule (cm)
d_w	diameter of water molecule (cm)
$\mathcal{E}$	electrostatic field (statvolt/cm)
F_{ij}	frictional force exerted on a mole of i by the relative motion of i and j (atm-cm^2/mole)
F(d/D)	function in hydrodynamic theory of relative diameters of particle and pore from Eq. (21)
f	external force per unit volume (dyn/cm^3)
f_{ij}	molar frictional coefficient of i with j (atm-cm-sec/mole)
G(d/D)	function in hydrodynamic theory of relative diameters of particle and pore from Eq. (21)
J(0)	flow at zero pressure (mole/sec)
J_D	differential rate of transport of solute and solvent (cm/sec)
J_s	flow of solute (mole/cm^2-sec)
J_v	rate of total volume transport (cm/sec)
J_w	flow of solvent (mole/cm^2-sec)
j	local flow per unit pore area and time (mole/cm^2-sec)
K_D	distribution coefficient of solute between the water in the membrane and in the solution
k	Boltzmann constant (erg/deg)
L	thickness of homogeneous phase or membrane (cm)
L*	effective pore length (cm)
L_D, L_p, L_{Dp}, L_{pD}	phenomenological coefficients for a given membrane, solute and solvent (cm/atm-sec)

M	molarity
m	number of solvent molecules in a pore
N	number of pores
N_A	Avagadro's number
P	pressure (dyn/cm^2) or (atm)
P*	characteristic pressure from $\Delta P^* = 32\eta D_o/D^2$
$P(\epsilon_2/\epsilon_1)$	function of dielectric constants in medium and in cavity from Eq. (72)
Q	volume flow (cm^3/sec)
q	ionic charge (esu)
R	molar gas constant (erg/mole-deg)
R^{-1}	conductance (Ω^{-1})
Re	Reynolds number
r	radial position in pore (cm)
T	absolute temperature ($^\circ$K)
V	velocity (cm/sec)
V*	characteristic velocity from $V^* = D_o/L$ (cm/sec)
V_M	molecular volume (cm^3/molecule)
$\bar{V}_s$	molar volume of solute (cm^3/mole)
$\bar{V}_w$	molar volume of solvent (cm^3/mole)
v	velocity (cm/sec)
v^*	characteristic velocity from $v^* = D_o/L$ (cm/sec)
v_{max}	velocity at center of tube far from sphere (cm/sec)

v_i velocity of component i (cm/sec)

v_j velocity of component j (cm/sec)

v_m velocity of membrane (cm/sec)

v_s velocity of solute (cm/sec)

v_w velocity of solvent (cm/sec)

w energy in electrostatic field (erg)

X_s negative local gradient in chemical potential of solute (atm-cm^2/mole)

X_w negative local gradient in chemical potential of solvent (atm-cm^2/mole)

α relative diameters of ball and tube (α=d/D)

ϵ dielectric constant

ϵ_1 dielectric constant of medium in cavity

ϵ_2 dielectric constant of outside medium

η effective viscosity (poise)

η_o viscosity of medium (poise)

π osmotic pressure (atm)

ρ density (gm/cm^3)

σ reflection coefficient

σ_{s-d} reflection coefficient for solution-diffusion model

σ_o conductivity (Ω^{-1}-cm^{-1})

Φ volume fraction of spheres

Φ_w volume fraction of water in the membrane

ω solute permeability coefficient (mole/atm-cm^2-sec)

ω_{HTO} permeability coefficient of tritiated water (mole/atm-cm^2-sec)

ω_s solute permeability coefficient (mole/atm-cm^2-sec)

REFERENCES

1. R. J. Forbes, in A History of Technology (C. Singer, E. J. Holmy, and A. P. Hall, eds.), Vol. 1, Clarendon Press, Oxford, 1954, p. 277.
2. For example: H. D. Smyth, Atomic Energy for Military Purposes, Princeton University Press, Princeton, N.J., 1945, p. 172.
3. J. Charpin, P. Plurien, and S. Mommejac, in Proceedings of the Second United Nations International Conference on the Peaceful Uses of Atomic Energy, Vol. 4, United Nations, Geneva, 1958, p. 380.
4. J. W. Gibbs, Trans. Conn. Acad., 3, 108 (1873), p. 138 et seq.
5. J. D. Ferry, Chem. Rev., 18, 373 (1936).
6. A. K. Solomon, J. Gen. Physiol., 51, 335s (1968).
7. R. Schlögl, Stofftransport durch Membranen, Steinkopff, Darmstadt, 1964.
8. N. Lakshminarayanaiah, Transport Phenomena in Membranes, Academic, New York, 1969.
9. N. Lakshminarayanaiah, Chem. Rev., 65, 491 (1965).
10. L. Onsager, Phys. Rev., 37, 405 (1931); 38, 2265 (1931).
11. A. J. Staverman, Rec. Trav. Chim., 70, 344 (1951).
12. J. G. Kirkwood, in Ion Transport across Membranes (H. T. Clark, ed.), Academic, New York, 1954, p. 119.
13. O. Kedem and A. Katchalsky, Biochim. Biophys. Acta, 27, 229 (1958).
14. A. Katchalsky and O. Kedem, Biophys. J., 2, 53s (1962).
15. A. Katchalsky and P. F. Curran, Nonequilibrium Thermodynamics in Biophysics, Harvard University Press, Cambridge, Mass., 1965.
16. J. Dainty, Adv. Bot. Res., 1, 279 (1963).
17. T. G. Kaufmann and E. F. Leonard, A.I.Ch.E. J., 14, 110 (1968).
18. K. S. Spiegler, Trans. Faraday Soc., 54, 1408 (1958).
19. For example: A. Sommerfeld, Mechanics of Deformable Bodies, Academic, New York, 1950, p. 113.
20. L. M. Milne-Thompson, Theoretical Hydrodynamics, 2nd ed., MacMillan, New York, 1950, p. 515.
21. R. Ladenburg, Ann. d. Phys., 23, 447 (1907).
22. H. Faxén, Arkiv. Mat. Astron. Fys., 17, No. 27 (1923).
23. T. Bohlin, Trans. Roy. Inst. Technol. (Stockholm), No. 155 (1959).

24. H. Faxén, Kolloid Z., 167, 146 (1959).
25. V. Fidleris and R. L. Whitmore, Brit. J. Appl. Phys., 12, 490 (1961).
26. J. Famularo, D. Eng. Sci. Thesis, New York University, New York; 1962; quoted in J. Happel and H. Brenner, Low Reynolds Number Hydrodynamics, Prentice-Hall, Englewood Cliffs, N.J., 1965, p. 309.
27. H. L. Goldsmith and S. G. Mason, J. Coll. Sci., 17, 448 (1962).
28. G. Segré and A. Siberberg, J. Fluid Mech., 11, 447 (1962).
29. A. Einstein, Ann Physik, 19, 289 (1906); with correction 34, 591 (1911).
30. G. H. Higginbotham, D. R. Oliver, and S. G. Ward, Brit. J. Appl. Phys., 9, 372 (1958).
31. J. L. Kavanau, Water and Solute-Water Interactions, Holden Day Inc., San Francisco, Calif., 1964, p. 8 et seq.
32. A. Holtzer and M. F. Emerson, J. Phys. Chem., 73, 26 (1969).
33. L. Korson, W. Drost-Hansen, and F. J. Millero, J. Phys. Chem, 73, 34 (1969).
34. J. H. Wang, C. B. Robinson, and I. S. Edelman, J. Am. Chem. Soc., 75, 466 (1953).
35. H. J. V. Tyrell, Diffusion and Heat Flow in Liquids, Butterworths, London, 1961 p. 159.
36. J. Morgan and B. E. Warren, J. Chem. Phys., 6, 666 (1938).
37. J. L. Kavanau, in Ref. 31, p. 34 et seq.
38. E. Forslind, in Proceedings of the Second International Congress of Rheology, Butterworths, London, 1953, p. 50.
39. P. Low, Advan. Agronomy, 13, 269 (1961).
40. B. V. Derjaguin, in The State and Movement of Water in Living Organisms, Academic, New York, 1965, p. 55.
41. U. B. Bazaron, B. V. Derjaguin, and A. V. Bulgadaeu, Soviet Physics JETP, 24, 645 (1967).
42. C. P. Bean and M. V. Doyle, unpublished work.
43. P. B. Price and R. M. Walker, Phys. Rev. Letters, 8, 217 (1962).
44. C. P. Bean, M. V. Doyle, and G. Entine, J. Appl. Phys., 41, 1454 (1970).
45. For example: W. Jost, Diffusion in Solids, Liquids and Gases, Academic, New York, 1952, p. 46.
46. G. Hertz, Physik. Z., 23, 433 (1922).
47. E. Manegold and K. Solf, Kolloid Z., 59, 179 (1932).
48. For example: A. S. Michaels, H. J. Bixler, and R. M. Hodges, Jr., J. Colloid Sci., 20, 1034 (1965).
49. L. Garby, Nature, 173, 444 (1954).
50. B. Andersen and H. H. Ussing, Acta Physiol. Scand., 39, 228 (1957).
51. A. R. Overman and R. J. Miller, J. Phys. Chem., 72, 155 (1968).

52. C. P. Bean, Characterization of Cellulose Acetate Membranes and Ultrathin Films for Reverse Osmosis (Final Report, Contract No. 14-01-0001-1480), Office of Saline Water, United States Department of the Interior, Washington, D.C., 1969.
53. K. J. Mysels, J. Gen. Physiol., 52, 120s (1968).
54. N. Bjerrum and E. Manegold, Kolloid-Z., 43, 5 (1927).
55. J. R. Pappenheimer, E. M. Renkin, and L. M. Borrero, Am. J. Physiol., 167, 13 (1951).
56. V. Koefoed-Johnsen and H. H. Ussing, Acta Physiol. Scand., 28, 60 (1953).
57. E. M. Renkin, J. Gen. Physiol., 38, 225 (1954).
58. A. H. Nevis, J. Gen. Physiol., 41, 927 (1958)
59. E. Robbins and A. Mauro, J. Gen. Physiol., 43, 523 (1960).
60. Quoted in L. B. Ticknor, J. Phys. Chem., 62, 1483 (1958).
61. G. Thau, R. Bloch, and O. Kedem, Desalination; 1, 66 (1966).
62. A. Cass and A. Finkelstein, J. Gen. Physiol., 50, 1765 (1967).
63. C. Huang and T. E. Thompson, J. Mol. Biol., 15, 539 (1966).
64. R. Holtz and A. Finkelstein, J. Gen. Physiol., 56, 123 (1970).
65. N. Lakshminarayanaiah and M. S. White, J. Polym. Sci., A-1, 7, 2235 (1969).
66. R. W. DeBlois and C. P. Bean, Rev. Sci. Instr., 41, 909 (1970).
67. J. D. Ferry, J. Gen. Physiol., 20, 95 (1936).
68. J. Dainty and B. Z. Ginzburg, J. Theoret. Biol., 5, 256 (1963).
69. Handbook of Chemistry and Physics, 44th ed., Chemical Rubber Company, Cleveland, Ohio, 1962.
70. S. G. Schultz and A. K. Solomon, J. Gen. Physiol., 44, 1189 (1961).
71. J. Dainty and B. Z. Ginsburg, Biochim. Biophys. Acta, 79, 102, 112, 122, 129 (1964).
72. Quoted in Ref. 15, p. 123.
73. B. Z. Ginsburg and A. Katchalsky, J. Gen. Physiol, 47, 403 (1963).
74. F. Auerbach, in Handbuch der Elektrizität und des Magnetismus (L. Graetz, ed.) Johann Ambrosius Barth, Leipzig, 1921, p. 110.
75. For example: R. W. Gurney, Ions in Solution, Dover Publications, New York, 1962.
76. C. Wagner, Physik Z., 25, 474 (1924).
77. A. Parsegian, Nature, 221, 844 (1969).
78. U. Merten, in Desalination by Reverse Osmosis (U. Merten, ed.), The MIT Press, Cambridge, Mass., 1966, p. 15.
79. S. Sourirajan, Indl. and Eng. Chem. (Fund.), 2, 51 (1963).
80. E. Glueckauf, in Proceedings of the First International Symposium on Water Desalination, 1965, p. 1.
81. E. J. Breton, Water and Ion Flow Through Imperfect Osmotic Membranes, Office of Saline Water, Research and Development Progress Report No. 16, PB 161391 (1957).
82. P. Meares, European Polymer J. 2, 241 (1966).

83. H. Kawabe, H. Jacobson, I. F. Miller, and H. P. Gregor, J. Colloid Sci., 21, 79 (1966).
84. H. C. Longuet-Higgins and G. Austin, Biophysic. J., 6, 217 (1966).
85. C. V. Paganelli and A. K. Solomon, J. Gen. Physiol., 41, 259 (1957).
86. G. T. Rich, R. I. Shàafi, T. C. Barton, and A. K. Solomon, J. Gen. Physiol., 50, 2391 (1967).

Chapter 2

TRANSPORT ACROSS ION-EXCHANGE RESIN MEMBRANES: THE FRICTIONAL MODEL OF TRANSPORT

P. Meares

Chemistry Department
The University
Old Aberdeen, Scotland

J.F. Thain

School of Biological Sciences
University of East Anglia
Norwich, England

D.G. Dawson*

Chemistry Department
The University
Old Aberdeen, Scotland

---

*Present address: I.C.I. Paints Division, Slough.

I. INTRODUCTION

In recent years transport processes across organic resin membranes, particularly ion-exchange membranes, have been widely studied. These materials are important in their own right with applications in ion exchange, desalination, and fuel-cell technology. They are frequently used also as models for biological membrane systems and as experimental systems for the testing of general theories of membrane transport. With regard to this last use, organic resin membranes possess many useful properties: They are usually robust and easy to handle; their chemical and physical properties such as composition, structure, and fixed charge concentration are relatively easy to study; and the membranes can often be synthesized with desired properties to a fair degree of accuracy and reproducibility.

Progress in the field of ion-exchange membranes has been covered in several symposia (1,2) and reviews (3,4). Accordingly it is not intended here to produce another general review of their transport properties.* Instead attention will be directed to one aspect of the study of transport, namely the application of nonequilibrium thermodynamics. This has aroused a considerable amount of interest since the papers of Staverman (5) and Kirkwood (6) were published. In particular the frictional model of membrane transport proposed by Spiegler (7) is discussed and a detailed experimental test of this model with an organic cation-exchange membrane is described.

---

*Literature available at the beginning of 1969 was taken into account when writing this article. It is not intended as a literature survey and no references have been added to cover literature which has appeared during the compilation of this volume.

II. THEORIES OF MEMBRANE TRANSPORT

A. Nernst-Planck and Pseudo-thermodynamic Treatments

Many of the earlier treatments of membrane transport (e.g., 8,9) used the Nernst-Planck equations to describe the relationships between the flows of the permeating species and the forces acting on the system. According to these equations the flux J_i of species i at any point is equal to the product of the local concentration c_i of i, the absolute mobility u_i of i, and the force acting on i. This force has been identified with the negative of the local gradient of the electrochemical potential μ_i of i. Thus, at a distance x from a reference plane at right angles to the direction of unidimensional flow through a membrane

$$J_i = -c_i u_i \, d\mu_i/dx \tag{1}$$

The electrochemical potential of i can be divided into its constituent parts giving in place of Eq. (1)

$$J_i = -c_i u_i (RTd \, \ln c_i/dx + RTd \, \ln \gamma_i/dx + \bar{V}_i dp/dx + z_i F d\psi/dx) \tag{2}$$

where γ_i, $\bar{V}_i$, z_i, p, and ψ represent the activity coefficient, the partial molar volume, the valence charge on i, the hydrostatic pressure, and the electrical potential, respectively. R is the gas constant, T the absolute temperature, and F the Faraday. It is apparent from Eq. (2) that the Nernst-Planck equations make use of the Nernst-Einstein relation between the absolute mobility u_i and the diffusion coefficient D_i of species i. This is

$$D_i = u_i RT \tag{3}$$

On replacing the electrochemical mobility in Eq. (2) by the diffusion coefficient, the more usual form of the Nernst-Planck flux equation is obtained.

$$J_i = -D_i \left(\frac{dc_i}{dx} + c_i \frac{d \ln \gamma_i}{dx} + \frac{c_i \bar{V}_i}{RT} \frac{dp}{dx} + \frac{c_i z_i F}{RT} \frac{d\psi}{dx} \right) \tag{4}$$

On the basis of the Nernst-Planck equations, the flow of species i is regarded as unaffected by the presence of any other permeating species except in so far as the other species either influences the force acting on i by, for example, affecting the values of γ_i or ψ, or alters the state of the membrane and hence alters the value of D_i.

To obtain relationships between the flows of the permeating species and the observable macroscopic differences in concentration, electrical potential, and hydrostatic pressure between the solutions on the two sides of the membrane, it is necessary to integrate the Nernst-Planck equation (4) for each mobile component across the membrane and the membrane/solution boundaries. In order to carry out this integration an additional assumption has to be made. The differences between the various treatments derived from the Nernst-Planck equations lie in the different assumptions used. For example, in the theory of Goldman (9), which is widely applied to biological membranes, it is assumed that the gradient of electrical potential $d\psi/dx$ is constant throughout the membrane. It is usually assumed also that thermodynamic equilibrium holds across the membrane/solution interfaces and that the system is in a steady state so that the flows J_i are constant throughout the membrane. Generally these integrations do not lead to linear relationships between the flows and the macroscopic differences of electrochemical potential between the two bathing solutions.

The main disadvantage of the Nernst-Planck approach is that it fails to allow for interactions between the flows of different permeating species. Such interactions are most obvious when a substantial flow of solvent, usually water, occurs at the same time as a flow of solute. For example, during the passage of an electric current across a cation-exchange membrane, the permeating cations and anions both impart momentum to the water molecules with which they collide. Since the number of cations is greater than the number of anions, the momentum imparted to the water by the cations is normally greater than the momentum imparted by the anions and an electroösmotic flow of water is set up in the direction of the cation current. The resultant bulk flow of the water has the effect of reducing the resistance to the flow of cations and increasing the resistance to the flow of anions. This flow of water occurs under the difference of electrical potential and in the absence of a concentration gradient of water. The appropriate Nernst-Planck equation would predict no flow of water under these conditions. Furthermore the flows of cations and anions differ from those which would be predicted from the respective Nernst-Planck equations on account of the effect of the water flow on the resistances to ionic flow.

This effect of solvent flow on the flows of solute molecules or ions can be allowed for by adding a correction term to the Nernst-Planck equations. Thus we may write

$$J_i = -c_i u_i \, d\mu_i/dx + c_i v \qquad (5)$$

where v is the velocity of the local center of mass of all the species (10, 11). The term $c_i v$ is often called the convective contribution to the flow of i and some authors have preferred to define v as the velocity of the local center of volume.

The addition of this convection term to the Nernst-Planck equation for the flow of a solute is probably a sufficient correction in most cases involving only the transport of a solvent and nonelectrolyte solutes across a membrane in which the solvent is driven by osmotic or hydrostatic pressure. The situation is much more complex when electrolyte solutes are considered. Even at low concentrations the flows of cations and anions may interact strongly with each other. Interactions between the different ion flows may be of similar size to their interactions with the solvent flow. Under these circumstances the convection-corrected Nernst-Planck equations may still not give a good description of the experimental situation regarding the ion flows.

The theoretical difficulties arising from interacting flows can be formally overcome by the use of theories of transport based on nonequilibrium thermodynamics. Such theories are described in the next section.

B. Treatments Based on Nonequilibrium Thermodynamics

Since the original papers of Staverman (5) and Kirkwood (6), many papers have appeared on the application of nonequilibrium thermodynamics to transport across synthetic and biological membranes. In particular, major contributions have been made by Katchalsky, Kedem, and co-workers. In view of the appearance of recent and extensive texts (4, 12), this account is intended only as a brief summary of the general principles.

1. The Phenomenological Equations

The theory of nonequilibrium thermodynamics allows that, in a system where a number of flows are occurring and a number of forces are operating, each flow may depend upon every force. Also, if the system is not too far from equilibrium, the relationships between the flows and forces are linear. Thus the flow J_i may be written as follows

$$J_i = \sum_{\text{all } k} L_{ik} X_k \tag{6}$$

where the X_k are the various forces acting on the system and the L_{ik} are the phenomenological coefficients which do not depend on the sizes of the fluxes or forces. The flow J_i may be a flow of a chemical species, a volume flow, a flow of electric current, or a flow of heat. The forces X_k may be expressed in the form of local gradients or macroscopic differences across the membrane of the chemical potentials, electric potential, hydrostatic pressure, or temperature. If a discontinuous formulation is used so that the macroscopic differences in these quantities across the membrane are chosen as the forces, then the L_{ik} coefficients in Eqs. (6) are average values over the membrane interposed between a particular pair of solutions.

Equations (6) imply, for example, that the flow of a chemical species i is dependent not only on its conjugate force X_i, i.e., the difference or negative gradient of its own chemical or electrochemical potential, but also on the gradients or differences of the electrochemical potentials of the other permeating species. Hence Eqs. (6) imply that a difference of electrical potential may cause a flow of an uncharged species, a fact which, as we saw previously, the Nernst-Planck equations do not recognize. In general, Eqs. (6) allow that any type of vectorial force can, under suitable conditions, give rise to any type of vectorial flow.

In a system where n flows are occurring and n forces are operating, a total of n^2 phenomenological coefficients L_{ik} are required to describe fully the transport properties of the system. This must be compared with the n mobilities used in the Nernst-Planck description of the system. A corresponding number n^2 experimental transport measurements would have to be made to permit the evaluation of all the L_{ik} coefficients.

Fortunately a simplification can be made with the help of Onsager's reciprocal relationship (13). This states that under certain conditions

$$L_{ik} = L_{ki} \tag{7}$$

The conditions required for Eq. (7) to be valid are that the flows be linearly related to the forces and that the flows and forces be chosen such that

$$T\sigma = \sum_i J_i X_i \tag{8}$$

where σ is the local rate of production of entropy in the system when the X_i are the local potential gradients. The quantity $T\sigma$ is often represented by the symbol Φ and called the dissipation function because it represents the rate at which free energy is dissipated by the irreversible processes. In fact there is no completely general proof of Eq. (7) but its validity has been shown for a large number of situations (14).

With the help of the reciprocal relationship the number of separate L_{ik} coefficients required to describe a system of n flows and n forces is reduced from n^2 to $\frac{1}{2}n(n + 1)$.

This nonequilibrium thermodynamic theory holds only close to thermodynamic equilibrium. The size of the departure from equilibrium for which the linear relationship between flow and force, Eqs. (6), and the reciprocal relationship, Eq. (7), are valid depends upon the type of flow considered. Strictly, the range of validity must be tested experimentally for each type of flow process. In the case of molecular flow processes, electronic conduction, and heat conduction the linear and reciprocal relationships have been found to be valid for flows of the order of magnitude commonly encountered in membranes (14). In describing the progress of chemical reactions the relationships are valid only very close to equilibrium. We exclude from the scope of this article systems in which chemical reactions are taking place.

2. The Choice of Flows and Forces

In an isothermal membrane system the most obvious choice of flows is the set of flows of the permeating species--solvent, nonelectrolyte solutes, and ions. The conjugate forces are then the differences or local gradients of the electrochemical potentials of these species. To accord with Eq. (8), in which $T\sigma$ must be positive, increasing potentials in the direction of positive fluxes constitute negative forces. A set of phenomenological equations corresponding to Eqs. (6) can be written relating the flows to the forces. The values of the L_{ik} coefficients appearing in these equations depend on the interactions occurring in the membrane, i.e., on the chemical nature of the permeating species and of the membrane, on the detailed microstructure of the membrane, and on the local concentrations of the permeating species.

In principle it should be possible to obtain values for the $\frac{1}{2}n(n + 1)$ L_{ik} coefficients by carrying out a suitable set of $\frac{1}{2}n(n + 1)$ independent experiments. For example, if all the forces except one, X_a, were held at zero and the flows J_i, J_j, etc. of all the n species were measured, then the values of the coefficients L_{ia}, L_{ja} etc. could be obtained directly. Similar experiments would give the values for the remaining L_{ik} coefficients. Other sets of experiments may be used, and one may combine experiments where some of the forces are kept at zero, experiments where some of the flows are kept at zero, and experiments where some forces and some flows are kept at zero (15,16).

Although the set of flows and conjugate forces outlined above may seem to be convenient for the molecular interpretation of the interactions occurring in a membrane system, the equations written in terms of these flows and forces are not convenient for the design of experiments for the evaluation of the L_{ik} coefficients. For example the forces which are usually controlled experimentally are not differences of electrochemical potential, but differences of concentration, electrical potential, and hydrostatic pressure. Also it may be more convenient to measure the total volume of the flows across a membrane rather than the flow of solvent, or to measure the electric current and one ionic flow rather than two ionic flows. For these reasons, sets of practical flows and forces are often chosen to describe membrane transport (16,17). These practical sets of flows and their conjugate forces must satisfy the relationship of Eq. (8), which gives the dissipation function.

A system involving the transport of water and a nonelectrolyte solute across a membrane can be described by giving the flows of water J_w and of solute J_s. The conjugate forces are then the differences, or the local gradients, of the chemical potentials of water μ_w and solute μ_s. The transport properties of this system are described by the following equations:

$$J_w = L_w \Delta\mu_w + L_{ws} \Delta\mu_s$$
$$J_s = L_{sw} \Delta\mu_w + L_s \Delta\mu_s \tag{9}$$

where according to the reciprocal relationship $L_{sw} = L_{ws}$ and the dissipation function of the system is given by the expression

$$\Phi = J_w \Delta\mu_w + J_s \Delta\mu_s \tag{10}$$

When considering ideal external solutions the forces $\Delta\mu_w$ and $\Delta\mu_s$ are often expanded into separate terms giving the contributions of the concentration differences and pressure difference to the total driving forces. Thus

$$\Delta\mu_w = (RT/\bar{c}_w)\Delta c_w + \bar{V}_w \Delta p$$

Here $\bar{V}_w$ is an average partial molar volume of water and $\bar{c}_w$ is an average concentration of water. When $\Delta\mu_w$ and $\Delta\mu_s$ in Eq. (10) are expanded in this way and the resulting concentration and pressure terms are grouped separately the expression for the dissipation function becomes (15)

$$\Phi = J_V \Delta p + J_D RT\Delta c_s \tag{11}$$

where J_V the total volume flow is equal to $(\bar{V}_w J_w + \bar{V}_s J_s)$ and J_D is equal to $(J_s/\bar{c}_s - J_w/\bar{c}_w)$. J_D is sometimes called the exchange flow and represents the apparent mean velocity of the solute relative to the water. According to Eq. (11) the system can be described in terms of J_V and J_D as flows and Δp and $RT\Delta c_s$ (or $\Delta\pi_s$) as their conjugate forces. Thus

$$J_V = L_p \Delta p + L_{pD} \Delta\pi_s \tag{12}$$
$$J_D = L_{Dp} \Delta p + L_D \Delta\pi_s$$

where L_{Dp} equals L_{pD} and $\Delta\pi_s$ is the difference in osmotic pressure between the solutions. Experimentally it is easier to control the values of the forces appearing in Eq. (12) than those appearing in Eqs. (9).

Similarly a system involving flows of water and a salt dissociated into a cationic species and an anionic species can be described (16) in terms of the flows J_w, J_1, and J_2 of these molecular species or by the set comprising the total volume flow, the electric current, and the defined flow of salt, i.e., J_v, I and J_s. In the former case the conjugate forces are the differences of the electrochemical potentials of the species across the membrane, in the latter case the conjugate forces are the pressure difference minus the osmotic pressure difference, the electrical potential difference, and the difference of the pressure-independent part of the chemical potential of the salt. Care must be taken in the precise definition of these forces, particularly of the electrical potential difference (17).

Since the choice of flows and forces is to some extent open as long as the flows and forces satisfy Eq. (8) a set can be chosen primarily for ease of theoretical interpretation of L_{ik} coefficients or for ease of experimental evaluation of the L_{ik} coefficients. Furthermore, given values of the L_{ik} coefficients relevant to one set of flows and forces, it is a straightforward operation to calculate the values of L_{ik} coefficients relevant to another set of flows and forces (17).

It is of course possible and often convenient to describe the transport properties of a system in terms of flows and forces which are not conjugate and which do not obey Eq. (8). The system where the membrane is permeated by a flow of water and a flow of a solute can be described in terms of the flow of water J_w, the flow of solute J_s, the pressure difference Δp, and the difference in concentration of the solute $RT\Delta c_s$ or $\Delta\pi_s$. These flows and forces are interrelated by the equations

$$\begin{aligned} J_v &= L_p \Delta p - \sigma L_p \Delta\pi_s \\ J_s &= \bar{c}_s(1 - \sigma)J_v + \omega\Delta\pi_s \end{aligned} \tag{13}$$

Here L_p has the same significance as in Eqs. (12). σ is called the reflection coefficient of the solute and is equal to $\Delta p/\Delta\pi_s$ at zero J_v, ω is the solute permeability $J_s/\Delta\pi_s$ at zero J_v, and $\bar{c}_s$ is the average concentration of the solute in the two solutions (15).

In practice Eqs. (13) may be easier to use than Eqs. (12) because the flows generally measured are J_v and J_s rather than J_v and J_D. However Eqs. (13) are not a proper set of phenomenological equations in the sense of Eqs. (6). Neither are σ and ω phenomenological coefficients in the sense used so far. They are related to the L_{ik} coefficients of Eqs. (12) by the relationships (15)

$$\sigma = -L_{pD}/L_p \text{ and } \omega = \bar{c}_s(L_p L_D - L_{pD}^2)/L_p$$

3. Uses and Limitations of the Theory

The theory of nonequilibrium thermodynamics has been applied to membranes in a number of papers where the aim has been to obtain general relationships between observable macroscopic flows and forces. Topics investigated in this way have included: isotopic tracer flows and flux ratios (18-22), electrokinetic phenomena (23,24), the transport properties of complex membranes (16,25,26), and the coupling of transport processes with chemical reactions, so-called active transport (12,27-31). However the main concern of these investigations has been the transport of nonelectrolyte solutes and ions across charged and uncharged membranes (4-6, 12,15,26,32-34).

The L_{ik} coefficients obtained from experimental measurements of transport phenomena under one set of conditions can either be used to predict values of flows and forces under other sets of conditions or they can be analyzed for the purpose of interpreting, at a molecular level, the various interactions which occur between the permeating molecules and ions and the membrane material. This second use of the L_{ik} coefficients is especially interesting but it is by no means simple.

Only a few such attempts have been made (15, 16, 26, 35-41) and the set of coefficients most commonly used for this purpose has been L_p, σ, and ω. The interpretation of L_{ik} coefficients at the molecular level gives rise to difficulties because they are complex quantities. Each one may reflect the combined effects of several different molecular interactions. For example, if all the forces except one, X_1 say, are kept at zero, measuring the flow J_1 of component 1 will give directly the value of the coefficient L_{11}. However it is obvious that this coefficient must include effects due to the interactions of species 1 with all other species present -- the other permeating species as well as the membrane material. Neither do the cross-coefficients L_{1k} ($k \neq 1$) represent only the direct molecular interaction between the flows of species 1 and k because the force X_1 may generate fluxes of all components. Thus, while J_1 and J_k may interact directly they interact also indirectly through their separate interactions with the other flows J_j.

An inspection of any of the sets of phenomenological equations [(6), (9), (12), and (13)] shows that nowhere is any direct reference made to the membrane or its properties. The L_{ik} coefficients relate the flows of the permeating species to the gross thermodynamic forces acting on these species and, in general, no particular coefficient represents only the interaction of a permeating species with the membrane. Instead the properties of the membrane material affect the values of each of the L_{ik} coefficients to a greater or lesser extent.

The physical interpretation of measurements of transport properties is made more straightforward by inverting the matrix of the phenomenological equations [Eqs. (6)] to give the set of Eqs. (14)

$$X_i = \sum_{\text{all } k} R_{ik} J_k \tag{14}$$

These represent the forces as linear functions of the flows. The R_{ik} and L_{ik} coefficients of Eqs. (14) and (6) are related by the expression

$$R_{ik} = A_{ik}/|L| \tag{15}$$

where A_{ik} is the minor of L_{ik} and $|L|$ is the determinant of the L_{ik} coefficients. If the reciprocal relation is valid for the L_{ik} coefficients, it is valid also for the R_{ik} coefficients. Whereas the L_{ik} coefficients have the dimensions of conductance (i.e., flow per unit force), the R_{ik} coefficients have the dimensions of resistance (i.e., force per unit flow) and are frequently called resistance coefficients.

The R_{ik} coefficients are easier to interpret at the molecular level than the L_{ik} coefficients. A nonzero R_{ik} $(i \neq k)$ implies a direct interaction between i and k, that is, the molecular flow of k directly causes a force to act on species i. On the other hand, a nonzero L_{ik} $(i \neq k)$ does not necessarily imply a direct molecular interaction between species i and k, it means that the force acting on k affects the flow of i, perhaps directly or indirectly.

In effect Eqs. (14) mean that, in the steady state, the gross thermodynamic force X_i acting on species i is balanced by the forces $R_{ik}J_k$ summed over all species k, including i. The term $R_{ii}J_i$ is the drag force per mole which would act on i when moving at a rate J_i/c_i through a medium where there was no net flow of any other species. Thus the R_{ii} coefficients are still complex quantities including contributions from the interactions between i and all other species present, including the membrane. However each R_{ik} $(i \neq k)$ coefficient represents only the single interaction between the flows of i and k. The R_{ii} coefficients, like the L_{ii}, must always be positive but R_{ik} $(i \neq k)$ and the L_{ik} coefficients may be positive, negative, or zero.

The resemblance of the R_{ik} ($i \neq k$) coefficients to macroscopic friction coefficients suggests a simple mechanistic interpretation of the interaction. This idea is the basis of Spiegler's frictional model of membrane transport (7) which is described in the following section.

C. The Frictional Model of Membrane Transport

The idea of describing steady-state transport processes in a membrane as balances between the gross thermodynamic forces acting on the system and frictional interactions between the components of the system is one of long standing (42). More recently (43), the term molecular friction coefficient has been applied to the coefficient which relates the frictional force between two components to the difference between their velocities. This approach has been used to describe transport processes across membranes by several authors (44-47). The precise treatment with which we are primarily concerned in this article is the frictional model as proposed by Spiegler (7).

The fundamental statement of the frictional model is that when the velocity of a permeating species has reached a constant value, the gross thermodynamic force X_i acting on one mole of that species must be balanced by the interactive forces, F_{ik}, acting between one mole of the same species and the other species present. Mathematically this is expressed by

$$X_i = - \sum_{k \neq i} F_{ik} \tag{16}$$

Furthermore these interactions are assumed to be frictional in character so that each force F_{ik} is equal to a friction coefficient f_{ik} multiplied by the difference between the velocities v_i and v_k of the two species. Thus

$$F_{ik} = -f_{ik}(v_i - v_k) \tag{17}$$

and

$$X_i = \sum_{k \neq i} f_{ik}(v_i - v_k) \tag{18}$$

It should be noted that f_{ik} is the force acting on one mole of i owing to its interaction with the amount of k normally in the environment of i and under unit difference between the mean velocities of i and k. In general the concentrations of i and k are not equal and consequently the coefficients f_{ik} and f_{ki} are not equal. When the balance of forces is taken over unit volume of the system it is readily seen that

$$c_i f_{ik} = c_k f_{ki} \tag{19}$$

The quantity f_{ik}/c_k or f_{ki}/c_i represents the force acting between one mole of i and one mole of k at unit velocity difference. Its value obviously depends on the chemical types of the two species.

Besides containing a term such as $f_{ik}(v_i - v_k)$ for the interactions between i and each of the other permeating species, the right-hand side of Eq. (18) also includes a term $f_{im}(v_i - v_m)$ which allows for the interaction between i and the membrane. Usually the membrane is taken as the velocity reference so that v_m is zero.

With the help of the relationship

$$J_i = c_i v_i \tag{20}$$

Eq. (18) can be rearranged to

$$X_i = (J_i/c_i) \sum_{k \neq i} f_{ik} - \sum_{k \neq i} (J_k f_{ik}/c_k) \tag{21}$$

Equation (21) has the same form as Eqs. (14) which relate the forces to the flows via the R_{ik} coefficients. Each R_{ii} coefficient can be equated to the corresponding $\Sigma f_{ik}/c_i$. This illustrates the complex nature of the R_{ii} coefficient. Each R_{ik} $(i \neq k)$ coefficient is equivalent to the corresponding $-f_{ik}/c_k$.

In a system with n flows, (n - 1) friction coefficients are required to describe the interactions of any one permeating species with the other permeating species. One further coefficient is required to describe its interaction with the membrane. A total of n^2 friction coefficients is thus required to describe the transport properties of the system but with the use of Eq. (19) this number is reduced to $\frac{1}{2}n(n + 1)$, i.e., the same as the minimum number of independent L_{ik} or R_{ik} coefficients. Hence the minimum number of experimental measurements required to characterize the system fully is the same whether it is described in terms of the L_{ik} coefficients, the R_{ik} coefficients, or the f_{ik} coefficients. The most convenient set of experimental parameters to be measured may depend on which set of coefficients is chosen to represent the properties of the system.

The choice of coefficients can be made mainly on the basis of experimental convenience because, having obtained values of one set of coefficients, it is no problem to obtain values for the other sets from these. The relationships between the R_{ik} and L_{ik} coefficients, and between these and the friction coefficients have already been given briefly above and are discussed in more detail elsewhere (7, 26, 35, 37, 39, 48). Direct relationships between the friction coefficients and experimentally measurable quantities have also been discussed in several papers (7,35,37,39,45-47, 49-55). The method of obtaining one such relationship is mentioned here as an illustration of Spiegler's treatment.

In a system consisting of a membrane, water, one species of univalent cation and one species of univalent anion the electrical conductivity k is given by the expression

$$k = F(J'_1 - J'_2) \tag{22}$$

where J'_1 and J'_2 are the flows of univalent cations and anions per unit area respectively under an electrical potential gradient of 1 V cm^{-1}. Under these conditions the forces acting on the cations, anions and water are F, -F, and 0 J cm^{-1} $mole^{-1}$, respectively. On substituting these forces into the set of Eqs. (21) describing the system, the equations can be solved for the flows J'_1 and J'_2 in terms of the friction coefficients and the concentrations of the ions and water. These expressions for J'_1 and J'_2 can then be substituted into Eq. (22) to give an expression for k in terms of the friction coefficients operating in the system and the concentrations of the permeating species.

It is possible to obtain expressions for other transport parameters, such as the electroösmotic permeability, transport numbers of the ions, and the self-diffusion coefficients of the permeating species in terms of the friction coefficients in a somewhat similar manner. A set of such expressions can then be solved to give the individual friction coefficients in terms of the transport parameters and the concentrations.

The procedure outlined above becomes rather tedious as the expressions giving the individual transport parameters in terms of the friction coefficients may be very complicated. Under certain circumstances a simpler procedure can be used to obtain values for the friction coefficients and this method will be described later.

The main advantage claimed for the use of the frictional model to describe transport processes in membranes is that each friction coefficient represents the interaction between a particular pair of flows. They are not complex combinations of several interactions as are the L_{ik} and R_{ii} coefficients. The model also permits a direct evaluation of the interactions between the various permeating species and the membrane, interactions which are hidden in treatments which use only the L_{ik} and R_{ik} coefficients.

It may be possible under favorable conditions to neglect some of the frictional interactions on the basis of previous knowledge of the properties of the membrane and permeants. A smaller number of experimental measurements is then necessary to describe the system. For example Spiegler (7) suggested that, in a system where a cation-exchange membrane is in equilibrium with a dilute electrolyte solution, the friction coefficient f_{12} (where 1 represents cations and 2 represents anions) can be set equal to zero because of the low concentration of diffusible anions.

Simplifications such as that described above should be made only with great care. It is possible that even through f_{ik} may be negligibly small f_{ki} may be quite large because the ratio c_k/c_i [c.f. Eq. (19)] may be large. In such a case the full number of experimental measurements must still be made.

The quantitative application of the frictional model to biological membrane systems is restricted by the difficulty of measuring or estimating values for the average or local concentrations of the permeating species in the membrane. These values are required for the calculation of the friction coefficients from the measured experimental parameters. Thus, although values for sets of L_{ik} coefficients (particularly L_p, σ, and ω) have been obtained for some biological systems, it has been possible to interpret these in terms of the friction coefficients in only a qualitative manner (15,26,35,37,48).

With homogeneous synthetic resin membranes the situation seems to be simpler. Some limited measurements of friction coefficients for such systems have been reported (49-55) over the past ten years. In the authors' opinion none of these sets of measurements has constituted a thorough test of the frictional model and the purpose of the experimental work, the results of which are to be presented and discussed in the remainder of this article, was principally to provide such a test.

III. EXPERIMENTAL MEASUREMENTS ON A CATION-EXCHANGE MEMBRANE

The aim of this experimental study has been to measure a large number of transport properties of a simple and well-characterized membrane system so that the friction, resistance, and conductance coefficients could be calculated with a minimum of assumptions. The data to be discussed here were obtained when using a homogeneous cation-exchange membrane in contact with solutions of single electrolytes at various concentrations.

The membrane was a polycondensate of formaldehyde, phenol, and a phenol sulfonic acid in which the active exchange groups were $-CH_2 \cdot SO_3^-$. This material was supplied by Permutit Ltd. under the name Zeo-Karb 315. It has been described in detail in several previous publications. In order to interpret its transport properties the equilibrium water, counterion, and co-ion contents and the linear dimensions of the membrane were required at each concentration studied in the three electrolytes NaBr, CsBr, and $SrBr_2$. A few of the more important characteristics are summarized in Table 1, while the detailed data and methods of measurement have been described elsewhere (56-58). The data refer to 25°C at which temperature the transport measurements were also made.

Data have already been published on the tracer-diffusion coefficients of the co-ions Br^- and the counterions Na^+, Cs^+, and Sr^{2+} (59, 60), ionic transport numbers (61), electrical conductance and electroösmotic permeability (51, 62). These properties were measured with the membrane in contact with solutions of equal concentration on its two sides. Many concentrations were studied in the range 0.01 to 1.00 eq/liter, and it was found that all the transport properties were strongly concentration dependent. The results of these experiments are not reproduced here, but they are used in the calculations which follow later.

In addition, data have now been obtained on the fluxes of the salts and water across the membrane when there is a difference in the concentrations of the solutions on its opposite sides (63). The measurements were made at 25°C and with zero electric current passing, i.e., under open circuit conditions.

The volume flux was determined by using a slightly improved version of the electroösmosis cell already described (51, 62). The cell volumes were sufficiently large for the changes in concentration due to salt diffusion during the time required to determine the volume flux to be disregarded. The plots of volume change versus time were always linear throughout the period of the experiment. In order to ensure these conditions the concentration difference across the membrane was kept small. Complete curves were built up of the volume flux J_V cm^3/cm^2 sec from the most dilute solution

TABLE 1

Properties of Zeo-Karb 315 Membranes Per Gram of Na^+ Form Fully Swollen in Water

Exchange capacity	Volume	Weight of water
0.486 meq	0.887 cm^3	0.665 g

studied into each of the more concentrated solutions by using the additivity principle which has been discussed and explained elsewhere (17). According to this principle, if J_V $(c_1 \rightarrow c_2)$, etc., represents the volume flux from concentration c_1 into c_2, etc., then

$$J_V (c_1 \rightarrow c_3) = J_V (c_1 \rightarrow c_2) + J_V (c_2 \rightarrow c_3) \tag{23}$$

The additivity principle holds when the linear flux equations [Eq. (6)] apply and the membrane is continuous and homogeneous on the macroscopic scale. It is necessary also that interfacial phenomena at the membrane/ solution boundaries do not affect the values of the fluxes under the conditions of the experiment.

The additivity principle has been exhaustively demonstrated to apply in the case of our volume flux measurements. A test of it is shown in Table 2 for NaBr. The composite volume flux curves for NaBr, CsBr, and $SrBr_2$ relative to a reference solution at 0.01 eq/liter are shown in Figs. 1 and 2. The actual fluxes are inversely proportional to the membrane thickness δ. The values of $J_V\delta$ which are plotted represent the volume fluxes normalized to a membrane of 1 cm thickness. The negative anomalous osmosis between $SrBr_2$ solutions at high concentrations is clearly seen (65).

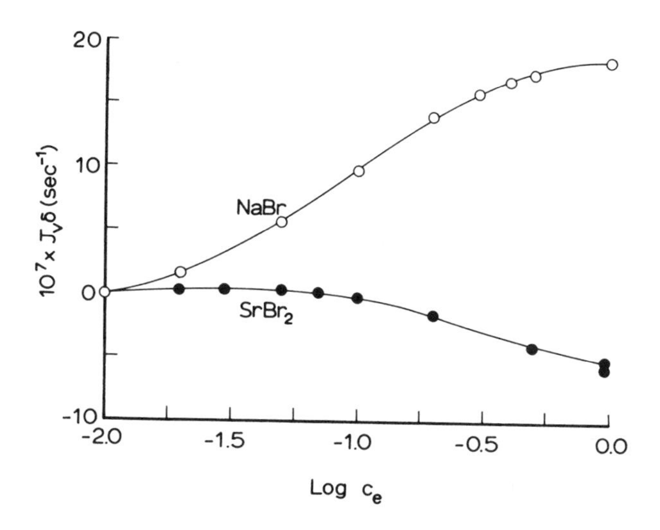

FIG. 1. Composite curves of normalized volume flux $J_V\delta$ (cm^2 sec^{-1}) against log c_e. (c_e = the external solution on the concentrated side). The dilute side is always at 0.01 eq/liter.

TABLE 2

Test of the Additivity Principle for the Volume Flux J_V of NaBr in Zeo-Karb 315 Membrane

c_1	c_2	c_3	$J_V(c_1 \rightarrow c_2)$	$J_V(c_2 \rightarrow c_3)$	$J_V(c_1 \rightarrow c_3)$	$[J_V(c_1 \rightarrow c_2) + J_V(c_2 \rightarrow c_3)]$
(eq/liter)			($cm^3\ cm^{-2}\ sec^{-1} \times 10^{-6}$)			
0.01	0.02	0.05	1.28	3.06	4.47	4.34
0.01	0.05	0.10	4.47	3.16	7.70	7.63
0.01	0.05	0.20	4.47	6.27	10.70	10.74
0.01	0.10	0.20	7.70	3.13	10.70	10.83
0.02	0.05	0.10	3.06	3.16	6.18	6.22
0.02	0.10	0.20	6.18	3.13	9.49	9.31
0.05	0.10	0.20	3.16	3.13	6.27	6.29
0.20	0.30	0.50	1.49	1.23	2.65	2.72
0.20	0.50	1.00	2.65	0.87	3.49	3.52
0.30	0.50	1.00	1.23	0.87	1.99	2.10

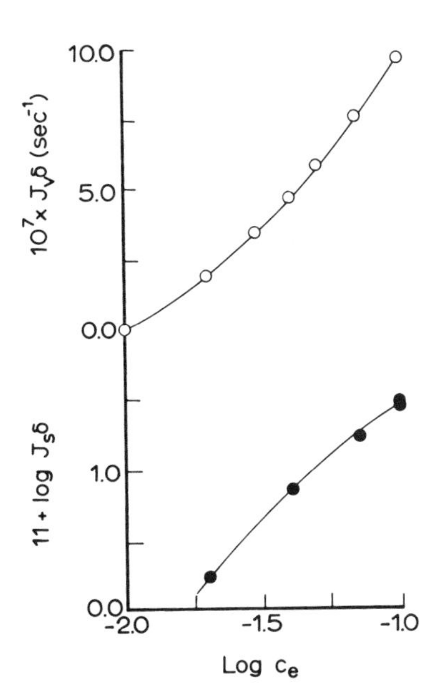

FIG. 2. Composite curves of normalized CsBr salt flux $J_s\delta$ (eq cm^{-1} sec^{-1}) (plotted logarithmically) and volume flux $J_V\delta$ (cm^2 sec^{-1}) against log c_e. (c_e = the external solution concentration on the concentrated side). The dilute side is always at 0.01 eq/liter.

The salt diffusion fluxes were obtained by measuring the radio-tracer flux of co-ions $^{82}Br^-$ up and down the concentration gradient in separate tracer flux experiments. If J_s is the flux of salt in moles per square centimeter per second, J_{Br} (net) the net flux of Br^-, and z_1 the counterion valence, while $\bar{J}_{Br}(f)$ and $\bar{J}_{Br}(b)$ are the total unidirectional fluxes of Br^- down the concentration difference and up the concentration difference, respectively, measured by using tracer $^{82}Br^-$, then

$$J_s = J_{Br}\,(\text{net})/z_1 = [\bar{J}_{Br}(f) - \bar{J}_{Br}(b)]/z_1 \tag{24}$$

The salt fluxes were measured by using the same apparatus and technique as were used to determine the tracer diffusion coefficients (59)*. The normalized fluxes $J_s\delta$ into a solution at 0.01 eq/liter for NaBr, CsBr, and $SrBr_2$ are shown in Figs. 2 and 3 as plots of log ($J_s\delta$) versus log c_e.

In order to compile these curves from the measured fluxes the additivity principle was used as for the volume fluxes. It is demonstrated in Table 3 that the salt fluxes obey the additivity principle. In this table the last two columns are used to compare the salt fluxes observed across particular concentration intervals with those obtained by summing the fluxes across two lesser and contiguous intervals.

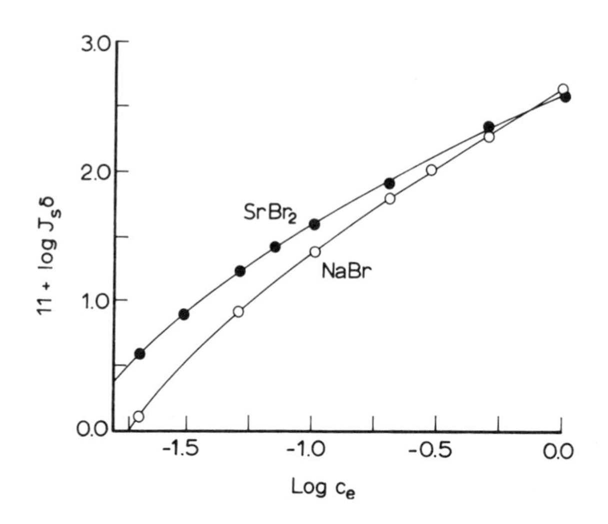

FIG. 3. Composite curves of normalized salt flux $J_s\delta$ (eq cm^{-1} sec^{-1}) plotted logarithmically against log c_e (c_e = the external concentration on the concentrated side in eq/liter.) The dilute side is always at 0.01 eq/liter.

*During this set of flux measurements, difficulty was experienced with a batch of membranes of abnormally high permeability. A scaling factor was determined carefully and in the work described here, use was made of this factor. Very recently extensive new measurements with normal membranes by T. Foley have confirmed the correctness of the values used here within the experimental uncertainty of ±2% except for NaBr above 0.50 eq/liter where the values given here appear to be a little too low.

TABLE 3

Test of the Additivity Principle for the Salt Flux J_s of $SrBr_2$ in Zeo-Karb 315 Membrane

c_1	c_2	c_3	$J_s(c_1 \to c_2)$	$J_s(c_2 \to c_3)$	$J_s(c_1 \to c_3)$	$[J_s(c_1 \to c_2) + J_s(c_2 \to c_3)]$
(eq/liter)			(eq cm^{-2} sec^{-1} x 10^{-8})			
0.01	0.05	0.10	0.124	0.165	0.284	0.289
0.01	0.05	0.20	0.124	0.467	0.590	0.591
0.05	0.10	0.20	0.165	0.297	0.467	0.462
0.10	0.50	1.00	1.332	1.455	2.760	2.787

Since the transport properties are functions of concentration the fluxes J_V and J_S are functions of integral or average values of the transport coefficients over the concentration ranges which were studied. In order to use J_V and J_S to evaluate friction and other coefficients of the membrane at a definite and known composition a differentiation procedure is necessary (17).

The flux of water J_W in moles per square centimeter second may be evaluated from J_V and J_S according to

$$J_W = (J_V - J_S \overline{V}_S)/\overline{V}_W \tag{25}$$

where $\overline{V}_S$ and $\overline{V}_W$ are the partial molar volumes of the salt and water respectively in the external solution into which J_V was measured.

The flux measurements described above when added to those determined previously give a total of seven pieces of independent information on fluxes. They are:

1. electrical conductance
2. ionic transport number
3. electroösmotic flux
4. salt diffusion flux
5. volume or water osmotic flux
6. counterion tracer diffusion coefficient
7. co-ion tracer diffusion coefficient

Eight phenomenological transport coefficients are needed to characterize this system because in addition to the six coefficients required for the system of three mobile components--counterions, co-ions, and water--the introduction of the tracer diffusion coefficients brings in two more phenomenological cross coefficients which are concerned with the interaction between each kind of ions and their radio-tracer isotopes.

An eighth transport coefficient can be obtained by making measurements in which a pressure difference is one of the driving forces (17). Often this measurement cannot be successfully achieved under clearly defined boundary conditions. Unless an eighth item of independent information is available it is necessary to make at least one assumption in order to estimate complete sets of phenomenological coefficients. Here we shall explore the consequences of four possible assumptions; all are based on a supposed understanding of the physical nature of interactions between fluxes.

IV. SPIEGLER'S APPROXIMATE FRICTIONAL TREATMENT (METHODS I AND II)

In view of the strong concentration dependence of the transport coefficients it would be very convenient if they could be evaluated from measurements made only at constant concentration. The electrical properties, 1-3 in the list above, and the tracer diffusion coefficients, 6 and 7, come into this category.

In conjunction with his mathematical treatment of the frictional model, Spiegler (7) suggested that the frictional coefficients might be evaluated approximately from these five items of data. In order to achieve this he proposed to neglect the frictional interactions between tracer and nontracer ions of the same ionic species and also the interaction between co-ions and counterions. This left five coefficients to be evaluated from the five items of data.

These assumptions may be expected to hold reasonably well when the membrane is in contact with very dilute solutions since the concentration of co-ions is then small. An alternative is to assume that the frictional interaction between the co-ions and the polymeric matrix of the ion-exchange membrane is negligible. This also should hold in dilute solutions and it might be expected to hold up to somewhat higher concentrations because the interaction between the co-ions and the matrix is governed by the electrostatic repulsion between them. Furthermore the co-ions are most concentrated locally in those microregions of the membrane where the fixed charge concentration and hence the matrix concentration are lowest (64). These two versions of Spiegler's theory will be examined and tested before some alternatives are considered.

The cations, anions, water, and matrix are distinguished in what follows by the subscripts 1, 2, 3, and 4 respectively. Concentrations in the membrane in moles per cubic centimeter of swollen membrane are denoted by c_i, and ionic valencies by z_1 and z_2, where z_2 is a negative integer. Symbols used to denote the transport properties are tracer diffusion coefficients D_1 and D_2, the electrical transport numbers t_1 and t_2, electroösmotic transference of water t_3 moles per Faraday, and specific conductance k. F, R, and T have their usual meanings.

Five parameters related to the experimental data are defined by

$$a = RT/D_1 \tag{26}$$

$$b = RT/D_2 \tag{27}$$

$$A = t_1 k/z_1 c_1 F^2 \tag{28}$$

$$B = t_2 k / z_2 c_2 F^2 \tag{29}$$

$$C = t_3 k / c_3 F^2 \tag{30}$$

A, B, and C are the velocities of the counterions, co-ions, and water, respectively, when a potential gradient of $-1/F$ V cm^{-1} is applied across the membrane. Under these circumstances the forces per mole on the components 1, 2, and 3 are z_1, z_2, and zero, respectively. In this case Eq. (18) becomes

$$z_1 = A(f_{12} + f_{13} + f_{14}) - Bf_{12} - Cf_{13} \tag{31}$$

$$z_2 = B(f_{21} + f_{23} + f_{24}) - Af_{21} - Cf_{23} \tag{32}$$

$$0 = C(f_{31} + f_{32} + f_{34}) - Af_{31} - Bf_{32} \tag{33}$$

since the velocity of the matrix v_4 is zero.

The friction coefficients which appear in Eq. (33) may conveniently be changed into those which appear in Eq. (31) and (32) by using Eq. (19). This gives, in place of (33)

$$0 = C(c_1 f_{13} + c_2 f_{23} + c_3 f_{34}) - Ac_1 f_{13} - Bc_2 f_{23} \tag{34}$$

The f_{ik} coefficients are related to the tracer diffusion coefficients (46) and hence to the parameters a and b by

$$a = f_{1^*1}(c_1 + c_{1^*})/c_1 + f_{12} + f_{13} + f_{14} \tag{35}$$

$$b = f_{2^*2}(c_2 + c_{2^*})/c_2 + f_{21} + f_{23} + f_{24} \tag{36}$$

since the kinetic similarity of tracer and nontracer ensures that $f_{i^*k} = f_{ik}$. Here c_{i^*} is the concentration of tracer i and f_{i^*i} is the friction coefficient between the tracer i^* and normal i. In usual tracer practice $(c_i + c_{i^*})/c_i$ may be replaced by unity.

A. Approximation I

In order to extract the f_{ik} coefficients from measurements made only at constant concentration, the approximations are made that

$$f_{1^*1} = f_{2^*2} = 0 \tag{37}$$

Spiegler (7) suggested the additional approximation

$$f_{12} = 0 \tag{38}$$

whence from Eq. (19)

$$f_{21} = 0 \tag{39}$$

With this set of approximations introduced, Eqs. (31)-(39) may be solved to give the following expressions for the nonzero f_{ik} coefficients.

$$f_{13} = (Aa - z_1)/C \tag{40}$$

$$f_{14} = a - f_{13} \tag{41}$$

$$f_{23} = (Bb - z_2)/C \tag{42}$$

$$f_{24} = b - f_{23} \tag{43}$$

$$f_{34} = [c_1 f_{13}(A - C) + c_2 f_{23}(B - C)]/c_3 c \tag{44}$$

The five relations of Eqs. (40)-(44) may be used to calculate the f_{ik} coefficients as functions of the external solution concentration by making use of the defining relations Eqs. (26)-(30), and the experimental data on the concentrations, transport numbers, diffusion coefficients, conductance, and electroösmosis referred to earlier.

Once the f_{ik} coefficients have been evaluated it is a straightforward matter to evaluate the R_{ik} $(i \neq k)$ from

$$R_{ik} = -f_{ik}/c_k \tag{45}$$

and the R_{ii} from

$$R_{ii} = \sum_{k \neq i} f_{ik}/c_i \tag{46}$$

The L_{ik} coefficients follow from inverting the matrix of Eqs. (14) whence

$$L_{ik} = B_{ik}/|R| \tag{47}$$

where $|R|$ is the determinant of the R_{ik} coefficients and B_{ik} the minor of R_{ik}.

In connection with Eqs. (6) it was indicated that the forces X_i in the nonequilibrium thermodynamic formulation could be represented by either the differences in the respective potentials between the two faces of a membrane or as the local gradients of these potentials. The former is used in the discontinuous formulation and the latter in the continuous formulation of the system. Since the flux densities and entropy production per square centimeter of membrane must be unaffected by the definition of the forces it follows that the dimensions and values of the L_{ik} and R_{ik} coefficients must be matched to the choice of forces. The formulation used above corresponds with choosing the gradients as forces and the tables of R_{ik} and L_{ik} are presented on this basis of regarding a homogeneous resin membrane as a continuous system in mechanical equilibrium.

Although setting f_{12} at zero may hold quite well in dilute solution, the reciprocal coefficient f_{21} is equal to $(c_1/c_2)f_{12}$ and the ratio c_1/c_2 increases rapidly as the external solution becomes more dilute because the co-ions are so effectively excluded from the membrane. Indeed, instead of tending to zero at zero concentration, as does f_{12}, it is likely that f_{21} increases somewhat.

B. Approximation II

A sensible alternative to Spiegler's approximation is, as explained above, to set

$$f_{24} = 0 \tag{48}$$

With the aid of Eq. (48), Eqs. (31)-(37) may be solved for the remaining f_{ik} coefficients, including f_{12}. The solutions are

$$f_{21} = c_1 f_{12}/c_2 = [b(B - C) - z_2]/(A - C) \tag{49}$$

$$f_{13} = (Aa - z_1 - Bf_{12})/C \tag{50}$$

$$f_{14} = a - f_{12} - f_{13} \tag{51}$$

$$f_{23} = b - f_{21} \tag{52}$$

$$f_{34} = [c_1 f_{13}(A - C) + c_2 f_{23}(B - C)]/c_3 C \tag{53}$$

Hence from Eqs. (37), (48), (49)-(53) and Eqs. (45)-(47) a further set of f_{ik}, R_{ik}, and L_{ik} coefficients may be evaluated by using the same experimental data. This permits a comparison to be made between the assumptions of $f_{12} = 0$ and $f_{24} = 0$.

C. The Transport Coefficients

Table 4 lists the f_{ik} coefficients calculated on the basis of both the above assumptions. In each case for NaBr at 1.00 eq/liter and $SrBr_2$ at 0.50 and 1.00 eq/liter the coefficient L_{33} derived from the f_{ik} coefficients was negative. This is in contradiction with the requirements of thermodynamics. Evidently both assumptions are seriously in error at these high concentrations and no coefficients have been listed in Table 4.

Examination of the columns headed f_{13} and f_{14} shows that the choice of $f_{12} = 0$ or $f_{24} = 0$ has no important effect on the calculated values of the counterion friction coefficients. Usually f_{23} is smaller when f_{12} is set at zero than when f_{24} is set at zero. In considering the meaning of this observation it must be remembered that the total co-ion friction is given by $(f_{21} + f_{23} + f_{24})$ and that $f_{21} = c_1 f_{12}/c_2$. Thus, although neglecting f_{12} in comparison with $(f_{13} + f_{14})$ may be permissible, f_{21} may be far more important since $c_1 \gg c_2$, especially in dilute external solutions. The wide divergences between f_{23} from Method I and from Method II at low c_e are probably due, in part at least, to the extreme sensitivity of the results to experimental errors in c_2 when f_{12} is set at zero. This belief is supported also by the apparently large values of f_{24} which arise at low concentrations. These do not accord with any reasonable interpretation of the physical situation of the co-ions when the membrane is in dilute external solutions.

When comparing the columns of f_{24} from Method I and f_{12} from Method II in Table 4 it is tempting to suppose that neglecting the latter introduces less error than neglecting the former, but the remark above about the relative values of f_{12} and f_{21} shows that no such simple conclusion may be drawn. Even the negative values of f_{24} may be real. Although the notion of negative friction between moving particles cannot be interpreted literally in a macroscopic sense, such values have been observed before in the case of interactions between the fluxes of ions of like charge (51).

The values of f_{34} in dilute solutions do not depend greatly on the assumption used. In more concentrated solutions a consequence of this method of calculation, which does not make use of any data from experiments in which the driving force is applied to the water, is that f_{34} varies erratically with errors in the other coefficients. Overall these data on the f_{ik} coefficients do not show any distinct relative advantage for either the assumption $f_{12} = 0$ or $f_{24} = 0$.

When the f_{ik} coefficients are combined according to Eqs. (45) and (46) to obtain the R_{ii} and R_{ik} resistance coefficients defined in Eqs. (14) it is found that the values of the straight R_{ii} coefficients scarcely depend upon whether f_{12} or f_{24} are set at zero (see Table 5). The same comment

TABLE 4

Frictional Coefficients Calculated by Spiegler's Method
I, Assuming $f_{12} = 0$; II, Assuming $f_{24} = 0$. (f_{ik} in J sec cm^{-2} $mole^{-1}$)

c_e	10^{-8} f_{13}		10^{-8} f_{23}		10^{-8} f_{14}		10^{-8} f_{24}		10^{-8} f_{12}		10^{-6} f_{34}	
eq/liter	I	II	I	II	I	II	I	II	I	II	I	II
					NaBr							
0.01	3.9	3.9	-0.3	11.0	4.2	4.3	4.0	0.0	0.0	0.0	1.6	1.5
0.02	3.5	3.5	1.3	9.0	4.1	4.2	2.1	0.0	0.0	0.0	1.7	1.6
0.03	3.2	3.1	2.0	6.5	4.1	4.2	1.3	0.0	0.0	0.0	1.6	1.5
0.05	2.8	2.8	2.6	4.4	4.1	4.1	0.6	0.0	0.0	0.0	1.7	1.5
0.10	2.8	2.8	3.2	3.0	3.4	3.4	-0.1	0.0	0.0	0.0	2.2	2.3
0.20	2.1	2.0	2.7	3.5	3.3	3.5	0.3	0.0	0.0	0.0	1.3	0.8
0.30	2.6	2.6	2.9	3.3	2.5	2.6	0.2	0.0	0.0	-0.1	2.1	1.7
0.50	2.1	2.0	2.7	3.3	2.6	2.9	0.3	0.0	0.0	0.0	0.1	1.6
					CsBr							
0.01	4.5	4.4	1.5	5.6	3.3	3.3	1.4	0.0	0.0	0.0	3.1	3.1
0.02	4.6	4.6	2.2	3.8	3.1	3.2	0.6	0.0	0.0	0.0	3.5	3.5
0.03	4.4	4.4	2.5	3.1	3.0	3.0	0.2	0.0	0.0	0.0	3.7	3.7
0.05	3.5	3.5	2.6	2.8	3.0	3.0	0.1	0.0	0.0	0.0	3.5	3.5
0.07	2.9	2.9	2.5	2.9	3.0	3.0	0.2	0.0	0.0	0.0	3.1	3.0
0.10	2.3	2.2	2.1	3.2	3.0	3.1	0.6	0.0	0.0	0.0	2.6	2.2
					$SrBr_2$							
0.01	10.2	9.8	0.0	8.9	18.7	19.2	2.9	0.0	0.0	-0.1	3.3	2.7
0.02	10.1	9.9	2.3	3.8	17.6	17.7	0.5	0.0	0.0	0.0	3.5	3.3
0.03	9.3	9.6	3.6	1.5	17.5	17.2	-0.8	0.0	0.0	0.1	3.0	3.5
0.05	8.6	9.6	4.5	0.3	16.7	15.6	-1.6	0.0	0.0	0.2	2.4	4.2
0.10	9.7	10.9	4.3	1.2	12.9	11.3	-1.4	0.0	0.0	0.3	3.2	6.3
0.20	6.9	8.4	3.9	1.9	11.7	9.9	-1.0	0.0	0.0	0.3	0.1	4.6
0.30	7.1	7.2	2.9	2.8	9.8	9.7	-0.1	0.0	0.0	0.0	1.5	2.0

applies to R_{13}, but the differences between the two sets of f_{23} are shown equally by the values of R_{23} in dilute solutions although they agree reasonably above 0.1 eq/liter. Since one expects $R_{23} \simeq R_{13}$, it appears that below 0.10 eq/liter $f_{12} = 0$ is the preferable assumption.

The L_{ik} coefficients may also be calculated from the f_{ik} coefficients in Table 4. Data on these are given in Table 6. The data for $f_{12} = 0$ only have been tabulated because this assumption appeared to give the better results with the R_{ik}. The columns in Table 6 give the values of L_{ii}/c_i and $L_{ik}/c_i c_k$ as these functions are not too dependent on concentration.

TABLE 5

R_{ii} and R_{ik} Coefficients in J cm sec mole^{-2} Calculated from the f_{ij} Coefficients in Table 4 for NaBr and $SrBr_2$; I Assuming $f_{12} = 0$; II Assuming $f_{24} = 0$, and; III from Model III[a]

c_e	$10^{-12} \times R_{11}$			$10^{-12} \times R_{22}$			$10^{-8} \times R_{33}$			$10^{-12} \times R_{12}$			$10^{-8} \times R_{13}$			$10^{-8} \times R_{23}$		
eq/liter	I	II	III	I	II	III	I	II	III	I	II	III	I	II	III	I	II	III
									NaBr									
0.01	1.50	1.50	1.24	556	556	509	1.62	1.60	0.85	0.00	2.37	0.00	-94	-94	-49	6	-399	- 8
0.02	1.40	1.40	1.19	151	151	149	1.54	1.50	0.83	0.00	1.04	0.00	-85	-85	-46	-31	-218	-32
0.05	1.23	1.23	1.12	31	31	32	1.34	1.31	0.91	0.00	0.22	0.00	-69	-68	-47	-63	-107	-57
0.10	1.07	1.07	0.99	10.1	10.1	10.0	1.54	1.55	1.09	0.00	-0.02	0.00	-67	-67	-49	-78	-72	-79
0.20	0.86	0.86	0.97	3.7	3.7	4.3	1.24	1.12	2.15	0.00	0.06	0.00	-52	-49	-81	-67	-84	-46
0.50	0.58	0.58	0.70	1.1	1.1	1.2	1.48	1.12	3.94	0.00	0.04	0.00	-52	-48	-106	-66	-82	-52
									$SrBr_2$									
0.01	10.4	10.4	10.5	88	88	122	2.47	2.27	2.59	0.00	2.18	0.00	-248	-238	-253	1.5	-215	154
0.02	9.9	9.9	9.2	36	36	39	2.51	2.44	1.87	0.00	0.35	0.00	-244	-241	-181	-56	-93	-30
0.05	8.8	8.8	7.5	11.4	11.4	11.9	2.12	2.64	0.45	0.00	-0.90	0.00	-209	-232	-61	-109	-6	-97
0.10	7.4	7.4	9.3	5.0	4.9	6.9	2.68	3.55	7.16	0.00	-0.53	0.00	-236	-268	-509	-104	-29	10
0.20	5.5	5.5	6.0	2.3	2.3	2.5	1.70	2.95	3.70	0.00	-0.26	0.00	-169	-204	-257	-94	-47	-60

[a]See Sec. V.C.

TABLE 6

L_{ii}/c_i in cm^2 mole J^{-1} sec^{-1} and L_{ik}/c_ic_k in cm^5 J^{-1} sec^{-1} Calculated; I, from Friction Coefficients in Table 4 with $f_{12} = 0$; and III, from Model III[a]

ce	$10^9 \times L_{11}/c_1$		$10^9 \times L_{22}/c_2$		$10^7 \times L_{33}/c_3$		$10^6 \times L_{12}/c_1c_2$		$10^6 \times L_{13}/c_1c_3$		$10^6 \times L_{23}/c_2c_3$	
eq/liter	I	III	I	III	I	III	I	III	I	III	I	III
						NaBr						
0.01	1.94	1.94	2.71	2.95	2.35	3.71	-0.20	0.26	2.73	2.73	-0.40	0.85
0.02	1.97	1.97	3.00	3.04	2.38	3.72	1.01	1.08	2.65	2.66	2.20	3.66
0.05	2.04	2.03	3.19	3.05	2.55	3.43	2.08	1.82	2.54	2.56	5.05	5.91
0.10	2.26	2.26	3.40	3.49	2.29	3.06	2.56	2.73	2.48	2.59	5.74	7.84
0.20	2.54	2.43	3.73	2.90	3.00	1.72	2.54	1.23	2.85	2.29	6.50	2.23
0.50	3.74	3.04	5.38	3.26	3.99	1.15	3.91	1.33	4.40	2.12	8.68	1.77
						CsBr						
0.01	2.07	2.07	3.48	4.34	1.74	4.56	1.28	2.84	2.48	2.50	2.23	12.91
0.02	2.11	2.11	3.61	3.45	1.65	4.35	1.95	1.68	2.45	2.47	3.28	7.36
0.05	2.25	2.24	3.76	3.11	1.74	3.10	2.31	1.15	2.34	2.34	4.26	3.82
0.07	2.35	2.34	3.82	3.51	1.92	3.12	2.20	1.66	2.33	2.37	4.53	5.48
0.10	2.49	2.48	3.87	3.62	2.18	3.77	1.89	1.46	2.35	2.35	4.34	4.29
						$SrBr_2$						
0.01	0.45	0.45	3.51	2.56	1.29	1.24	-0.02	-1.73	1.10	1.07	-0.07	-4.83
0.02	0.48	0.48	3.53	3.26	1.28	1.60	0.91	0.44	1.13	1.11	2.51	1.54
0.05	0.52	0.53	3.75	4.18	1.60	7.50	2.08	2.83	1.32	2.11	6.09	24.4
0.10	0.64	0.57	3.96	2.53	1.42	0.55	2.23	-0.11	1.48	0.98	5.18	-0.15
0.20	0.89	0.71	5.24	3.34	3.07	0.99	3.75	0.97	2.77	1.24	10.10	1.86
0.30	0.90	0.69	4.37	2.62	1.94	-1.19	2.03	-0.38	1.98	-0.24	4.82	-4.5

[a]See Sec. V.C.

L_{ii}/c_i may be thought of as an intrinsic mobility coefficient of i which would be observed if all other fluxes and forces other than those on i were held at zero. L_{ik}/c_ic_k is a measure of the flow interaction per molecule of i and k. It should be noted that setting $f_{12} = 0$ does not result in $L_{12} = 0$ since the ion flows interact through coulombic as well as frictional forces.

The physical interpretation of the trends in these transport coefficients will be considered at the end of the chapter.

D. A Test of Spiegler's Treatment

The data on the fluxes of salt and water driven across the membrane by a concentration difference were not used in deriving the sets of phenomenological coefficients. Consequently some aspects of Spiegler's treatment may be tested by using the coefficients to predict the salt and water fluxes and then comparing these predictions with the experimental data. These calculations may be carried out most readily by expressing the salt and water fluxes in terms of the L_{ik} coefficients.

The linear flux equations [Eqs. (6)] may be written out for the set of particle fluxes J_1, J_2, and J_3 expressed in mole per square centimeter second by using the corresponding negative electrochemical potential gradients $(-\partial\mu_i/\partial x)_T$ as the correctly conjugated forces. The flux equations in this form are

$$J_1 = -L_{11}\frac{\partial\mu_1}{\partial x} - L_{12}\frac{\partial\mu_2}{\partial x} - L_{13}\frac{\partial\mu_3}{\partial x} \tag{54}$$

$$J_2 = -L_{21}\frac{\partial\mu_1}{\partial x} - L_{22}\frac{\partial\mu_2}{\partial x} - L_{23}\frac{\partial\mu_3}{\partial x} \tag{55}$$

$$J_3 = -L_{31}\frac{\partial\mu_1}{\partial x} - L_{32}\frac{\partial\mu_2}{\partial x} - L_{33}\frac{\partial\mu_3}{\partial x} \tag{56}$$

In what follows Onsager's reciprocal relation [Eq. (7)] will be used without further comment. If subscript s refers to the neutral salt then

$$\partial\mu_s = \nu_1\partial\mu_1 + \nu_2\partial\mu_2 \tag{57}$$

where ν_1 and ν_2 are the numbers of cations and anions respectively per molecule of salt. Electroneutrality requires

$$z_1\nu_1 + z_2\nu_2 = 0 \tag{58}$$

When salt is transported in the absence of a net electric current

$$z_1 J_1 + z_2 J_2 = 0 \tag{59}$$

must hold.

The salt flux J_s is given by

$$J_s = J_1/\nu_1 = J_2/\nu_2 \tag{60}$$

Equation (56) may be used to eliminate $\partial\mu_2$ from Eqs. (54) and (55). The resulting equations may be combined to eliminate $\partial\mu_1$ and when use is made of Eqs. (58)-(60) the following expression is obtained for J_s

$$J_s = -\left(\frac{z_1}{\nu_2}\right)^2 \left[\frac{L_{11}L_{22} - L_{12}^2}{z_1^2 L_{11} + 2z_1 z_2 L_{12} + z_2^2 L_{22}}\right]\left(\frac{\partial\mu_s}{\partial x}\right) - \left(\frac{z_1}{\nu_2}\right)\left[\frac{L_{23}(z_1 L_{11} + z_2 L_{12}) - L_{13}(z_2 L_{22} + z_1 L_{12})}{z_1^2 L_{11} + 2z_1 z_2 L_{12} + z_2^2 L_{22}}\right]\left(\frac{\partial\mu_3}{\partial x}\right) \tag{61}$$

Equation (1) will be written for brevity in the form

$$J_s = -\theta\left(\frac{\partial\mu_s}{\partial x}\right) - \phi\left(\frac{\partial\mu_3}{\partial x}\right) \tag{62}$$

where θ and ϕ are the coefficients of $(\partial\mu_s/\partial x)$ and $(\partial\mu_3/\partial x)$ in Eq. (61).

Equation (62) is now multiplied through by dx and integrated across the membrane, whose thickness is δ, to give

$$J_s\delta = -\int_{x=0}^{x=\delta} \theta\left(\frac{\partial\mu_s}{\partial x}\right)dx - \int_{x=0}^{x=\delta} \phi\left(\frac{\partial\mu_3}{\partial x}\right)dx \tag{63}$$

Provided the membrane is homogeneous--and hence the local L_{ik} coefficients are functions only of the local composition, i.e., of μ_s and μ_3, and not explicitly of position x--Eq. (63) may be written

$$J_s\delta = -\int_{\mu_s(x=0)}^{\mu_s(x=\delta)} \theta\, d\mu_s - \int_{\mu_3(x=0)}^{\mu_3(x=\delta)} \phi\, d\mu_3 \tag{64}$$

On making the usual assumption (which is expected to hold in our well-stirred membrane cell) that thermodynamic equilibrium holds at the membrane/solution interfaces; the limits of the integrations in Eq. (64) are the chemical potentials in the external solutions, i.e., $\mu_s(1)$, $\mu_s(2)$, $\mu_3(1)$, and $\mu_3(2)$, where (1) and (2) refer to the external solutions on the sides $x = 0$ and $x = \delta$, respectively. Furthermore the Gibbs-Duhem relation holds in these solutions, i.e.,

$$m_s d\mu_s + 55.51\, d\mu_3 = 0 \tag{65}$$

where m_s is the molal concentration, and

$$d\mu_s = RT\, d\, \ell n\, a_s \tag{66}$$

where a_s is the activity of the salt.

When $(\partial\mu_1/\partial x)$ and $(\partial\mu_2/\partial x)$ are eliminated from Eq. (56)by a procedure similar to that used to obtain Eq. (61), the following expression is obtained for the water flux J_3, which will in what follows be written J_W to conform with the notation used earlier

$$J_W = \left(\frac{z_1}{2}\right)\left[\frac{L_{13}(z_2L_{22} + z_1L_{12}) - L_{23}(z_1L_{11} + z_2L_{12})}{z_1^2L_{11} + 2z_1z_2L_{12} + z_2^2L_{22}}\right]\left(\frac{\partial\mu_s}{\partial x}\right) + \left[\frac{(z_1L_{13} + z_2L_{23})^2}{(z_1^2L_{11} + 2z_1z_2L_{12} + z_2^2L_{22})} - L_{33}\right]\left(\frac{\partial\mu_3}{\partial x}\right) \tag{67}$$

Equation (67) may, like Eq. (61), be integrated across the membrane on the assumptions of its homogeneity and membrane/solution equilibrium at the boundaries. The Gibbs-Duhem equation is used again to introduce the external salt activity difference as the sole driving force.

In order to predict J_S and J_W the functions of the L_{ik} coefficients which appear in Eqs. (61) and (67) have been evaluated at each value of the external concentration m_S for which data are available in Table 4. These functions of the L_{ik} coefficients have been plotted against ℓna_S and integrated graphically between the appropriate values of ℓna_S, which correspond with the values of m_S for which the prediction of flux is described. The evaluation of J_S and J_W is then straightforward. Data on the activity coefficients which connect a_S and m_S are to be found in many standard compilations.

The results of these calculations appear in Table 7. The values of the L_{ik} coefficients derived with $f_{12} = 0$ have been used in the calculations which

TABLE 7

Comparison Between Normalized Fluxes of Salt $J_s\delta$ and Water $J_w\delta$, Calculated from Spiegler's Approximation Method I and Observed.

c_e	$J_s\delta$		$J_w\delta$	
(eq/liter)	(mole cm^{-1} sec^{-1} x 10^{-10}) Calculated	Observed	(mole cm^{-1} sec^{-1} x 10^{-9} Calculated	Observed
		NaBr		
0.02	0.12	0.13	8	9
0.03	0.29	0.31	16	17
0.05	0.82	0.81	29	32
0.10	2.54	2.27	50	54
0.20	6.48	6.35	82	78
0.30	11.38	10.40	118	89
0.50	19.06	19.20	179	100
		CsBr		
0.02	0.17	0.18	5	10
0.03	0.44	0.41	8	19
0.05	1.14	0.99	14	33
0.07	2.00	1.68	19	43
0.10	3.42	2.97	27	55
		$SrBr_2$		
0.02	0.20	0.20	0.7	1.6
0.03	0.42	0.41	0.0	1.8
0.05	0.91	0.85	-2.8	1.5
0.10	2.24	1.94	-13.1	-1.4
0.20	4.86	3.98	-33.7	-9.0

are tabulated but closely similar values of J_s and J_w are predicted when the L_{ik} coefficients estimated with $f_{24} = 0$ are used.

The observed and calculated salt and water fluxes are compared in the table for a series of concentrations c_e on one side of the membrane and with a constant concentration of 0.01 eq/liter on the other. The table actually lists $J_s\delta$ and $J_w\delta$ i.e. the fluxes are scaled to a membrane of 1 cm thickness. The experimental flux of water J_w was obtained from the measured volume and salt fluxes by using the relation

$$J_w = (J_v - J_s\bar{V}_s)/\bar{V}_w \tag{68}$$

It may be seen from Table 7 that the observed and calculated salt fluxes agree excellently at low concentrations although the calculated values appear to be somewhat too high at the highest concentrations. Such an agreement might reasonably have been expected because Spiegler's method of evaluating the phenomenological coefficients depends heavily on the various measurements of ionic mobility. These are also the most important parameters which influence the flux of salt down a concentration gradient.

The comparison between the observed and calculated water fluxes is more perplexing. For NaBr at low concentrations the agreement is good but the calculated fluxes are too large at higher concentrations. For example $J_w\delta$, when calculated for an experiment with 0.30 eq/liter on one side of the membrane and 0.50 eq/liter on the other, is 61×10^{-9} while the observed value is 11×10^{-9}. Data on CsBr are available only at low concentrations and the calculated flux is consistently about a half of that observed.

The case of $SrBr_2$ is particularly interesting. In this system anomalous negative osmosis is observed (65). Table 7 shows that the water flux J_w also becomes positive in the direction in which its chemical potential increases at higher concentrations. It is encouraging to note that the calculated fluxes also predict that anomalous negative osmosis will take place at about the same concentrations at which it is actually observed. The quantitative agreement is not good; the calculated negative flows between 0.05 and 0.10 eq/liter and 0.05 and 0.20 are about three times those observed.

It is felt that the lack of quantitative agreement in the water fluxes between theory and experiment may be due to the lack of any data on flows driven by a force applied to the water being used in Spiegler's approximate scheme for evaluating the phenomenological coefficients. This scheme results in the major uncertainties in the coefficients accumulating principally in f_{34} and in L_{33}. We now describe three alternative methods of estimating the coefficients from the experimental data which we have available. It is hoped that these alternative methods may give more reliable values of L_{33}.

V. OTHER METHODS OF ESTIMATING THE L_{ik} COEFFICIENTS

We have available five independent types of transport data which may be related unequivocally to the L_{ik} coefficients. They are the electric conductance, the ionic transport number, the electroösmotic flow, the salt flux, and the water flux driven by a concentration gradient. All have been determined as functions of the external solution concentration and at constant temperature and pressure. To determine six L_{ik} coefficients from these data requires at least one assumption to be made.

In order to proceed one has to face first a practical problem of algebra. The experimental information can be expressed in simple equations containing the L_{ik} coefficients. Much more complex equations result when the data are related to the R_{ik} coefficients. It has to be remembered that an explicit solution of the set of equations is ultimately required.

When the question of choosing an assumption about the transport properties which is likely to be obeyed in practice is considered, one thinks naturally in terms of resistance or frictional coefficients because these have a relatively direct physical meaning. When examining Spiegler's treatment f_{1^*1} and f_{2^*2} were always taken as zero, in order to make use of the tracer diffusion coefficients, and either f_{12} or f_{24} was also neglected.

Of the former assumptions the one more likely to hold well is $f_{2^*2} = 0$ because the total concentration of co-ions is much lower than that of the counterions, especially when in dilute external solutions. Of the latter assumptions $f_{12} = 0$ appeared to give a better representation of the facts than $f_{24} = 0$

In the following section we describe three new methods of estimating the L_{ik} coefficients which make use of the assumptions $f_{2^*2} = 0$ or $f_{12} = 0$ or both.

A. Assumptions and the L_{ik} Coefficients

The first step is to convert these assumptions about f_{ik} into equations involving L_{ik} coefficients.

In his derivation of the frictional model equations Spiegler (7) showed that for the case under consideration, a single salt and water as the only mobile components, when $f_{12} = 0$.

$$L_{12} L_{33} = L_{13} L_{23} \tag{69}$$

necessarily holds

The assumption $f_{2*2} = 0$ holds provided $R_{22}* = 0$. It will now be shown that when $R_{22}* = 0$ then

$$L_{22} = D_2 c_2/RT \tag{70}$$

holds to a good approximation, where D_2 is the tracer diffusion coefficient as normally measured, and c_2 is the total concentration (tracer + nontracer) of component 2 in the membrane.

The true relationship for the tracer diffusion coefficient of the ions in a membrane of unit thickness is (61)

$$L_{22} - c_2 L_{22}*/c_2* = D_2 c_2/RT \tag{71}$$

Hence in order to substantiate Eq. (70) it is necessary to show that

$$L_{22} \gg c_2 L_{22}*/c_2* \tag{72}$$

holds when

$$R_{22}* = 0 \tag{73}$$

The exact connection between $L_{22}*$ and $R_{22}*$ may be obtained by appropriate application of Eq. (15) but the result is too complex to enable the deviation of L_{22*} from zero to be assessed clearly. We shall instead adopt a procedure suggested by Wills and Lightfoot (54, 66).

The fundamental equation of the frictional model Eq. (18) may be re-written in the form

$$X_i = - v_i \sum_{j \neq i} c_j R_{ij} + \sum_{j \neq i} c_j R_{ij} v_j \tag{74}$$

A mobility $\bar{U}_i$ of each species is defined by

$$\bar{U}_i = - 1/ \sum_{j \neq i} c_j R_{ij} \tag{75}$$

where the summations extend over each component capable of exerting friction and hence include the membrane matrix. Multiplication of Eq. (74) by $\bar{U}_i c_i$ gives

$$\bar{U}_i c_i X_i = v_i c_i + \bar{U}_i c_i \sum_{j \neq i} c_j R_{ij} v_j \tag{76}$$

By making use of Eq. (20) the fluxes can be introduced. Whence

$$J_i = \bar{U}_i c_i X_i - \bar{U}_i c_i \sum_{j \neq i} J_j R_{ij} \tag{77}$$

An equation analagous to Eq. (77) but written for J_j may now be substituted into Eq. (77) to give

$$J_i = \bar{U}_i c_i X_i - \bar{U}_i c_i \sum_{j \neq i} \left[\bar{U}_j c_j X_j R_{ij} - \bar{U}_j c_j R_{ij} \sum_{k \neq j} J_k R_{jk} \right] \tag{78}$$

Clearly a similar substitution can be made in Eq. (78) for J_k and a series developed in increasing powers of R_{jk}.

Ideal fluxes $\bar{J}_i$ are defined by

$$\bar{J}_i = \bar{U}_i c_i X_i \tag{79}$$

Wills and Lightfoot showed that the series in Eq. (78) converged rapidly provided

$$\left| (\bar{J}_i - J_i)/J_i \right| << 1 \tag{80}$$

i.e., provided the directly driven component of the flux is large compared with that due to the coupling phenomena. This must certainly hold in a tracer diffusion experiment and we may safely make the approximation of replacing J_k by $\bar{J}_k$ in the third term of the series in Eq. (78).

In the case of tracer diffusion of the co-ions at constant total composition all the forces are zero except X_2 and X_{2*}. In this case Eq. (78), with $\bar{J}_k$ introduced in place of J_k, may be fully written out and rearranged to give

$$J_2 = X_2 \bar{U}_2 c_2 [1 + \bar{U}_2 c_2 (\bar{U}_1 c_1 R_{12}^2 + \bar{U}_{2*} c_{2*} R_{22*}^2 + \bar{U}_3 c_3 R_{23}^2 + \bar{U}_4 c_4 R_{24}^2]$$
$$- X_{2*} \bar{U}_2 c_2 \bar{U}_{2*} c_{2*} (R_{22*} - \bar{U}_1 c_1 R_{12} R_{12*} - \bar{U}_3 c_3 R_{23} R_{2*3} - \bar{U}_4 c_4 R_{24} R_{2*4})$$

(81)

A comparison of Eq. (81) with the linear flux Eqs. (6) serves to identify the following relations

$$L_{22}/c_2 = \bar{U}_2 + \bar{U}_2^{\,2} c_2 G \tag{82}$$

and

$$L_{22*}/c_{2*} = \bar{U}_2^{\,2} c_2 G \tag{83}$$

where

$$G = (\bar{U}_1 c_1 R_{12}^2 + \bar{U}_3 c_3 R_{23}^2 + \bar{U}_4 c_4 R_{24}^2) \tag{84}$$

In order to obtain Eqs. (82), (83), and (84) from Eq. (81) use has been made of the assumption $R_{22*} = 0$ and of the Onsager reciprocal relations which, also bearing in mind the kinetic similarity of isotopes of the same element, led to the identities

$$R_{2k} = R_{k2} = R_{k2*} = R_{2*k} \tag{85}$$

It may be seen that, although $R_{22*} = 0$, $L_{22*} \neq 0$ in this approximation. From Eqs. (75) and (84) we obtain

$$\bar{U}_2 c_2 G = \frac{c_2}{\sum\limits_{j\neq 2} c_j R_{2j}} \sum_{k\neq 2} \left[\frac{c_k R_{2k}^2}{\sum\limits_{\ell\neq k} c_\ell R_{k\ell}} \right] \tag{86}$$

The right-hand side of Eq. (86) may be expanded to a sum of three terms. Each term consists of a denominator which is a sum of nine terms and a numerator which, since c_2 is a small concentration, is also the smallest term in the denominator. Thus we may conclude that

$$\bar{U}_2 c_2 G \ll 1$$

holds, or from Eqs. (82) and (83)

$$L_{22*}/c_{2*} \ll L_{22}/c_2$$

Returning to Eq. (71) we find therefore

$$L_{22} \simeq D_2 c_2/RT$$

in conformity with Eq. (70).

It may be noted that if we had adopted the next lower approximation of Wills and Lightfoot and had replaced J_j by $\bar{J}_j$ in Eq. (77) we would have obtained $L_{22}* = 0$ directly on writing out in full the equation for tracer diffusion of the co-ions and introducing the assumption $R_{22}* = 0$.

B. Experimental Data and the L_{ik} Coefficients

In order to develop methods for determining the L_{ik} coefficients the experimental data items have to be expressed in terms of the L_{ik}. The salt and water fluxes have already been put into this form in Eqs. (61) and (67), respectively. Because of the strong dependence of the L_{ik} coefficients on concentration, the salt and water flux data have to be expressed for the present purposes in a differential form. Thus if the flux across a membrane of unit thickness--when the solute activities in the two external solutions differ by only a differential amount $d(\ln a_s)$--is written $d(J_i\delta)$ then the additivity principle for fluxes across a homogeneous membrane, demonstrated earlier in Tables 2 and 3, ensures that

$$d(J_s\delta)/d(\ln a_s) = d(J_s\delta)/d(\Delta \ln a_s) \tag{87}$$

and

$$d(J_w\delta)/d(\ln a_s) = d(J_w\delta)/d(\Delta \ln a_s) \tag{88}$$

where the differentials on the left hand sides refer to the particular external salt activity a_s at which the differential on the right-hand side is evaluated from the experimental data.

A series of fluxes observed or obtained from the additivity principle when the concentration on one side of the membrane is held constant and that on the other is varied are plotted against $\ln a_s$ on the side which is varied. Slopes of these plots are determined graphically at the chosen values of $\ln a_s$ in order to evaluate the required differentials. Then from Eqs. (61)-(67) we have

$$\frac{d(J_s\delta)}{d(\ln a_s)} = \left(\frac{RT}{z_1 \nu_1}\right)\left[\frac{1}{z_2\nu_2}(z_1{}^2 L_{11}t_2 - z_1 z_2 L_{12}t_1) + \frac{m_s}{55.51}(z_1 L_{13}t_2 - z_2 L_{23}t_1)\right] \tag{89}$$

and

$$\frac{d(J_w \delta)}{d(\ln a_s)} = \left[RT \frac{1}{z_2 \upsilon_2} (z_1 L_{13} t_2 - z_2 L_{23} t_1) + \frac{m_s}{55.51} (L_{33} - t_3{}^2 k/F^2) \right] \quad (90)$$

In order to express Eqs. (89) and (90) in a compact form, use has been made of the well-known relations between the electrical flow coefficient and the L_{ik} coefficients. These are (7, 12)

$$t_1 k/z_1 F^2 = z_1 L_{11} + z_2 L_{12} \quad (91)$$

$$t_2 k/z_2 F^2 = z_2 L_{22} + z_1 L_{12} \quad (92)$$

$$t_3 k/F^2 = z_1 L_{13} + z_2 L_{23} \quad (93)$$

C. Method III, $f_{12} = 0$

The five operational Eqs. (89)-(93) may be combined with Eq. (69), which derives from the assumption $f_{12} = 0$. The resulting set of six equations may be solved for the L_{ik} coefficients. The algebra is lengthy and the explicit solutions appear clumsy. The L_{ik} coefficients may be evaluated from the experimental data by making use, in sequence, of Eqs. (94)-(99) together with the definitions of the parameters L, M and N given in Eq. (100)-(102)

$$L_{23} = z_2 \upsilon_2 F^2 LN/(z_2 F^2 L - z_2 \upsilon_2 F^2 MN + t_3 kM) \quad (94)$$

$$L_{13} = t_3 k/z_1 F^2 - z_2 L_{23}/z_1 \quad (95)$$

$$L_{12} = L + ML_{23} \quad (96)$$

$$L_{11} = t_1 k/z_1{}^2 F^2 - z_2 L_{12}/z_1 \quad (97)$$

$$L_{22} = t_2 k/z_2{}^2 F^2 - z_1 L_{12}/z_2 \quad (98)$$

$$L_{33} = L_{13} L_{23}/L_{12} \quad (99)$$

where

$$L = \left(\frac{t_2 k}{z_2 F^2}\right)\left(\frac{t_1}{z_1} - Mt_3\right) - \left(\frac{\nu_1 \nu_2}{RT}\right) \cdot \frac{d(J_s \delta)}{d(\ln a_s)} \tag{100}$$

$$M = \nu_1 m_s / 55.51 \tag{101}$$

$$N = -\left(\frac{t_3 k}{\nu_1 F^2}\right)\left(\frac{t_2}{z_1} + Mt_3\right) - \left(\frac{1}{RT}\right) \cdot \frac{d(J_w \delta)}{d(\ln a_s)} \tag{102}$$

Sample results of these calculations are included in Table 6 where the values may be compared with those obtained from Spiegler's method in which f_{12} was assumed zero.

It can be seen that L_{11}/c_1 and L_{22}/c_2 increase with increasing c_e on both models but the extent of the increase appears to be less according to Method III.

The L_{ik} cross-coefficients vary somewhat irregularly on both models. Apart from $SrBr_2$ at 0.05 eq/liter, the new method tends to give the lower values for the cross-coefficients.

The major differences are to be found in the water flow straight coefficient L_{33}/c_3. This was to be expected in view of the failure of the coefficients calculated by Spiegler's method to predict J_w satisfactorily. According to Spiegler's method L_{33}/c_3 increases with increasing c_e whereas according to Method III it decreases. (The negative value of L_{33} with $SrBr_2$ at 0.30 eq/liter is thermodynamically impossible.)

A similar trend is found with L_{33}/c_3 for CsBr but here the values estimated by Method III are all larger than those from Spiegler's method. This finding is consistent with J_w observed in Table 7 being larger than J_w calculated from the coefficients of Method I.

Table 5 shows a similar comparison between the R_{ik} coefficients calculated by methods I, II, and III. The sets of R_{11} and R_{22} do not differ very significantly. R_{12} is assumed zero in I and III, and the other cross-coefficients vary too irregularly to permit any definite conclusions. R_{33} from Spiegler's Method I and from Method II was relatively constant for each salt but on the new Method III the values of R_{33} were found to vary to an improbably large extent with variations in c_e.

There seemed to be two possible reasons why Method III might give physically implausible values of some R_{ik} coefficients. Either the assumption $f_{12} = 0$ might be very unsatisfactory or some of the experimental data might be in serious error.

The assumption of $f_{12} = 0$ did not lead to implausible results at reasonable concentrations with Method I. The least accurate item of experimental data was $d(J_S\delta)/d(\ln a_S)$ because it is obtained by graphical differentiation of the net salt flux measurements each of which is the difference between two measured tracer fluxes. By making two assumptions instead of one it has been possible to devise another method of calculating the L_{ik} coefficients without having recourse to the salt flux data.

D. Method IV, $f_{12} = 0$ and $f_{2^*2} = 0$

By assuming that f_{12} and f_{2^*2} are both zero Eqs. (69) and (70) may be combined with the four experimental data equations (90)-(93) to give a set of six equations which may be solved for the L_{ik} coefficients.

The solutions may be expressed conveniently in the form:

$$L_{11} = \frac{1}{z_1^2}\left[\frac{(t_1 - t_2)k}{F^2} + \frac{z_2 D_2 c_2}{RT}\right] \tag{103}$$

$$L_{22} = D_2 c_2/RT$$

$$L_{12} = \frac{1}{z_1}\left[\frac{t_2 k}{z_2 F^2} - \frac{z_2 D_2 c_2}{RT}\right] \tag{104}$$

$$L_{13}^2 + PL_{13} + Q = 0 \tag{105}$$

where

$$P = \left[\frac{55.51 L_{12}}{\upsilon_2 m_S} + \frac{t_3 k}{z_1 F^2}\right] \tag{106}$$

and

$$Q = \left(\frac{55.51 z_2}{m_S z_1 F^2}\right)\left[\frac{F^2}{RT} \cdot \frac{d(J_W\delta)}{d(\ln a_S)} - \frac{t_1 t_3 k}{z_1 \upsilon_1} + \frac{m_S t_3^2 k}{55.51}\right] L_{12} \tag{107}$$

$$L_{23} = (t_3 k/F^2 - z_1 L_{13})/z_2 \tag{108}$$

$$L_{33} = L_{13} L_{23}/L_{12} \tag{109}$$

The correct solution of Eq. (105) is found by taking the positive square root of $(P^2 - 4Q)$.

The results of Methods III and IV are compared in Table 8 which lists the R_{ik} coefficients for NaBr and $SrBr_2$. When applied to CsBr the two methods gave results which agreed closely.

It may be noted that Method IV gave unsatisfactory values of the R_{ik} for NaBr and $SrBr_2$ at $c_s = 0.01$ eq/liter but gave satisfactory values with CsBr. The reason for this failure is not known. The slope $d(J_w\delta)/d(\ell n\ a_s)$ is most uncertain at the end of the experimental curve but this difficulty should affect Methods III and IV equally.

It is seen in Table 8 that Methods III and IV give values which are in essential agreement for R_{11}, R_{22}, and (by assumption) R_{12}. For NaBr, the coefficients R_{13}, R_{23}, and R_{33} from both methods agree as well as can be expected. For $SrBr_2$ Method III tends to give the larger values of R_{33} and R_{13} and very erratic values of R_{23}. There is no large systematic divergence between the results of the two methods, remembering also that there is good agreement with CsBr.

Method IV gives the more smoothly varying values of of the R_{ik} and L_{ik} coefficients, and the straight coefficients remain positive up to higher concentrations. It appears that the sequence of calculation used in Method IV makes the values of the coefficients less sensitive to small errors in the experimental data. Thus, in spite of requiring two assumptions, it appears to be a more satisfactory source of the thermodynamic coefficients than Method III. The results obtained with Method IV are given in more detail in Tables 9 and 10. These will be discussed after the final method of calculation has been described.

E. Method V, $f_{2*2} = 0$

The comparison of Methods III and IV having supported our confidence in the salt flux data used in method III, another method of evaluating the L_{ik} coefficients can be devised.

In principle this entails solving the set of six equations: (70) and (89)-(93). This approach rests only lightly on the frictional model in order to justify neglecting L_{22*}/c_{2*} and makes direct use of six items of experimental

TABLE 8

Comparison of R_{ii} and R_{ik} from Models III and IV.
The units are J cm sec mole^{-2}

c_e eq/liter	$10^{-12} R_{11}$ III	$10^{-12} R_{11}$ IV	$10^{-12} R_{22}$ III	$10^{-12} R_{22}$ IV	$10^{-8} R_{33}$ III	$10^{-8} R_{33}$ IV	$10^{-12} R_{12}$ III	$10^{-12} R_{12}$ IV	$10^{-8} R_{13}$ III	$10^{-8} R_{13}$ IV	$10^{-8} R_{23}$ III	$10^{-8} R_{23}$ IV
						NaBr						
0.02	1.19	1.19	149	151	0.83	0.84	0.0	0.0	-46	-47	-32	-31
0.05	1.12	1.12	32	31	0.91	0.89	0.0	0.0	-47	-46	-57	-61
0.10	0.99	1.01	10	11	1.09	1.21	0.0	0.0	-49	-53	-79	-72
0.20	0.97	0.93	4.3	4.0	2.15	1.85	0.0	0.0	-81	-72	-46	-58
0.50	0.70	0.70	1.2	1.2	3.94	3.89	0.0	0.0	-106	-105	-52	-52
						$SrBr_2$						
0.02	9.3	8.8	38.9	36.4	1.87	1.35	0.0	0.0	-181	-133	-30	-54
0.03	8.8	8.4	23.4	22.4	1.59	1.03	0.0	0.0	-156	-106	-59	-78
0.05	7.5	7.8	11.9	12.4	0.45	0.93	0.0	0.0	-61	-100	-97	-86
0.10	9.3	6.6	6.9	5.5	7.16	1.34	0.0	0.0	-509	-128	11	-74
0.20	6.0	5.7	2.5	2.5	3.70	2.86	0.0	0.0	-257	-213	-60	-66

data: D_2, k, t_1, t_3, J_S, and J_W. In principle it appears to be a more satisfactory route to the L_{ik} coefficients than any of the methods described so far.

It turns out that when the six items of data are expressed in terms of the L_{ik} coefficients the algebraic solution of the resultant set of equations is extremely awkward. A far simpler route can be found by choosing a different set of conjugated fluxes and forces (5,16). Full discussions of these fluxes and forces and of the $\mathcal{L}_{\alpha\beta}$ phenomenological coefficients which interconnect them have been given recently (17). The interested reader should refer to these papers for details. Only an outline will be given here of the principles involved and of the conversion of the $\mathcal{L}_{\alpha\beta}$ to the L_{ik} coefficients.

TABLE 9

L_{ii}/c_i in cm^2 mole J^{-1} sec^{-1} and L_{ik}/c_ic_k in cm^5 J^{-1} sec^{-1} Calculated from Spiegler's Approximation Model I, Model IV[a] and Model V[b]

c_e equiv/liter	$10^6 L_{11}^*/c_1c_1^*$ IV and V	$10^9 L_{11}/c_1$ I	IV	V	$10^9 L_{22}/c_2$ I	IV	V	$10^7 L_{33}/c_3$ I	IV	V	$10^6 L_{12}/c_1c_2$ I	IV	V	$10^6 L_{13}/c_1c_3$ I	IV	V	$10^6 L_{23}/c_2c_3$ I	IV	V
NaBr																			
0.02	1.22	1.97	1.97	1.97	3.00	3.00	3.01	2.38	3.70	3.55	1.01	1.01	1.04	2.65	2.66	2.65	2.20	3.40	0.95
0.05	1.05	2.04	2.04	2.04	3.19	3.15	3.14	2.55	3.50	3.72	2.08	2.00	2.00	2.54	2.57	2.61	5.05	6.60	8.49
0.10	1.09	2.26	2.25	2.25	3.40	3.21	3.21	2.29	2.76	2.41	2.56	2.25	2.24	2.48	2.49	2.39	5.74	6.03	3.99
0.20	1.01	2.54	2.48	2.48	3.73	3.26	3.26	3.00	2.01	2.24	2.54	1.78	1.78	2.85	2.46	2.59	6.50	3.54	4.51
0.50	1.14	3.74	3.05	3.05	5.38	3.29	3.29	3.99	1.17	1.15	3.91	1.36	1.36	4.40	2.14	2.13	8.68	1.83	1.79
1.00	0.96	0.77	3.37	3.37	-1.62	3.32	3.32	-2.91	0.73	0.55	-3.19	1.04	1.04	-3.04	1.61	1.33	-7.66	1.18	0.64
CsBr																			
0.02	1.46	2.11	2.11	2.11	3.61	3.60	3.62	1.65	4.45	5.08	1.95	1.94	1.95	2.45	2.48	2.52	3.28	8.66	17.05
0.05	1.58	2.25	2.25	2.25	3.76	3.71	3.70	1.74	3.74	5.34	2.31	2.21	2.21	2.35	2.44	2.70	4.26	8.47	20.24
0.10	1.00	2.49	2.49	2.49	3.87	3.74	3.74	2.18	2.92	3.05	1.87	1.66	1.66	2.35	2.40	2.44	4.34	5.07	5.78
$SrBr_2$																			
0.02	0.41	0.48	0.48	0.48	3.53	3.51	3.51	1.28	2.13	5.22	0.91	0.89	0.88	1.13	1.15	1.35	2.51	3.98	18.12
0.05	0.43	0.52	0.52	0.52	3.75	3.51	3.51	1.60	3.28	2.47	2.08	1.66	1.66	1.32	1.45	1.32	6.09	9.12	6.21
0.10	0.57	0.64	0.62	0.62	3.96	3.51	3.51	1.42	2.44	4.76	2.23	1.50	1.50	1.48	1.54	2.22	5.18	5.76	13.07
0.20	0.56	0.89	0.73	0.73	5.24	3.51	3.51	3.07	1.28	1.37	3.75	1.22	1.22	2.77	1.40	1.45	10.10	2.71	2.96
0.50	0.26	0.70	0.82	0.82	2.89	3.51	3.51	-0.73	0.60	0.12	-0.01	0.70	0.70	-0.01	1.04	0.55	-1.85	0.98	-0.33
1.00	-0.16	0.71	0.72	0.72	3.51	3.51	3.51	-0.34	0.18	0.07	-0.05	0.05	0.05	0.37	0.45	0.28	-0.11	0.05	-0.27

[a] See Sec. V.D.
[b] See Sec. V.E.

TABLE 10

R_{ik} Coefficients in J cm sec mole^{-2} Calculated from the L_{ik} Derived from Model I[a], Model IV[b], and Model V[c]

c_e eq/liter	$10^{-12}R_{11}$			$10^{-12}R_{22}$			$10^{-8}R_{33}$			$10^{-11}R_{11}^*$		$10^{-11}R_{12}$	$10^{-10}R_{13}$			$10^{-10}R_{23}$		
	I	IV	V	I	IV	V	I	IV	V	IV	V	V	I	IV	V	I	IV	V
									NaBr									
0.02	1.40	1.19	1.20	151	151	150	1.54	0.84	0.88	-2.17	-2.05	-1.63	-0.85	-0.47	-0.49	-0.31	-0.31	-0.22
0.05	1.23	1.12	1.11	30.9	31.5	31.9	1.34	0.89	0.84	-1.15	-1.24	+0.94	-0.69	-0.46	-0.44	-0.63	-0.61	-0.78
0.10	1.07	1.01	1.03	10.1	10.7	10.4	1.54	1.21	1.36	-0.58	-0.44	-1.18	-0.67	-0.53	-0.58	-0.78	-0.72	-0.46
0.20	0.86	0.93	0.93	3.71	3.97	4.11	1.24	1.85	1.72	+0.71	+0.71	+0.74	-0.52	-0.72	-0.69	-0.67	-0.58	-0.74
0.50	0.58	0.70	0.71	1.11	1.24	1.24	1.48	3.89	3.99	+1.24	+1.29	-0.02	-0.52	-1.05	-1.07	-0.66	-0.52	-0.52
1.00	0.38	0.50	0.47	0.49	0.57	0.53	0.98	7.35	8.22	+1.21	+0.84	-0.74	-0.46	-1.30	-1.26	-0.79	-0.57	-0.17
									CsBr									
0.02	1.39	1.00	0.98	102	103	103	2.45	0.66	0.58	-3.86	-4.02	+2.22	-1.14	-0.31	-0.27	-0.55	-0.54	-1.01
0.05	1.15	0.94	0.91	21.9	22.5	24.5	2.13	0.81	0.60	-0.74	-0.98	+2.93	-0.88	-0.35	-0.27	-0.66	-0.63	-1.23
0.10	0.89	0.84	0.84	7.51	7.78	7.84	1.55	1.09	1.05	-0.57	-0.59	+0.25	-0.57	-0.41	-0.40	-0.53	-0.48	-0.54
									$SrBr_2$									
0.02	9.85	8.79	8.17	36.1	36.4	38.5	2.51	1.35	0.54	-10.5	-16.7	+7.29	-2.44	-1.33	-0.61	-0.56	-0.54	-1.11
0.05	8.75	7.80	7.95	11.4	12.4	12.0	2.12	0.93	1.22	-9.5	-8.0	-1.99	-2.09	-0.99	-1.22	-1.09	-0.86	-0.72
0.10	7.38	6.60	6.82	4.95	5.45	6.62	2.68	1.34	0.85	-7.7	-5.6	+8.00	-2.36	-1.28	-1.10	-1.04	-0.74	-1,16
0.20	5.47	5.73	5.74	2.25	2.46	2.49	1.70	2.86	2.73	+2.5	+2.6	+0.47	-1.69	-2.13	-2.09	-0.94	-0.66	-0.70
0.50	3.22	4.60	8.44	0.87	0.92	1.16	-2.7	7.05	65.1	+13.7	+52.1	-14.76	-0.02	-3.52	-19.1	-0.72	-0.50	+4.18
1.00	2.14	3.82	4.12	0.48	0.48	0.54	-6.4	21.6	65.4	+16.8	+19.8	-3.67	+1.35	-5.50	-10.6	-0.09	-0.08	+2.14

[a] $R_{11}^* = 0$, $R_{12} = 0$, $R_{22}^* = 0$.
[b] $R_{12} = 0$, $R_{22}^* = 0$.
[c] $R_{22}^* = 0$.

In place of the three molar particle fluxes J_1, J_2, and J_3, or J_W used so far, we take instead the three fluxes J_1, the cation flux density, J_V, the volume flux density defined by

$$J_V = \bar{V}_S J_1/\upsilon_1 + \bar{V}_W J_W \tag{110}$$

and I, the electric current density given by

$$I = F(z_1 J_1 + z_2 J_2) \tag{111}$$

It may be shown (16) that the three forces which conjugate with these fluxes so as to give the dissipation function are

$$X_1 = \pi_S/\upsilon_1 \bar{c}_S \tag{112}$$

$$X_V = p - \pi_S \tag{113}$$

$$X_I = E \tag{114}$$

where $\bar{c}_S$ is the mean concentration of salt defined by

$$\bar{c}_S = \pi_S/RT\Delta \underline{\ell n}\, a_S \tag{115}$$

Here p is the hydrostatic pressure difference and E is the electric potential difference, expressed with respect to electrodes reversible to the anions, between the solutions on opposite sides of the membrane.

The linear relations between these fluxes and forces are

$$J_1 = L_\pi (\pi_S/\upsilon_1 \bar{c}_S) + L_{\pi p} (p - \pi_S) + L_{\pi E} E \tag{116}$$

$$J_V = L_{p\pi} (\pi_S/\upsilon_1 \bar{c}_S) + L_p (p - \pi_S) + L_{pE} E \tag{117}$$

$$I = L_{E\pi} (\pi_S/\upsilon_1 \bar{c}_S) + L_{Ep} (p - \pi_S) + L_E E \tag{118}$$

These equations refer to fluxes across macroscopic differences in the salt activity Δa_S. The $L_{\alpha\beta}$ coefficients are therefore average values over this activity interval. Like the L_{ik} coefficients they are functions of concentration. Nevertheless the average $L_{\alpha\beta}$ coefficients also obey the Onsager reciprocal relations

$$L_{\alpha\beta} = L_{\beta\alpha} \tag{119}$$

Differential flux coefficients $\mathcal{L}_{\alpha\beta}$ may be defined by relations of the kind

$$\mathcal{L}_{\alpha\beta} = (\partial J_\alpha / \partial X_\beta)_{X_{\gamma=0,\ \gamma\neq\beta}} \tag{120}$$

For example, from Eq. (117)

$$\mathcal{L}_{pE} = (\partial J_V / \partial E)_{p=0,\ \pi_s=0} \tag{121}$$

These differential $\mathcal{L}_{\alpha\beta}$ coefficients refer to the transport properties of the membrane at a particular composition, i.e., in equilibrium with a solution of a particular salt activity a_s.

They may be related to the L_{ik} coefficients which we have used previously by the expression

$$(L_{ik}/\delta) = \Gamma^{-1} (\mathcal{L}_{\alpha\beta}) \Gamma^{-1T} \tag{122}$$

where (L_{ik}/δ) and $(\mathcal{L}_{\alpha\beta})$ are the matrices of the L_{ik}/δ and $\mathcal{L}_{\alpha\beta}$ coefficients and

$$\Gamma = \begin{pmatrix} 1 & 0 & 0 \\ \bar{V}_s/\upsilon_1 & 0 & \bar{V}_w \\ z_1 F & z_2 F & 0 \end{pmatrix} \tag{123}$$

Γ is the matrix which connects the fluxes J_1, J_V, and I with J_1, J_2, and J_w, i.e.

$$\begin{pmatrix} J_1 \\ J_V \\ I \end{pmatrix} = \Gamma \begin{pmatrix} J_1 \\ J_2 \\ J_w \end{pmatrix} \tag{124}$$

Γ^{-1} and Γ^{-1T} are the reciprocal and transposed reciprocal of Γ. Thus a determination of the $\mathcal{L}_{\alpha\beta}$ coefficients permits the L_{ik} coefficients to be evaluated.

It may be noted that the relations we have given are between $\mathcal{L}_{\alpha\beta}$ and L_{ik}/δ, where δ is the membrane thickness. This is because the L_{ik} coefficients are defined in terms of potential gradients as forces, whereas X_1, X_v, and X_I in Eqs. (112)-(114) are formulated as potential differences.

The following relations exist between the electrical quantities and the $\mathcal{L}_{\alpha\beta}$ coefficients (17)

$$\mathcal{L}_E = k/\delta \tag{125}$$

$$\mathcal{L}_{\pi E} = \mathcal{L}_{E\pi} = t_1 k/z_1 \delta F \tag{126}$$

$$\mathcal{L}_{pE} = \mathcal{L}_{Ep} = kW/\delta \tag{127}$$

where the electroösmotic flow in cubic centimeters per coulomb W is given by

$$W = (t_3 \overline{V}_w + t_1 \overline{V}_s/z_1)/F \tag{128}$$

Also, for the ion and volume fluxes

$$\left(\frac{\partial J_1}{\partial \pi_s}\right)_{p=0}^{I=0} = \frac{t_1 Wk}{z_1 \delta F} - \frac{t_1{}^2 k}{z_1{}^2 F^2 v_1 c_s \delta} + \frac{\mathcal{L}_\pi}{v_1 c_s} - \mathcal{L}_{\pi p} \tag{129}$$

and

$$\left(\frac{\partial J_v}{\partial \pi_s}\right)_{p=0}^{I=0} = \frac{kW^2}{\delta} - \frac{t_1 Wk}{z_1 \delta F v_1 c_s} + \frac{\mathcal{L}_{p\pi}}{v_1 c_s} - \mathcal{L}_p \tag{130}$$

where

$$\partial \pi_s = c_s RT\, \partial \ln a_s \tag{131}$$

and

$$\mathcal{L}_{p\pi} = \mathcal{L}_{\pi p} \tag{132}$$

From Eq. (122) one can obtain the following equation for L_{22}

$$L_{22} = \delta(\mathcal{L}_E/F - 2z_1 \mathcal{L}_{E\pi} + z_1^2 F \mathcal{L}_\pi)/z_2^2 F \tag{133}$$

which may be transformed with the help of Eqs. (125) and (126) to

$$L_{22} = (k/F - 2t_1 k/F + z_1^2 \delta F \mathcal{L}_\pi)/z_2^2 F \tag{134}$$

When this equation is combined with Eq. (70) for L_{22} and rearranged it is found that

$$\mathcal{L}_\pi = \left(\frac{z_2}{z_1}\right)^2 \frac{D_2 c_2}{\delta RT} + \frac{(t_1 - t_2)k}{z_1^2 F^2 \delta} \tag{135}$$

Hence the experimental data permit $\mathcal{L}_E$, $\mathcal{L}_{\pi E}$, $\mathcal{L}_{pE}$, and $\mathcal{L}_\pi$ to be evaluated without difficulty from Eqs. (125)-(127) and (135). Once $\mathcal{L}_\pi$ is known, $\mathcal{L}_{\pi p}$ follows from Eq. (129). $\mathcal{L}_p$ is then obtained from Eqs. (130) and (132).

Having evaluated the $\mathcal{L}_{\alpha\beta}$ coefficients, the L_{ik} coefficients follow from Eq. (122). These calculations have been carried out and some results from Method V are compared with those from Methods I and IV in Tables 9 and 10 (see pp. 100 and 101).

It may be seen immediately from Table 9 that Methods IV and V give identical values of L_{11}, L_{22} and L_{12} for all salts and concentrations. There is a fair agreement for L_{13}, except at the highest $SrBr_2$ concentrations, and also for L_{33} except for some scattering of the values with Method V when applied to $SrBr_2$.

The situation with regard to L_{23} is less satisfactory; on both models there are fluctuations in L_{23}/c_2c_3 which it is hard to believe are physically real. It seems that L_{23}/c_2c_3 is somewhat greater than L_{13}/c_1c_3 and decreases as the external concentration is increased.

Method I agrees reasonably with Methods IV and V at concentrations below about 0.2 eq/liter for L_{11}, L_{22}, L_{12}, and L_{13} although Method I gives slightly higher values with $SrBr_2$. Method I gives values of L_{33}/c_3 which increase with increasing concentration, in contrast to Methods IV and V. It gives negative, and therefore impossible values, at the highest concentrations, thus demonstrating the failure of some of the underlying assumptions of Method I at these concentrations. Below 0.2 eq/liter the values of L_{23} from Method I are a little smaller than, but not very different from, those given by Method IV.

Summarizing these comments it appears that either of Methods IV and V gives a thermodynamically plausible set of L_{ik} coefficients over a wide range of concentrations. L_{11}, L_{22}, L_{12}, and L_{13} are probably determined with satisfactory precision. By plotting L_{33}/c_3 against c_e and putting a reasonable curve through the points, L_{33} could probably be estimated to within ±10% in most cases. L_{23} is rather small and on all methods of calculation which we have tried, the values obtained are very sensitive to errors in the experimental data as well as to the approximations of the theory.

It is worth emphasizing that the assumption of $R_{12} = 0$ in Method IV does not lead, of course, to the result $L_{12} = 0$.

VI. THE INTERACTIONS BETWEEN ISOTOPES

The use of radio-isotopes as tracers in transport studies has been widely practiced in chemistry and biology. For a long time it was supposed that the behavior of tracers gave directly a measure of the behavior of the parent substance in the absence of tracer. More recently, with the introduction of the nonequilibrium thermodynamic formulation of transport equations, it has become apparent that, when writing an expression for the flux of tracer, terms have to be included for the interaction of the tracer with every other component including the nontracer particles of the same chemical species.

Thus if the thermodynamic and kinetic similarity of the isotopes is fully accepted and minute contributions to the entropy production resulting from the difference in the atomic masses between tracer and nontracer are ignored, one expects to find

$$R_{ik} = R_{i^*k}, \quad k \neq i, \ i^* \tag{136}$$

and

$$L_{ik}/c_i = L_{i^*k}/c_{i^*}, \quad k \neq i, \ i^* \tag{137}$$

where i* is a tracer isotope of i. c_{i*} is known only very rarely in absolute terms. The normal way of analyzing the results of tracer experiments involves only relative specific activities which are proportional to $c_{i*}/(c_i + c_{i*}) \simeq c_{i*}/c_i$ (because $c_{i*} \lll c_i$), and so L_{ik} is obtained directly, rather than L_{i*k}, from Eq. (137).

The problem is that adding tracer to the system introduces the term R_{ii*} or L_{ii*}, which does not arise in the absence of tracer. In most cases lack of information on these coefficients has left the users of tracers with no alternative other than to ignore them. Alternatively some workers have tried to design their experiments so that the effects of tracer interaction could be eliminated from the final results (61).

Whenever it has been possible to estimate tracer interaction coefficients, it has transpired from the results that they are negligible only in very dilute solutions and may be important at the concentrations commonly met inside ion-exchange membranes (22, 53). It may be remarked that the minute amount of tracer has no detectable effect on the average behavior of the nontracer particles most of which probably never meet a tracer particle.

The tracer particles are immensely outnumbered by the nontracer particles and may meet them frequently and be affected by these encounters. It is, of course, the average behavior of the tracer particles which is observed in the experiment. Only in the case that the labeled species is dilute in the system as a whole will tracer/nontracer encounters be sufficiently rare for their effect on the behavior of the tracer to be ignored.

These considerations form the basis of our assumption that R_{22*} might be neglected because the co-ions 2 are as a whole dilute in the membrane while R_{11*} cannot be neglected because the counterions are a concentrated component.

The fixed charge concentration in the membrane we have studied was rather low, about 0.55 M, consequently the concentrations of Br^- co-ions sorbed from the more concentrated external solutions are quite considerable (approximately 0.3 M from 0.5 eq/liter and approximately 0.6 M from 1.0 eq/liter external concentrations). The assumption $R_{22*} = 0$ would not be expected to hold precisely at these concentrations. The data we have obtained permits us to study this problem at least semiquantitatively.

The tracer diffusion coefficient of the counterions D_1 is related to the L_{ik} coefficients by an expression analagous to Eq. (71) for D_2. Hence

$$L_{11} = D_1 c_1/RT + (c_1/c_1*)L_{11*} \qquad (138)$$

Thus the experimental data on D_1, which were used in Models I and II where R_{11*} was neglected, may be combined with L_{11} from Models IV and V (which give the same values for L_{11}) in order to calculate $L_{11*}/c_1 c_1*$ which is tabulated in the second column of Table 9.

It can be seen that $L_{11*}/c_1 c_{1*}$ is not far from constant for each salt and is rather smaller than $L_{12}/c_1 c_2$ because there is less coupling between the flows of particles that electrostatically repel one another than between those that attract (the value in $SrBr_2$ at 1.0 eq/liter may not be reliable because L_{33} appears to fall to an improbably low value at this concentration).

When L_{11*}/c_{1*} is compared with L_{11}/c_1 it is found to be about 30% of L_{11}/c_1 for all salts and concentrations. Thus it would have been quite incorrect to neglect L_{11*}/c_1* and to estimate L_{11} directly from D_1 in the way that L_{22} has been evaluated from D_2.

It is not unreasonable to suppose that for these flow interaction parameters

$$L_{11*}/c_1c_{1*} \simeq L_{22*}/c_2c_2* \tag{139}$$

may hold for simple univalent inorganic cations 1 and anions 2. Thus a rough estimate of L_{22*}/c_{2*} may be made from the average value of L_{11*}/c_1c_{1*} when in the NaBr and CsBr solutions.

Table 11 lists this estimate of L_{22*}/c_{2*} and compares it with L_{22}/c_2. At external concentrations below about 0.2 eq/liter it is clear that in Models IV and V little error is introduced by ignoring L_{22*}/c_{2*} but at higher concentrations this is not true. The listed values of L_{22}/c_2 are, at the highest concentrations, up to 10% too small on account of the neglect of L_{22*}/c_{2*}.

This error results, in turn, in L_{12} being given too low a value. The value of L_{12} then affects the value of L_{13}, L_{23} and L_{33} and so the consequences of the error in L_{22} are passed on to all the other L_{ik} coefficients.

A more direct understanding of tracer interaction may be had from the friction and resistance coeffficients f_{1*1} and R_{11*}. These may be evaluated from the relation

$$RT/D_1 = c_1(R_{11} - R_{11*}) = c_1R_{11} + f_{1*1} \tag{140}$$

Values of R_{11*} appear in Table 10. They show that R_{11*} is negative at low concentrations, becomes less negative as concentration is increased and, in the cases of NaBr and $SrBr_2$, changes sign between 0.1 and 0.2 eq/liter external concentration and then becomes increasingly positive with further increase in concentration.

If the frictional interpretation of the R_{ik} is adopted then a positive value of R_{ik} implies a negative friction coefficient which has no counterpart in macroscopic phenomena. It means that a flux of k in one direction produces a force on i in the opposite direction.

TABLE 11

Estimated Tracer Interaction Coefficient for Br^- Co-ions and L_{22*}/c_{2*} in cm^2 mole J^{-1} sec^{-1} Compared with the Straight Coefficient L_{22}/c_2

c_e (eq/liter)	10^9 L_{22*}/c_{2*}	10^9 L_{22}/c_2
	NaBr	
0.02	0.003	3.01
0.05	0.012	3.14
0.10	0.036	3.21
0.20	0.097	3.26
0.50	0.318	3.29
1.00	0.717	3.32
	CsBr	
0.02	0.003	3.62
0.05	0.014	3.70
0.10	0.042	3.74
	$SrBr_2$	
0.02	0.009	3.51
0.05	0.029	3.51
0.10	0.067	3.51
0.20	0.148	3.51
0.50	0.384	3.51
1.00	0.702	3.51

It is not difficult to imagine such an effect occurring in a system of electrostatically repelling components i and k at high concentrations where the movement of particles requires the rearrangement of ionic atmospheres. It is interesting that positive values of R_{ik} were obtained also in a similar membrane for the mutual resistance of Na^+ and Sr^{2+} ions (52). In that work the external concentration was 0.10 eq/liter and R_{ik} lay in the range 2 to 5×10^{11} J sec cm mole^{-2} which is comparable with R_{11*} for Sr^{2+} at 0.2 eq/liter in Table 10. Positive resistance coefficients seem therefore not to be abnormal in the interaction of flows of particles of like charge.

In Spiegler's approximate treatment which formed the basis of Models I and II R_{11*} was neglected in comparison with R_{11}. Table 10 shows that the error in doing this changes from positive to negative over the concentration range studied and is at a minimum in the region of 0.10-0.20 eq/liter. At the highest and lowest concentrations the error in R_{11} from Models I and II is probably in the region of at least 20 percent. At low concentrations R_{11} is too large and at high concentrations too small.

VII. TRANSPORT MECHANISMS IN A HOMOGENEOUS MEMBRANE

A primary objective in evaluating the various sets of transport coefficients is to gain information from them on the physical mechanisms by which particles of various kinds are transported in an almost homogeneous cation-exchange membrane. As mentioned previously, the complex nature of the L_{ik} coefficients restricts their usefulness for this purpose. A few features in Table 9 (page 100) deserve comment. The coefficients L_{1k}/c_1c_k ($k \neq 1$) vary remarkably little with concentration when k = 1*, 2, or 3 and when 1 = Na^+, Cs^+, or Sr^{2+}. This observation suggests that all flux interactions take place through a basically similar mechanism and that the concept of a frictional mechanism may be reasonable since this would predict similar interactions as to order of magnitude, irrespective of whether the interacting flows were of similarly charged, oppositely charged, or uncharged particles.

L_{23}/c_2c_3 appears on the whole to be somewhat larger than L_{13}/c_1c_3 This is a reflection of the tendency of the sorbed co-ions to be more concentrated in regions where the fixed charge and matrix concentrations are below average and c_3 is above average. The counterions, on the other hand, are most concentrated in the densest regions where c_3 is below average.

The most notable feature regarding the L_{ii}/c_i coefficients is that, whereas the values for the counterions and co-ions are of the same order of magnitude--although somewhat smaller for the counterions because their mobilities are lower--L_{33}/c_3 is larger by two orders of magnitude. This is not a consequence of a greater diffusional mobility of water in the resin (67); it shows that in osmotic and electroösmotic flow the transport of the solvent has mainly the characteristics of viscous flow in fine pores.

A more readily comprehensible picture of the interactions between the flowing particles is obtained by an examination of the resistance coefficients R_{ik} listed in Table 10. An immediate and obvious conclusion from this table is that in no case is R_{ii} or R_{ik} independent of concentration. Simplified theories of membrane processes which ignore this variation cannot provide a reliable basis for quantitative predictions.

Straight coefficients R_{ii} measure the force on i per mole required to generate unit flux density of i when all other fluxes are zero. It is to be expected that R_{ii} will tend to vary inversely with the concentration of i. This is observed in the case of R_{11} and R_{22}.

The concentration of water in the resin scarcely changes as the external concentration of salt is increased in the range examined here and one might have expected R_{33} to be almost constant. Discounting the values at the two highest concentrations of NaBr and $SrBr_2$, where the assumptions in all the models are unreliable, it can be seen that R_{33} appears to decrease slightly with increasing c_s on Spiegler's Model I whereas Models IV and V (and also Model III, for which data have not been listed in order to avoid overloading the tables) show R_{33} at least doubling over the range 0.02 to 0.20 eq/liter. In this range c_3, the concentration of water in the membrane, decreases by only about 0.5%.

It is felt that Models III, IV, and V are more reliable than Models I and II, especially where data pertaining to water in the membrane are concerned, because they take into consideration data on the osmotic as well as on the electroösmotic flow of water. It appears therefore that the introduction of a small amount of sorbed electrolyte and a shrinkage of less than 1% by volume of a membrane which contains about 75% by volume of imbibed solution results in a large increase in the resistance to flow of water. The physical origin of this behavior is not easily deduced from observations on R_{33} alone because this includes friction of the water with cations, anions, and matrix in a single term. The discussion of the f_{3k} coefficients, which follows in the next section will make the behavior of R_{33} more understandable.

The discussion of the direct interactions measured by the R_{ik} $(i \neq k)$ will be clearer if expressed in terms of the f_{ik} coefficients and this will be done shortly. Before leaving the R_{ik} coefficients it is worth noting that, although the values of R_{12} from model V (R_{12} is assumed zero in Models I, III, and IV) are frequently irregular and positive (this holds also for Model

II), they are of an order ten times larger than R_{13} and R_{23}. This is a consequence of the strong interaction between the flows of charged particles.

VIII. EVALUATION AND INTERPRETATION OF FRICTION COEFFICIENTS

The f_{ik} coefficients are evaluated from relations of the types

$$f_{ik} = -c_k R_{ik}; \quad i, k = 1, 2, 3 \tag{141}$$

and

$$f_{i4} = c_i R_{ii} - f_{ij} - f_{ik}; \quad i, j, k, = 1, 2, 3 \quad i \neq j \neq k \neq 4 \tag{142}$$

These calculations have been carried out for Models I-V starting from the L_{ik} coefficients as calculated at each individual concentration and also from L_{ik} coefficients interpolated from smooth curves drawn through plots of the $L_{ik}/c_i c_k$ and L_{ii}/c_i versus log c_e. The f_{ik} coefficients are more sensitive to errors in the experimental data than are the L_{ik}.

The development of theory and experiment has not yet reached a stage at which a precise quantitative account of the f_{ik} coefficients can be attempted. Our data have been presented graphically in Figs. 4-10.

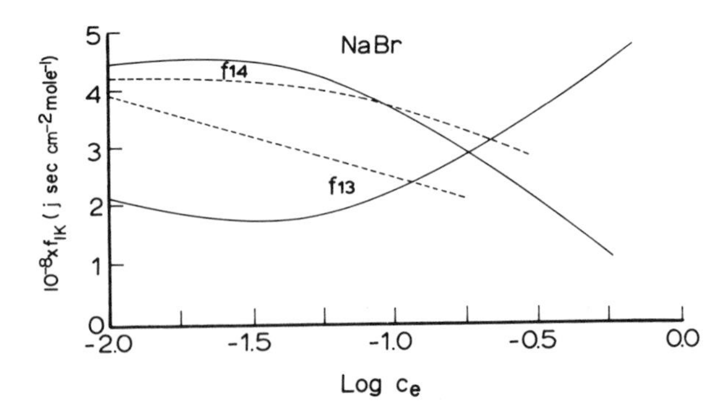

FIG. 4. Friction coefficients of cations with water f_{13} and cations with matrix f_{14} plotted against log c_e. (c_e = the external concentration in eq/liter). The dashed curves are from Model I and the full lines are the combined results of Models III, IV, and V (NaBr system).

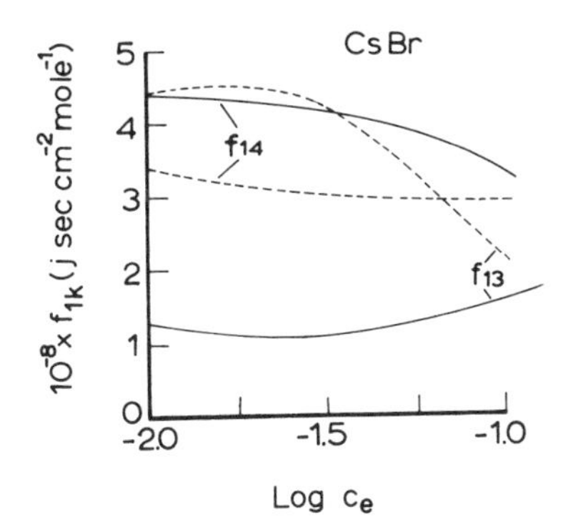

FIG. 5. Same as Fig. 4, CsBr.

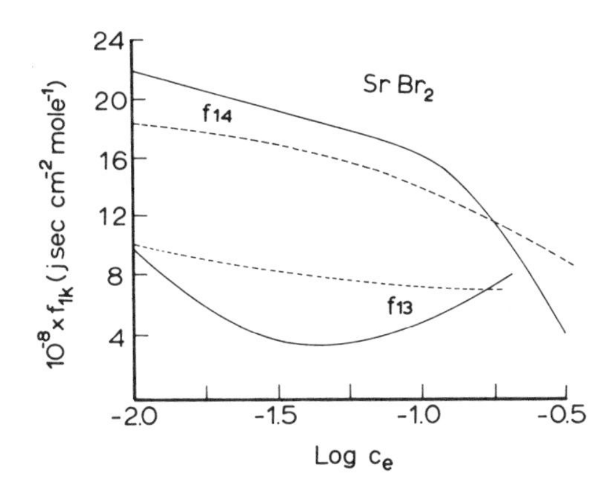

FIG. 6. Same as Fig. 4, $SrBr_2$

When drawing these graphs it was found that single compromise curves could be drawn (the solid lines) which represented the results of Models III, IV, and V within the general limits of certainty of the whole scheme. The dashed curves represent the results of Model I (Spiegler's approximation). For f_{13}, f_{14}, and f_{34} Model II did not differ greatly from Model I.

No graphs have been drawn for f_{12}. This coefficient is assumed zero in Models I, III, and IV. On models II and V it turned out to be very small and negative in nearly all cases where it differed significantly from zero.

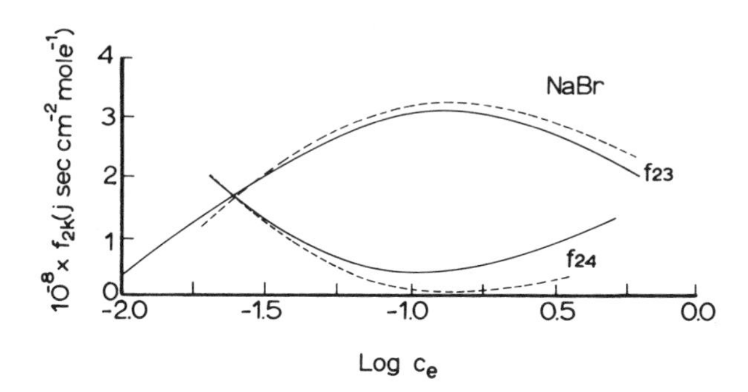

FIG. 7. Friction coefficients of anions Br^- with water f_{23} and anions Br^- with matrix f_{24} plotted against log c_e. (c_e = the external concentration in eq/liter.) The dashed curves are from Model I and the full lines are the combined results of Models III, IV, and V (NaBr system).

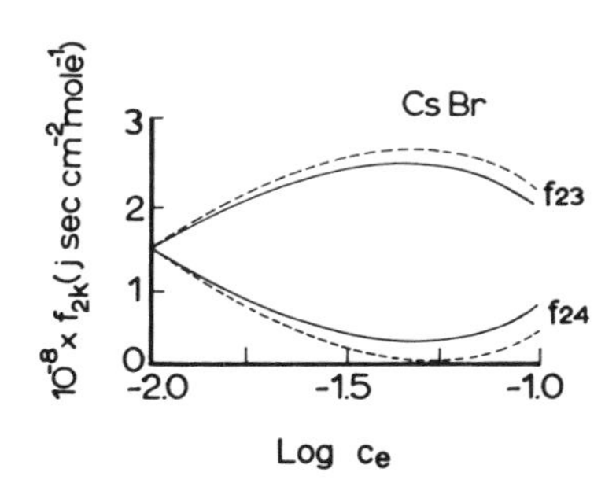

FIG. 8. Same as Fig. 7, CsBr.

When attributing a physical significance to the friction coefficients it must be recalled that they are calculated by equating the thermodynamic forces to a sum of products of friction coefficients and velocities. If friction is to have any meaning when applied to molecular encounters, which may take any form between the extremes of head-on and tangential, the relevant velocity is the local average velocity of the colliding particles. However the velocities used in the frictional model calculations are macroscopic averages obtained by dividing the flux density, which is an average over a macroscopic surface element of the membrane, by the concentration,

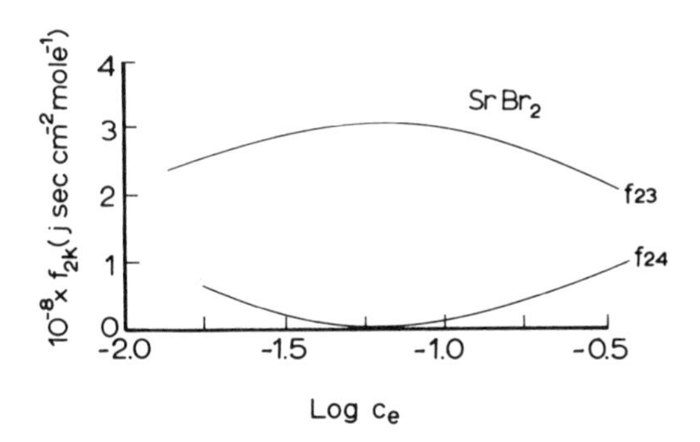

FIG. 9. Same as Fig. 7 ($SrBr_2$) except that Model I gave no satisfactory results.

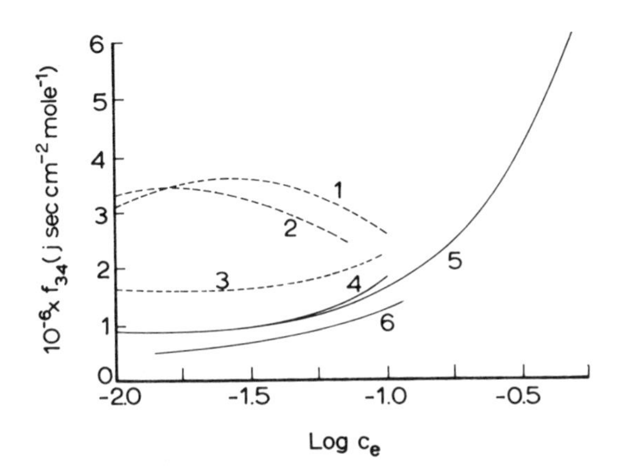

FIG. 10. Friction coefficient of water with matrix f_{34} plotted against log c_e. (c_e = the external concentration in eq/liter.) Curve 1: in CsBr from Model I; Curve 2: in $SrBr_2$ from Model I; Curve 3: in NaBr from Model I; Curve 4: in CsBr from Models III, IV, and V; Curve 5: in NaBr from Models III, IV, and V; Curve 6: in $SrBr_2$ from Models III, IV, and V.

which is an average given by the analytical total content of the membrane divided by its total volume. If, as is almost certainly the case even in the most homogeneous of membranes, the mobile components are not uniformly distributed on a local time average basis then these macroscopic average velocities are not equal to the local averages which determine the nature of the collisions between particles of different types. Consequently, when any changes in composition and concentration in the membrane result in large changes in the f_{ik} coefficients calculated as described above, it is to be anticipated that the changes in the f_{ik} coefficients will indicate changes in the distributions of the particles and their fluxes rather than changes in the intermolecular forces which would determine the true microscopic friction coefficients if we could evaluate them. Indeed, for relatively small changes in composition we may suppose that the microscopic friction coefficients remain virtually unaltered.

The principal frictions experienced by the counterions are illustrated in Figs. 4, 5, and 6. In each case in dilute solutions, friction with the matrix f_{14} is greater than with the water f_{13} despite the fact that water occupies about three-quarters of the total volume. The friction coefficients obtained by Method I show the same feature for NaBr and $SrBr_2$, although they underestimate the difference between f_{13} and f_{14}. For CsBr, Method I makes f_{13} appear larger than f_{14}. This is improbable because Cs^+ is thermodynamically preferred by the resin to Na^+ and also Cs^+ ions are less strongly hydrated than Na^+. Thus one expects f_{14} to exceed f_{13} for Cs^+ by a greater margin than for Na^+ and this is the result found from Methods III, IV, and V.

f_{14} has about the same value for Na^+ and Cs^+ despite the thermodynamic preference for the latter. In the case of Sr^{2+}, f_{14} is about five times larger than for the univalent ions, a reflection of the stronger coulombic attraction between the bivalent ions and the matrix.

f_{13} increases in the order $Cs^+ < Na^+ < Sr^{2+}$ which is also the order of increasing hydration. In the definition of components used throughout the thermodynamic treatment given here, the unhydrated ions have been regarded as one component and the total water present as another. This approach avoids the ambiguities associated with assigning hydration numbers to the ions. As a result the hydration forces, which result in the ions dragging some closely associated water molecules along with them, appear in the values of f_{13}.

The variations in the friction coefficients as the external solution concentration is increased are clearly seen in Figs. 4, 5, and 6. f_{13} and f_{14} are functions of the conditions inside the membrane and the changes in f_{13} and f_{14} must result most directly from the changes in the concentration of sorbed electrolyte.

f_{14} decreases in each case with increasing concentration. When f_{14} is plotted against the concentration of sorbed electrolyte the graph is not far from linear in each case. Two effects may contribute to this decrease. In the presence of sorbed electrolyte the total charge on the cations exceeds that on the matrix and the average interaction with the matrix per cation must be lowered by the presence of this excess. The observed reduction in f_{14} is too large to be explained entirely in this way. The sorbed co-ions can act as electrostatic bridges (60) and permit the counterions to escape more readily from the electrical environment of the matrix during their thermal motions.

Of course, when the cations are less bound to the matrix they should spend more time diffusing among the water molecules and so one expects and finds that f_{13} increases with increasing concentration. The shallow minima in f_{13} with Na^+ and Cs^+ are scarcely significant but in the case of Sr^{2+} the effect appears to be definite although its explanation is obscure; perhaps it indicates a decrease in ionic hydration as a_w is decreased.

At sufficiently large concentrations the effect of the matrix charge is effectively swamped by the other ions present. The cations will then interact with matrix and water roughly in proportion to their relative volumes and f_{13} will exceed f_{14}. It is seen that this trend exists in each case and occurs in NaBr within the concentration range examined although the exact values of f_{13} and f_{14} are least reliable at the highest concentrations.

Figures 7, 8, and 9 show the frictional coefficients involving the Br^- ions. f_{23} and f_{24} are very sensitive to assumptions which affect f_{21} and so they are less securely known than f_{13} and f_{14}. As anticipated, the repulsion between the co-ions and the matrix leads to f_{24} being small and less than f_{23}, except in low concentrations of NaBr. The pattern of the data is quite similar for each counterion, with f_{23} lying between 2 and 3 x 10^8 J sec cm^{-2} $mole^{-1}$ and f_{24} between 0 and 1 x 10^8 J sec cm^{-2} $mole^{-1}$.

The maximum in f_{23} at an external concentration of about 0.05 eq/liter may result from two conflicting influences. The evidence from the co-ion sorption and diffusion data discussed in another paper (60) was that at very low concentrations the co-ions were confined mainly to regions of abnormally low fixed charge, and hence counterion, concentration. As the external concentration is increased the co-ions must enter more concentrated regions of the membrane and meet the counterions more frequently. This would result in an increasing interaction between the co-ions and the loosely bound hydration water dragged in the opposite direction by the counterions. This is analogous to the electrophoretic effect which is observed in the conductance of electrolytes in free solution. It would cause f_{23} to increase with increasing concentration when the external solution was dilute.

At higher concentrations the swamping of the effect of the fixed charges enables the co-ions to penetrate the membrane volume more uniformly. Thus at higher external concentrations f_{24} must increase and f_{23} decrease as the concentration is increased.

The frictional coefficient of the water with the matrix f_{34} in all the cationic states is shown in Fig. 10. The solid curves (the combined results of Methods III, IV, and V) are well grouped together. f_{34} would not be expected to vary greatly with cationic state, provided that, as in this case, the water content of the membrane was scarcely affected by the exchange of counterions. The dashed curves from Method I are scattered and give much larger values for f_{34}. Only in the case of NaBr is there an agreement between Method I and Methods III, IV, and V as to trend with changing concentration.

The gentle increase in f_{34} with increasing concentration is consistent with the progressive loosening of counterions from the matrix which would thus become more exposed to interaction with the water. A similar picture accounts for f_{34} being a little less in the Sr^{2+} state than in Na^+ or Cs^+ because the strongly attracted Sr^{2+} ions shield the matrix more effectively.

f_{34} is lower than f_{13} or f_{23} by a factor of about 100. The more comparable coefficients f_{31} and f_{32} are easily calculated from

$$f_{31} = c_1 f_{13}/c_3 \tag{143}$$

and

$$f_{32} = c_2 f_{23}/c_3 \tag{144}$$

One finds that at or below 0.1 eq/liter f_{32} is small enough to be ignored in comparison with $(f_{31} + f_{34})$ but f_{31} is always greater than f_{34}. In the Na^+-state up to 0.1 eq/liter, f_{31} is about 3×10^6 J sec cm^{-2} mole^{-1} and increases at higher concentrations roughly parallel to f_{34}. In the Sr^{2+}-state f_{31} appears to fall sharply from about 7×10^6 in 0.01 eq/liter to a minimum of about 3×10^6 at 0.05 eq/liter and then increase again to 7×10^6 at 0.20 eq/liter. It is clear that the main resistance to water flow through the membrane in dilute solutions under a hydrostatic or osmotic pressure gradient is friction with the counterions which is two to three times greater than friction with the matrix. One presumes that this phenomenon is a consequence of the relatively hydrophobic nature of the membrane matrix.

Our general conclusion from the foregoing attempt to apply the frictional model of transport to one particular membrane is that, although it would be desirable to make a sixth independent flux measurement under a pressure gradient so as to avoid the need for any approximations, either of the approximate Methods IV and V gives reliable values of the membrane transport coefficients over a fair concentration range. They require a substantial quantity of good quality experimental data but of a kind that is not too difficult to determine. Once the transport coefficients have been obtained they can be used to evaluate the molecular friction coefficients and from these many useful inferences may be drawn regarding the distribution and flow of the various components in the membrane.

If a body of results were built up on the properties of a range of membranes of different types, the significances of several aspects of the frictional coefficients and of their variations with concentration and composition might become clearer and permit a theory of membrane transport to be developed which took into quantitative account the structural parameters of the membrane in a more detailed way than is possible at present.

ACKNOWLEDGMENTS

D.G. Dawson's participation in this work was made possible by the award of a Science Research Council Studentship and, later, by financial support under U.S. Office of Saline Water Research Grant 14-01-0001-638.

The authors gratefully acknowledge the extensive help of Dr. J. Klinowski in the computational work.

LIST OF PRINCIPAL SYMBOLS

A	parameter of frictional Models I and II
A_{ik}	minor of L_{ik} in $\lvert L\rvert$
a	parameter of frictional Models I and II
$a_{s,w}$	activity of salt s or water w
B	parameter of frictional Models I and II
B_{ik}	minor of R_{ik} in $\lvert R\rvert$
b	parameter of frictional Models I and II
C	parameter of frictional Models I and II
c_e	external concentration (eq. liter^{-1})
c_i	concentration of i (mole cm^{-3})

c_s	external concentration (mole cm^{-3})
D_i	diffusion coefficient of i determined with a tracer
E	electrical potential with respect to electrode reversible to anions
F	Faraday's number
F_{ik}	frictional force of k on i per mole of i
f_{ik}	molar frictional coefficient of i with k
I	electric current density
J_D	exchange flow in Eq. (11)
J_i	flux density of i (mole cm^{-2} sec^{-1})
J_v	total volume flux density (cm sec^{-1})
$\overleftarrow{J}_{Br}$	unidirectional tracer flux of Br^-
k	specific electrical conductance
L_D	phenomenological coefficient in Eqs. (12)
L_{ik}	phenomenological conductance coefficient
L_p	phenomenological coefficient in Eqs. (12)
L_{pD}	phenomenological coefficient in Eqs. (12)
$\|L\|$	determinant of L_{ik}
$\mathcal{L}_{\alpha\beta}$	differential practical phenomenological coefficient
m_s	molal concentration of solution
n	number of components
p	pressure
R	gas constant
R_{ik}	phenomenological resistance coefficient

$\|R\|$	determinant of R_{ik}
T	absolute temperature
t_i	electrical transference number of i
$\bar{U}_i$	defined mobility of i
u_i	absolute mobility of i
$\bar{V}_i$	partial molar volume of i
v	velocity of local center of mass
v_i	mean velocity of i
W	electroösmotic flow (cm^3 C^{-1})
X_i	force on i per mole
x	distance from reference plane in membrane
z_i	number of positive charges per ion i (i.e., valency with sign)
Γ	transformation matrix [Eq. (122)]
γ_i	activity coefficient of i
δ	membrane thickness
μ_i	electrochemical potential of i
υ_i	number of ions i per mole of salt
π_s	osmotic pressure
σ	rate of entropy production, reflection coefficient
Φ	dissipation function
Ψ	electrical potential
ω	solute permeability

REFERENCES

1. "Membrane Phenomena," Discussions Faraday Soc., 21 (1956).
2. "Stofftransport durch Membranen in Chemie und Biologie," Ber. Bunsenges f. physik. Chem., 71, 750-911 (1967).
3. N. Laksminarayanaiah, Chem. Rev., 65, 491 (1965); P. Lauger, Angew. Chem., 81, 56 (1969).
4. S. R. Caplan and D. C. Mikulecky, in Ion Exchange, (J.A. Marinsky, ed.), Dekker, New York, 1966, Chap. 1.
5. A. J. Staverman, Trans. Faraday Soc., 48, 176 (1952).
6. J. G. Kirkwood, in Ion Transport across Membranes, (H. T. Clarke, ed.), Academic, New York, 1954, p. 119.
7. K. S. Spiegler, Trans. Faraday Soc., 54, 1408 (1958).
8. T. Teorell, Progr. Biophys. Biophys. Chem., 3, 305 (1953).
9. D. E. Goldman, J. Gen. Physiol., 27, 37 (1943).
10. R. Schlögl, Discussions Faraday Soc., 21, 46 (1956).
11. F. Helfferich, Ion Exchange, McGraw-Hill, New York, 1962, p. 344.
12. A. Katchalsky and P. F. Curran, Nonequilibrium Thermodynamics in Biophysics, Harvard University Press, Cambridge Mass., 1965
13. L. Onsager, Phys. Rev., 37, 405 (1931).
14. D. G. Miller, Chem. Rev., 60, 15 (1960).
15. O. Kedem and A. Katchalsky, Biochim. Biophys. Acta, 27, 229 (1958).
16. O. Kedem and A. Katchalsky, Trans. Faraday Soc., 59, 1918, 1931, 1941 (1963).
17. H. Krämer and P. Meares, Biophys. J., 9, 1006 (1969).
18. T. Hoshiko and B. D. Lindley, Biochim. Biophys. Acta, 79, 301 (1964).
19. O. Kedem and A. Essig, J. Gen. Physiol., 48, 1047 (1965).
20. A. Essig, J. Theoret. Biol., 13, 63 (1966).
21. H.G. L. Coster and E. P. George, Biophys. J., 8, 457 (1968).
22. P. F. Curran, A. E. Taylor, and A. K. Solomon, Biophys. J., 7, 879 (1967).
23. P. Mazur and J. T. Overbeek, Rec. Trav. Chim., 70, 83 (1951).
24. J. Dainty, P. C. Croghan, and D. S. Fensom, Can. J. Botany, 41, 953 (1963).
25. F. M. Snell and B. Stein, J. Theoret. Biol., 10, 177 (1966).
26. A. Katchalsky and O. Kedem, Biophys. J., 2, No. 2, Part 2, Suppl. 53 (1962).
27. O. Kedem, in Membrane Transport and Metabolism, (A. Kleinzeller and A. Kotyk, eds.), Czech. Acad. Sci., distributed by Academic, New York, 1961, p. 87.
28. B. T. Scheer, Bull. Math. Biophys., 20, 231 (1958); 22, 269 (1960).
29. R. Blumenthal, S. R. Caplan, and O. Kedem, Biophys. J., 7, 735 (1967).

30. O. Kedem, in The State and Movement of Water in Living Organisms, Fourteenth Symposium of the Society for Experimental Biology, Cambridge Univ. Press, 1965, p. 61.
31. T. L. Hill and O. Kedem, J. Theoret. Biol., 10, 399 (1966).
32. H. Kimizuka and K. Koketsu, J. Theoret. Biol., 6, 290 (1964).
33. A. Essig. O. Kedem, and T. L. Hill, J. Theoret. Biol., 13, 72 (1966).
34. O. Kedem and A. Leaf, J. Gen. Physiol., 49, 655 (1966).
35. O. Kedem and A. Katchalsky, J. Gen. Physiol., 45, 143 (1961).
36. B. Andersen and H. H. Ussing, Acta Physiol. Scand., 39, 228 (1957).
37. J. Dainty and B. Z. Ginzburg, Biochim. Biophys. Acta, 79, 102, 112, 122, 129 (1964).
38. V. W. Sidel and A. K. Solomon, J. Gen. Physiol., 41, 243 (1957); D. A. Goldstein and A. K. Solomon, ibid, 44, 1 (1960).
39. B. Z. Ginzburg and A. Katchalsky, J. Gen. Physiol., 47, 403 (1963).
40. W. Dorst, P. L. Polak, R. Caramazza, and A. J. Staverman, Gazz. Chim. Ital., 92, 1241 (1962).
41. J. Dainty, Advan. Botan. Res., 1, 279 (1963).
42. Lord Rayleigh, Proc. Math. Soc. (London), 4, 363 (1873).
43. R. W. Zwanzig, J. G. Kirkwood, I. Oppenheim, and B. J. Alder, J. Chem.. Phys., 22, 783 (1954).
44. L. Onsager, Ann. N.Y. Acad. Sci., 46, 241 (1945).
45. A. Klemm, Z. Naturforsch., 8a, 397 (1953).
46. R. W. Laity, J. Phys. Chem., 63, 80 (1959).
47. R. W. Laity, J. Chem. Phys., 30, 682 (1959).
48. J. Dainty and B. Z. Ginzburg, J. Theoret. Biol., 5, 256 (1963).
49. R. Caramazza, W. Dorst, A.J.C. Hoeve, and A.J. Staverman, Trans. Faraday Soc., 59, 2415 (1963).
50. W. Dorst and A. J. Staverman, Rec. Trav. Chim., 86, 61 (1967).
51. D. Mackay and P. Meares, Trans. Faraday Soc., 55, 1221 (1959).
52. P. Meares, D. G. Dawson, A. H. Sutton, and J. F. Thain, Ber. Bunsenges f. Physik. Chem., 71, 765 (1967).
53. E. M. Scattergood and E. N. Lightfoot, Trans. Faraday Soc., 64, 1135 (1968).
54. G. B. Wills and E. N. Lightfoot, Ind. Eng. Chem. Fundamentals, 5, 114 (1966).
55. A. S. Tombalakian, M. Worsley, and W. F. Graydon, J. Am. Chem. Soc., 88, 661 (1966).
56. J. S. Mackie and P. Meares, Proc. Roy. Soc., A. 232, 485 (1955).
57. P. Meares and J. F. Thain, J. Phys. Chem., 72, 2789 (1968).
58. W. J. McHardy, P. Meares, and K. R. Page, in preparation.
59. P. Meares, in Diffusion in Polymers, (J. Crank and G. S. Park, eds.), Academic, London, 1968, Chap. 10.
60. W. J. McHardy, P. Meares, and J. F. Thain, J. Electrochem. Soc., 116, 920 (1969).

61. P. Meares and A. H. Sutton, J. Colloid Interface Sci., 28, 118 (1968).
62. W. J. McHardy, P. Meares, A. H. Sutton, and J. F. Thain, J. Colloid Interface Sci., 29, 116 (1969).
63. D. G. Dawson, Ph. D. Thesis, Aberdeen University, 1966.
64. E. Glueckauf and D. C. Watts, Proc. Roy. Soc., A, 268, 339 (1962).
65. D. G. Dawson, W. Dorst, and P. Meares, J. Polymer Sci., C, 22, 901 (1969).
66. G. B. Wills, Ind. Eng. Chem. Fundamentals, 6, 142 (1967).
67. D. Mackay, J. Phys. Chem., 64, 1718 (1960).

Chapter 3

LIQUID MEMBRANES AS ELECTRODES AND BIOLOGICAL MODELS

John Sandblom

Departments of Physiology and Medical Biophysics
Biomedical Centre, University of Uppsala
Box 572, S-751 23 Uppsala
Sweden

Frank Orme
Department of Physiology
University of California
Berkeley, California 94720

I. INTRODUCTION

The design of artificial membranes for simulation of biological transport phenomena was started around the turn of the century, (1,2) and some of the first membranes to be used as biological models consisted of water immiscible liquids. Most of the work in this field, however, was later carried out on solid membranes--which have the advantage of being better defined structurally, both with respect to boundaries and bulk phase--and it was not until very recently that membranes composed of nonaqueous liquids once more became the subject of numerous investigations. The reason for this was twofold.

With the commercially available liquid ion exchangers (3) it has been possible to develop liquid electrodes with a high degree of specificity for inorganic (4,5) as well as organic ions (6,7). These electrodes are continuously being improved and are therefore becoming useful in an increasing variety of electrochemical applications.

The second reason for the interest in liquid membranes is their resemblence to biological membranes. There is plenty of evidence, for example, which indicates that a liquid-like state prevails in biological membranes. Such evidence has been obtained by spin labeling techniques (7a, 7b) and in a recent review, Stoeckenius and Engleman (7c) have summarized other results supporting the notion of a liquid-like hydrocarbon region of cell membranes.

Also very recently, the striking effect of certain neutral organic molecules in facilitating cation transport in lipid bilayers has been successfully explained by assuming that the transport of ions occurs through a liquid bulk phase with the ions chemically bound to the organic molecules (8). A carrier substance need not only be neutral, however, since a liquid ion exchanger having charged or dissociable groups will also function as a kind of carrier mechanism and will have many properties in common with the neutral carriers. In both cases, the carrier can be viewed as a mobile membrane component with which a set of physicochemical properties, specific as well as nonspecific, can be associated. It is the purpose of this chapter to summarize some of these properties of liquid membranes, in particular liquid ion-exchange membranes, which are not only interesting models of biological membranes but presently also constitute a rapidly expanding field in analytical chemistry.

II. DESCRIPTION AND CLASSIFICATION OF LIQUID MEMBRANES

A. Geometry

A system consisting of a poorly water miscible liquid membrane interposed between two aqueous solutions and containing a dissociable acid or base can be said to define a liquid ion-exchange membrane (3). This often-used description, however, does not define the system from a geometrical point of view. The bulk phase, for example, can be subjected to stirring (9) in which case the transport of dissolved components occurs primarily by means of convection. As a matter of fact, a very vigorous stirring of the interior membrane phase will confine the rate limiting step of the transport processes to the interphases and produce a state of pure surface diffusion (10, 11).

On the other hand, one may take suitable precautions to avoid convection and thereby obtain a system in which bulk diffusion will constitute the rate limiting process (12). The two systems will naturally have entirely different properties and it is important to bear this geometrical aspect in mind in attempting to analyze the properties of liquid membranes.

A second important geometrical aspect that must be considered in defining a liquid membrane is the number of liquid junctions and their locations within the system. The boundaries between aqueous and nonaqueous phases are obvious liquid junctions but experimental situations have also been realized in which liquid junctions have been located inside the bulk phase (13-16). If, for example, two nonaqueous liquids containing different solutes are brought into contact, a liquid junction will appear in the interior of the membrane as shown below.

Reversible	Aqueous	‖	Oil	|	Oil	‖	Aqueous	Reversible
Electrode	AX	‖	AX	|	BX	‖	BX	Electrode

Some interesting phenomena have been observed in cells of this type, where a moving boundary crosses the membrane: namely the appearance of rapid potential shifts as the moving boundary reaches one or the other interphase (13, 16). A theoretical analysis of systems with such a geometry is difficult and may explain why some workers attribute the potential shifts to changes in distribution equilibria (13), and others to reversal of surface charge (16).

It is also important in considering the geometrical aspects of liquid membranes to take into account the shape of their surfaces. For example, it was found by Bonner and Lunney (17) that, in order to obtain reproducible potentials in a liquid ion-exchange membrane system, they had to use a continuous flow of exchanger from carefully designed capillary tips into the aqueous solution. The drop rates as well as the formation of drops turned out to be an important factor in determining the potential response.

It must, therefore, be concluded that the experimental design as well as the theoretical analysis of liquid membranes must be based on a well defined geometry, specifying the liquid junctions as well as the natures of the surfaces and interior of the system.

B. Liquid Membranes with Oleophilic Compounds

An ion which moves across a liquid membrane can do this either by dissolving in the liquid phase and diffusing across the membrane, or it can combine with another ion or molecule to form a membrane-soluble complex which will diffuse across the membrane and dissociate at the other phase boundary. These different modes of transport can be used to classify liquid membranes, although to some extent both types will always be present simultaneously in every system. A classification based on these principles, however, was made by Beutner (18) who studied the properties of cells with and without the addition of oleophilic salts. An excellent review of Beutner's work was given by Kahlweit (19), who has carried this classification further and divided liquid membranes into three groups as follows.

1. In the first group the same electrolyte is present in all phases

Reversible	Aqueous	‖	Oil	‖	Aqueous	Reversible
Electrode	AX	‖	AX	‖	AX	Electrode

The first studies of liquid membranes were performed by Nernst and Risenfeld (1) on membranes belonging to this group, which were characterized with respect to diffusion of electrolyte in the bulk phase and distribution of electrolyte across the phase boundaries.

2. The second type of cell was primarily examined by Beutner (18). Its initial conditions can be represented as follows:

Reversible	Aqueous	Oil	Aqueous	Reversible
Electrode	AX	AX AY	AY	Electrode

In the systems used by Beutner, AX usually represented an electrolyte such as NaCl, whereas AY represented the salt of an oleophilic substance, for example, oleic acid. From the potential response of these systems Beutner drew some interesting conclusions regarding the phase boundaries and suggested that they responded as electrodes to changes in external solution conditions.

3. In the third type of cell, the oleophilic ion is entirely confined to the membrane phase and therefore acts as an ion exchanger.

Reversible	Aqueous	Oil	Aqueous	Reversible
Electrode	AX	AY	AX	Electrode

The current interest in liquid membranes is focused on systems belonging to this group, and most of this chapter will therefore be devoted to a description of the properties of these systems.

C. Liquid Ion-exchange Membranes

One can define an ideal liquid ion-exchange membrane as a membrane belonging to the third category of liquid membranes described in the previous section; since in essence the species Y does not take part in the transfer across the membrane boundaries. Instead it contains the functional groups which constitute the sites for exchange; and it is according to the structure of these groups that liquid ion exchangers can be classified as weak acids and bases (e.g., alkyl phosphoric acids or alklyl amines) or strong acids and bases (e.g., carboxylic or sulfonic acids and quarternary ammonium salts) (3).

In general, in a partition between oil and water, the exchanger becomes more oleophilic the longer the alkyl chains and the weaker the tendency to ionize. The strongly acidic and basic exchangers are, therefore, unsuitable for ideal liquid ion-exchange membranes, partly because of their increased water solubility but also because of their great surface activity and a strong tendency to emulsify. A greater branching of the residues, however, will reduce this tendency. For reviews of liquid ion exchangers see the articles by Högfeldt (20) and by Kunin and Winger (3).

III. BIOLOGICAL MODELS AND MEASUREMENTS

The early work on liquid membranes was primarily carried out in order to characterize transport processes in general, and to attempt to describe the properties of biological membranes.

In a classical treatment (18) mentioned in the previous section, Beutner studied the potential of cells with the addition of various oleophilic salts and postulated his theory of phase boundary potentials, according to which the potential is due to a partition equilibrium between oil and water. It had also been suggested by others (15, 16, 21, 22), however, that the phase boundary potential could be due to selective adsorption of the ions at the interfaces and the nature of the membrane potential became the subject of a controversy between Baur (21) and Beutner.

The difference between adsorption potentials and equilibrium distribution potentials was further explored by Ehrensvärd and Sillén (22) and recently by Colaccico (15, 16), who examined the influence on the interfacial potential of the polar head charges at the membrane surfaces and the way that the potential varied with electrolyte concentration.

Adsorption potentials will probably always arise at the membrane interfaces through the presence of surface charges and through specific interactions expressed by the Freundlich and Langmuir isotherms. The total boundary potential, however, is composed of at least two steps, namely, the double layer potentials in the aqueous and nonaqueous phases. Schematically we may represent the potential profile at an interface as shown in Fig. 1, where E'E" represent adsorption potentials and E_d is the total potential. Only if the two phases are in thermodynamic equilibrium with each other will the potential E_d become equal to the distribution potential. But whether local equilibrium at the interfaces exists or not, and whether the measured potential will be equal to a distribution potential or an

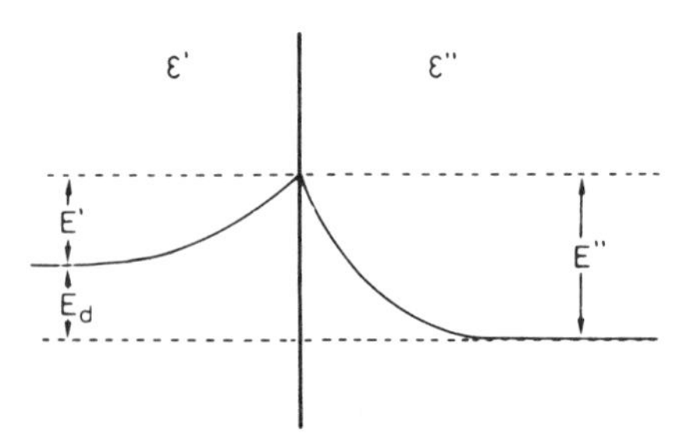

FIG. 1. The potential profile at a liquid-liquid interface. The surface carries an excess charge and the resulting double layer potentials are denoted by E' and E".

adsorption potential, depends in general on the rate of diffusion through the membrane and on the conductivity of the two phases. In most of the experimental work referred to in this chapter, however, the systems have had rate limiting bulk transport with rapid establishment of equilibrium across the interfaces and E_d is set equal to the distribution potential.

Undoubtedly, however, adsorption potentials may be important in many biological situations, particularly where electrokinetic phenomena play an important role, and liquid membranes of the proper design can therefore be used for these types of investigations.

Following the systematic studies of Beutner (18), Osterhout (24) turned his attention to the carrier properties of liquid membranes with oleophilic salts. He used guajacol as a membrane and measured the fluxes of potassium and sodium guajacolates and was able to demonstrate that the guajacolate ions acted as carriers and greatly enhanced the transport of cations. Also, by adjusting the pH on the two sides of the membrane it was possible to produce an uphill movement of cations across the membrane and this system could, therefore, also serve as a model for active transport. A theoretical analysis of the fluxes in the Osterhout and Stanley model (23, 24) was made by Longsworth (25).

The same system was also studied in some detail by Shedlovsky and Uhlig (26), who made very careful measurements of distribution coefficients, dissociation constants, and electrical conductances of potassium and sodium guajacolates in guajacol. These are measurements of parameters which are quite generally found to determine the permeability of liquid ion-exchange membranes and the contributions of Shedlovsky and Uhlig are therefore of fundamental importance in understanding the properties of these membranes.

The original work of Beutner was extended by several investigators (27-29), in particular by Bonhoeffer and his group (30-32), having as their purpose the analysis of the potential response of liquid membranes with oleophilic salts. In a series of papers (19, 30-32), they used the fixed charge concepts of the Teorell-Meyer-Sievers-theory (TMS) to predict the e.m.f. by taking into account the internal transport processes. As in the TMS-theory, the concept of membrane sites functioning as ion-exchange sites was introduced, and consequently a connection between Beutner cells of type 3 and modern ion-exchange membranes was made through the treatment of Bonhoeffer et al. (30). It was pointed out by Kahlweit (19), however, that two important distinctions must be made between solid and liquid ion-exchange membranes, namely that:

1. A liquid ion exchanger has a finite solubility in water and is therefore not entirely confined to the membrane phase. This is very likely an important factor influencing the potential response of ion specific electrodes.

2. The sites in a liquid membrane, although largely confined to the membrane phase, are not fixed but are free to rearrange within the membrane phase under the influence of electrochemical potential gradients.

These two factors were considered by several investigators in determining the potential response of various types of membranes which can be said to fulfill the criteria of liquid ion-exchange membranes (13,17,19).

Although the rearrangeability of the sites was recognized by previous investigators (30) to be one of the distinguishing features of liquid membranes, it was first explored in detail by Conti and Eisenman (33) and later by Ciani and Gliozzi (33a). They treated the ideal case of complete dissociation between sites and counterions and assumed at the same time completely trapped and mobile sites. Since dissociated liquid ion exchangers are usually highly soluble in water, the treatment of Conti and Eisenman is more applicable to electrode processes at solid-liquid interfaces than to liquid ion-exchange membranes; and in fact their theory was beautifully verified in a convection-free system of HCl trapped between two silver chloride electrode boundaries (34). Nevertheless, the theory of Conti and Eisenman demonstrated some interesting properties which are common to all mobile site membranes, particularly the presence of limiting currents.

In order to develop a theoretical formalism that would be more applicable to experimental situations involving liquid ion-exchange membranes, Sandblom, Eisenman, and Walker (35,36) also took into account the possibility for association between sites and counterions. They developed expressions for the electrical properties and transport processes in liquid ion-exchange membranes and related them to such physical parameters as distribution coefficients, transference numbers, and dissociation constants. Experimental tests of these predictions showed that, despite the relatively crude approximations of the theory, it was reasonably adequate in describing the properties of a commonly used liquid ion-exchange membrane (di-2-ethylhexylphosphoric acid in n-pentanol) (12).

Through the development of commercial liquid ion exchangers a variety of membranes have been made and studied with respect to their ionic permeability. Sollner and Shean (37) measured fluxes and potentials across these membranes and noted their very high selectivity, a property which makes them very useful as electrodes. They also pointed out the carrier properties of liquid membranes.

Bonner and Lunney (17), in their important paper measured potentiometrically the activity coefficients of p-toluene sulfonic acid and suggested that liquid ion-exchange membranes could be used for measuring activities of large ions, for which solid ion exchangers are extremely impermeable.

It was also discovered that alkylphosphoric acids had a high selectivity for divalent ions (38,39), and with the design of commercial electrodes (5) it became possible to measure the activity of calcium and other divalent cations with a high degree of accuracy. Numerous reports have since been presented on the properties and applications of liquid electrodes (40-48) and the literature on this subject is very rapidly expanding.

Investigations of the potential response of liquid ion-exchange membranes have been made by Rechnitz and coworkers (49, 50); in order to compare experimental results with theoretical predictions. They also measured kinetic properties by measuring impedance loci (50) and concluded that some rate process other than electrodiffusion must be present in these membranes. The nature of this other rate process was not clear, however, and their measurements did not suggest a physical explanation.

Despite the difficulties in attempting to make a quantitative formulation of transport processes in liquid ion-exchange membranes, in view of the many factors involved, it is informative to write expressions for the fundamental processes responsible for the transfer of ions across the membranes. From such a theoretical formulation, some useful concepts may be derived as will be shown in the next section.

IV. THEORETICAL FORMULATION

A. Transport Processes

An ion which is transported across a liquid membrane containing an exchanger may, as we pointed out, either diffuse through the liquid as a solvated ion or as a complex consisting of the ion and the exchanger sites.

Most exchangers will probably form bonds of primarily ionic nature, and if the association is a simple ion pair formation involving univalent electrolytes the reaction can be written as

$$I^{+} + S^{-} \rightleftarrows IS \tag{1}$$

where I^{+} is the counterion which is free to cross the boundary, S^{-} is the site, and IS the neutral ion pair. In writing equations for the fluxes we follow the procedure and keep the notations used in previous treatments (51). Since every mode of transport in the liquid phase must be taken into account, it is necessary to write flux equations for each of the species appearing in Eq. (1), or

$$J_{ii} = -L_{ii} \frac{d\bar{\mu}_i}{dx} \tag{2a}$$

$$J_{ss} = -L_{ss} \frac{d\bar{\mu}_s}{dx} \tag{2b}$$

$$J_{is} = -L_{is} \frac{d\bar{\mu}_{is}}{dx} \tag{2c}$$

where J denotes the flux, $\bar{\mu}$ the electrochemical potential, and L the product of mobility and concentration of a particular species.

It is very useful to introduce the total fluxes and concentrations, which are easily measured quantities--and indicating total quantities by asterisks they are expressed in terms of the partial quantities by

$$J_i^* = J_{ii} + J_{is} \tag{3a}$$

$$J_s^* = J_{ss} + \Sigma_j J_{js} \tag{3b}$$

$$C_i^* = C_{ii} + C_{is} \tag{3c}$$

$$C_s^* = C_{ss} + \Sigma_j C_{js} \tag{3d}$$

We shall also assume that a state of local chemical equilibrium exists between the individual species, the concentrations being related by a simple mass law. This assumption implies that the transport processes are much slower than the chemical reactions given by Eq. (1), which seems to be a reasonable approximation in electrodes and in most experimental arrangements described in the literature. Its validity in biological membranes, however, is uncertain in view of their thinness, and the effects which can be deduced by considering finite chemical reaction rates have been studied theoretically by Katchalsky and Spangler (52).

In the following, however, we shall only consider systems in which the condition of local equilibrium is fulfilled and which enable us to write a relationship between the electrochemical potentials corresponding to Eq. (1):

$$\bar{\mu}_i + \bar{\mu}_s = \bar{\mu}_{is} \tag{4}$$

Equations (1) through (4) can now be used to obtain some very useful expressions for the total fluxes J_i^*. Combining these equations and rearranging them yields:

$$J_i^* - \frac{L_{is}}{L_{ss} + \sum_j L_{js}} \cdot J_s^* = -\left[L_{ii} + L_{is} - \frac{L_{is}^2}{L_{ss} + \sum_j L_{js}}\right] \frac{d\bar{\mu}_i}{dx} + \sum_{j \neq i} \frac{L_{is} L_{js}}{L_{ss} + \sum_j L_{js}} \cdot \frac{d\bar{\mu}_j}{dx} \tag{5}$$

If we consider the particular case where the exchanger is entirely confined to the membrane phase, analogous to the situation in biological carrier systems, the total flow of sites J_s^* will be zero in the steady state. Under these circumstances Eqs. (5) reduce to a set of linear relationships between fluxes and forces. The corresponding phenomenological coefficients will be assigned asterisks to distinguish them from the coefficients appearing in Eq. (2), and it is now evident from Eqs. (5) that this set of phenomenological cross coefficients L^*_{ij} obey reciprocal relationships which can be written

$$L_{ij}^* = L_{ji}^* \tag{6}$$

Another interesting property of the coeffficient matrix is the negative sign of the cross coefficients, which is due to the nature of the chemical coupling between counterions, in this case a competition between the counterions for the sites. When cross coefficients arise from frictional interactions between the various ions or when they are due to a chemical coupling which is additive instead of competitive (e.g., triple ion formation), their sign will be positive.

The negative signs of the cross coefficients also imply that the flow of a species in one direction induces a coupled flow of species in the opposite direction. This property has been called counter transport and is often used as a criterium for carrier mediated transport in biological membranes (53).

To obtain the current-voltage relationship it is necessary to introduce the electric current, which is equal to the total flow of charge carrying species or

$$I = \sum_i z_i J_i + z_s J_s + zJ \tag{7}$$

where z is the valence and J the total flow of co-ions. In most experimental situations of interest, however, the co-ions are effectively excluded and their flux can be neglected. The mechanism by which this occurs is probably very similar to that of solid ion exchangers, namely through the electric field surrounding the sites -- which acts as a barrier to the entry of co-ions. In addition, it is possible to enhance co-ion exclusion in liquid membranes by the proper choice of solvent (see Sec. IV.B).

Neglecting the flow of co-ions we get--by inserting Eqs. (5) and Eqs. (3a) and (3b) into Eq. (7):

$$\frac{I}{L^*} + \frac{L_{ss}}{L_{ss} + \sum_j L_{js}} \cdot \frac{J_s^*}{L^*} = -\frac{d\psi}{dx} - \frac{1}{z_i F} \sum_j t_j^* \, d\mu_j \tag{8}$$

where the total conductance L^* and the transference numbers t_j^* are given by

$$L^* = F^2 \left(\sum_j L_{jj} + \frac{L_{ss} \sum_j L_{js}}{L_{ss} + \sum_j L_{js}} \right)$$

$$t_j^* = \frac{L_{ii} + \dfrac{L_{ss} \sum_j L_{js}}{L_{ss} + \sum_j L_{js}}}{\sum_j L_{jj} + \dfrac{L_{ss} \sum_j L_{js}}{L_{ss} + \sum_j L_{js}}}$$

Some interesting features of Eq. (8) become apparent if the transference numbers are rearranged in the following way

$$t_j = (1 - t) \frac{L_{ii}}{\sum_j L_{jj}} + t \frac{L_{js}}{\sum_j L_{js}} \tag{9}$$

where

$$t = \frac{L_{ss} \cdot \sum_j L_{js} / (L_{ss} + \sum_j L_{js})}{\sum_j L_{jj} + L_{ss} \cdot \sum_j L_{js} / (L_{ss} + \sum_j L_{js})} \tag{10}$$

Equation (8) can then be written as follows

$$\frac{Idx}{L^*} + \frac{L_{ss} \cdot dx}{L_{ss} + \sum_j L_{js}} \cdot \frac{J_s^*}{L^*} = -d\psi - \frac{1-t}{z_i F} \cdot \frac{\sum_j L_{jj}\, d\mu_j}{\sum_j L_{jj}} - \frac{t}{z_i F} \cdot \frac{\sum_j L_{js}\, d\mu_j}{\sum_j L_{js}} \tag{11}$$

and it is now easy to construct the equivalent circuit of a liquid ion-exchange membrane. This is shown in Fig. 2 for the case where $J_s^* = 0$, and the form of the equivalent circuit is easily verified by adding the two circuit branches shown in the figure, which will yield an expression for the current-voltage relationship identical to Eq. (11).

Figure 2 summarizes the electrical properties of liquid ion-exchange membranes, and the most striking feature is the electrical heterogeneity which distinguishes homogeneous solid and liquid ion-exchange membranes. This heterogeneity is of fundamental importance in determining the

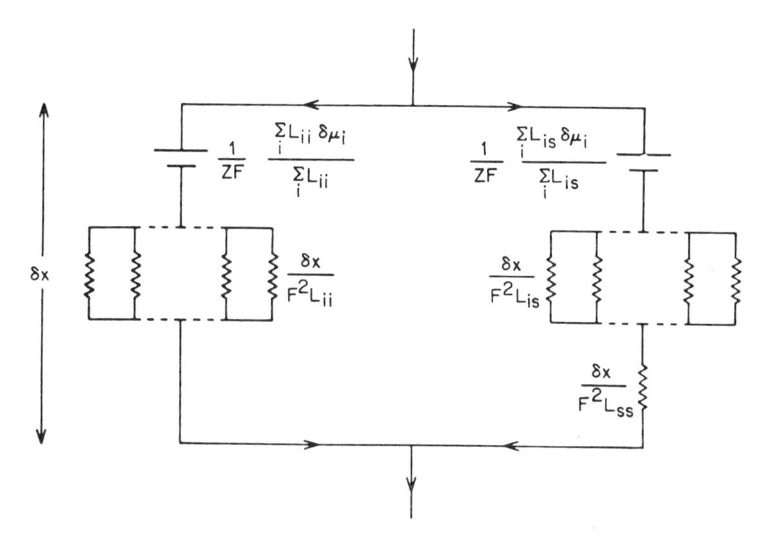

FIG. 2. Equivalent circuit element of a liquid ion-exchange membrane. [Reprinted from Sandblom (51).]

specificity of the membrane and it also leads to discrepancies between measurements of the tracer flux on one hand and measurements of the total flux or membrane resistance on the other hand (36).

The two branches shown in Fig. 2 represent the flow of dissociated and associated species, respectively, and it is therefore evident that each of these contribute to the potential and the resistance of the membrane.

Equation (11) can be integrated from 0 to d, where d is the membrane thickness, and with a knowledge of the boundary conditions a quantitative description of the electrical properties of the membrane can be given.

B. Boundary Conditions

As long as transport processes in the bulk phase are rate limiting the membrane boundaries may be treated as surfaces of discontinuity, with equilibrium distributions of ions across the boundaries. In many cases, however, it is not correct to treat the boundaries of liquid membranes as discontinuous surfaces, since the aqueous and nonaqueous phases may become structured at the area of contact. Such particular surface structures are known to give rise to high activation energies and will therefore act as barriers to the entry of ions. It is likely, however, that if an ion exchanger is dissolved in the membrane phase, it will be oriented at the surfaces in such a way that the polar groups will extend into the aqueous phases and facilitate the formation of ion pairs between sites and counterions. Such a process will tend to decrease the surface resistance and promote a rapid distribution of ions across the boundaries.

From the condition of equilibrium distributions across the boundaries, the following expressions for the potential across the membrane and the equilibrium distribution of univalent ions across each boundary can be derived (35).

$$V = \Delta\psi + \frac{RT}{z_i F} \ln \frac{C_s^0}{C_s^d} + \frac{RT}{z_i F} \ln \frac{\sum_j a_j' k_j}{\sum_j a_j'' k_j} \tag{12}$$

$$\frac{a_i'}{C_i^0} = \frac{\sum_j a_j' k_j}{C_s^0 \cdot k_i} \; ; \; \frac{a_i''}{C_i^d} = \frac{\sum_j a_j'' k_j}{C_s^d \cdot K_i} \tag{13}$$

The superscripts 0 and d refer to the membrane phase at the two boundaries and the superscripts ' and " refer to the external solutions. The constants k_j, called partition coefficients, are given by:

$$k_j = \frac{1}{\gamma_j} \exp \left(\frac{\Delta\mu_j^o}{RT}\right) \tag{14}$$

where γ_j is the activity coefficient of the dissociated ion in the liquid membrane phase and $\Delta\mu_j^o$ the difference in standard chemical potential between the two phases. If the same anion X is used in measuring the limiting distribution coefficients S_{ix} and S_{jx} of two salts, IX and JX, it is possible to relate the distribution coefficients to the partition coefficients k_j. This relationship is given by (54):

$$\frac{k_i}{k_j} = \frac{\gamma_j}{\gamma_i} \left(\frac{S_{ix}}{S_{jx}}\right)^2 \tag{15}$$

and provides a means for experimental determination of the partition coefficient ratio.

C. V_0, the Potential at Zero Current

Perhaps the most important expression describing the properties of membranes is that of V_0, the membrane potential at zero current, from which the specificity of the membrane for different ions can be derived. We shall deal with this subject of ion specificity in more detail in Sec. V after some useful expressions for V_0 have been derived.

Integrating Eq. (11) at zero current and inserting the boundary conditions from Eqs. (12) and (13) we get for V_0

$$\frac{z_i F V_0}{RT} = -\ln \frac{\sum_j u_j k_j a_j''}{\sum_j u_j k_j a_j'} - \int_0^d t \sum_j \left[\frac{L_{js}}{\sum_j L_{js}} - \frac{L_{jj}}{\sum_j L_{jj}}\right] d\mu_j \tag{16}$$

$$- \int_0^d \frac{L_{ss}}{L_{ss} + \sum_j L_{js}} \cdot \frac{J_s^*}{L^*} \cdot dx$$

where z_i is the valence of the counter ions and u_j the mobilities.

It has been assumed that mobility ratios and activity coefficient ratios are constant, which are valid assumptions if the Debye-Huckel and the Onsager limiting laws can be applied. This seems reasonable, however, for the low dielectric media of hydrocarbons, where the ionic strength is very low.

The first integral in Eq. (16) can be evaluated in a number of limiting cases--of particular interest is that of strong association, which is a characteristic property of most liquid ion-exchange membranes.

The evaluation of the last integral appearing in Eq. (16), on the other hand, is more difficult and its value depends on the solubility of the exchanger in the aqueous phase, the presence of moving boundaries, and possible stirring of the bulk phase, etc. A few interesting qualitative conclusions can nevertheless be drawn from the form of the last integral. If, namely, the membrane contains a moving boundary separating two parts of the membrane containing different concentrations of the exchanger, the last term will contribute a potential with a sign which depends on the direction of the moving boundary and on the valence of the exchanger. As the moving boundary reaches an interface there will be an abrupt change in J_s^* due to the fact that the exchanger is preferentially soluble in the nonaqueous phase. Accompanying this abrupt change in J_s^* will be an equally abrupt change in the measured membrane potential. In this way, the rapid potential shifts reported by Dupeyrat (13) and by Colacicco (16) in systems with moving boundaries (see Figs. 3 and 4) can be explained with respect to sign and at least qualitatively with respect to magnitude by the last term in Eq. (16).

The effect of the solubility of the exchanger can also be deduced from the last term in Eq. (16). A difference in solubility on the two sides of the membrane, due for instance to differences in salt concentration, will give rise to a net flow of exchanger across the membrane. The net effect of the potential will therefore be the same as a movement of co-ions and could explain the departure from an ideal Nernst behavior of the potential in many liquid ion-exchange membranes at low solution concentrations (17, 55).

Although Bonner and Lunney (17) have taken the solubility of the exchanger into account in a theoretical analysis of the potential response, they have neglected contributions from the last term. Instead they have explained the effect of the potential by a change in the external activities of the ions due to the partial solubility of the exchanger.

In some important circumstances, however, it is possible to neglect the second integral entirely. One such case is when the external solutions are changed without altering the membrane phase, a condition which is

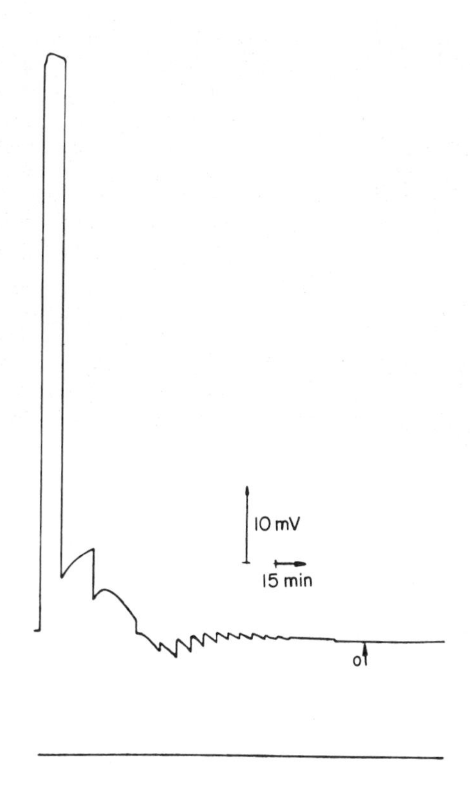

FIG. 3. The potential of a liquid membrane containing a moving boundary. The external solutions are identical and as the moving boundary reaches an interface the potential attains the value of zero. [Reprinted from Dupeyrat (13).]

usually met in an electrode measurement. If, furthermore, the exchanger is sparingly soluble in the aqueous phases, so that no appreciable flow of exchanger occurs across the boundaries, the total flow of sites can be set equal to zero.

It then remains to evaluate the first integral in Eq. (16). For strong association, i.e., when the concentrations of dissociated species are negligible compared to the concentrations of undissociated species, Eq. (16) yields after integration (two counterions) (35):

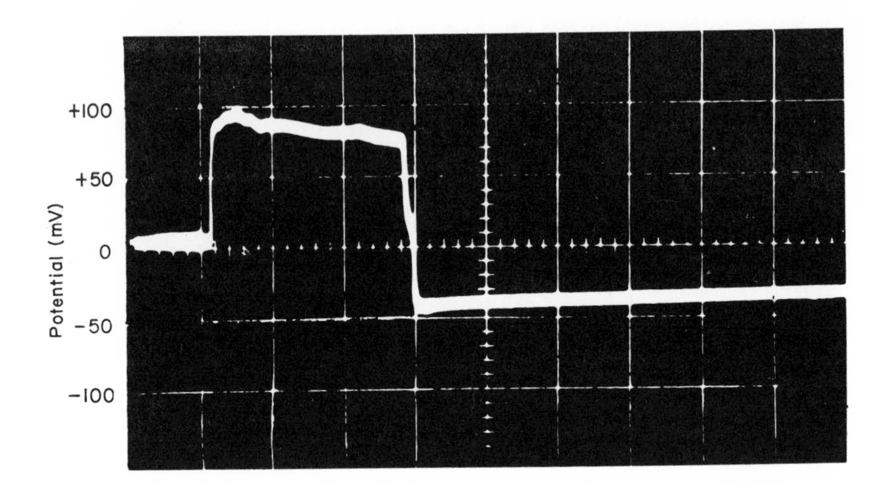

FIG. 4. Oscilloscope tracing of a reversal of potential across a pentanol membrane (membrane thickness 3 mm). Sodium decylsulfate was applied to the left side of the membrane. Note the reversal of potential as the moving boundary reaches the right membrane surface. The solution on the left side contained 10 mM KCl, the solution on the right side contained 1 mM KCl and was connected to ground potential. [Reprinted from Colacicco (16).]

$$\frac{z_i F V_0}{RT} = (1-\tau)\ln \frac{a_1' + \left(\frac{P_2}{P_1}\right)^s a_2'}{a_1'' + \left(\frac{P_2}{P_1}\right)^s a_2''} + \tau \ln \frac{a_1' + \left(\frac{P_2}{P_1}\right)^e a_2'}{a_1'' + \left(\frac{P_2}{P_1}\right)^e a_2''} \tag{17}$$

where

$$\tau = \frac{t_{1s}\left(\frac{P_2}{P_1}\right)^e - t_{2s}\left(\frac{P_2}{P_1}\right)^s}{\left(\frac{P_2}{P_1}\right)^e - \left(\frac{P_2}{P_1}\right)^s} \tag{18}$$

and $\left(\frac{P_2}{P_1}\right)^e = \frac{u_{2s}}{u_{1s}} K_{21}$ $\quad \left(\frac{P_2}{P_1}\right)^s = \frac{t_{1s}}{t_{2s}} \cdot \frac{k_2}{k_1}$

u_{2S}, u_{1S} are the mobilities of undissociated species and t_{1S}, t_{2S} are the transference numbers of the sites with respect to ions 1 and 2. K_{21} is the ion-exchange selectivity constant defined as

$$K_{21} = \frac{C_{2S}}{C_{1S}} \cdot \frac{a_1}{a_2} \tag{19}$$

Equation (17) for V_0 is seen to comprise two logarithmic terms, each defining a set of separate selectivities. Depending on the form of representation, however, Eq. (17) can be approximated by a single logarithmic term. Figure 5 shows an exponential representation of the potential as a function of concentration and the curves are seen to be very nearly straight lines. Figure 6 shows an experimental verification of this, as reported by Rechnitz (49). The experiment was performed with a chloride ion-selective electrode in the presence of iodide.

A different representation commonly used is to plot the potential versus the logarithm of the concentration of one of the ions. This is shown in Fig. 7, using the same selectivity values as those used in Fig. 5. One of the curves corresponds to a value of $\tau > 1$, which can occur for some values of the parameters (see Eq. (18)), and the curve is seen to exhibit a maximum. This may be compared to the potential curves describing the pH dependence of a calcium electrode (5) (Fig. 8), which exhibit minima.

The curves in Figs. 5 and 7 also demonstrate how the selectivities shift from one set of values to another with the quantity τ, and this quantity may therefore be regarded as a specificity controlling factor in liquid ion-exchange membranes.

D. Monovalent-Divalent Exchange

Equation (11), describing the current-voltage relationship as well as the equivalent circuit shown in Fig. 2, can easily be extended to higher valences and to mixtures of univalent and divalent electrolytes. The conclusions concerning the two pathways and the two sets of selectivities reached for univalent ions will remain valid for mixtures, but the form of the expressions will be slightly different.

An expression which has been reported to describe the data for mixtures of monovalent and divalent ions is the following (7,38,47):

$$V_0 = \text{const} + \frac{RT}{2F} \ln (a_{B^{2+}} + K_{AB} \cdot a^2_{A^+}) \tag{20}$$

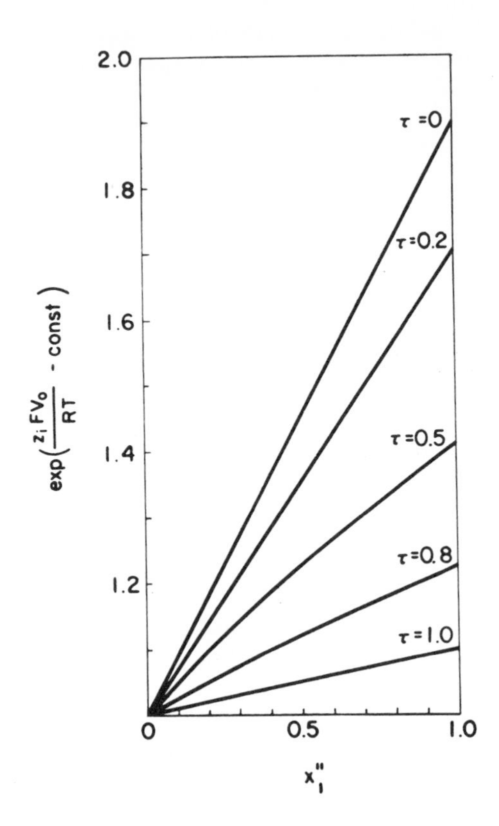

FIG. 5. Equation (16) is represented by plotting $\exp \frac{FZ_iV_0}{RT}$ - const as a function of mole fraction (x_1) of species 1 in the external solution. The parameters are chosen so that Eq. (16) may be written as $\exp \frac{FZ_iV_0}{RT}$ - const = exp $(1-\tau)\ln(1 + 0.9\, x_1'') + \tau \ln(1 + 0.1\, x_1'')$. This function is plotted for the following values of τ: 0, 0.2, 0.5, 0.8, and 1.0. [Reprinted from Sandblom et al. (35).]

which has been derived theoretically for solid ion-exchange membranes (56). In a solid ion exchanger, however, the exchange reaction can be written as

$$2A^+ + BX_2 \rightleftarrows A_2X_2 + B^{2+}$$

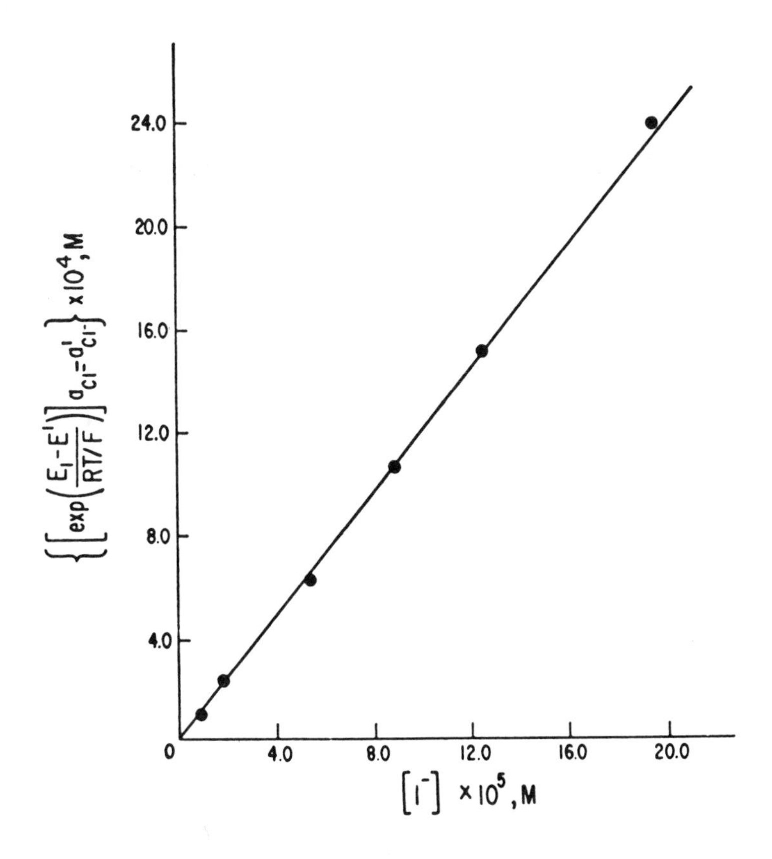

FIG. 6. Selectivity plot for iodide with a chloride ion-selective electrode. [Reprinted from U. Srinivasan and G.A. Rechnitz (49).]

and if a liquid ion exchanger is to take part in a similar reaction, a dimerization of the ion pairs is required.*

*A thermodynamic analysis of an ion exchange reaction involving two ions must take into account the fact that for a solid ion exchanger, the number of components is three (membrane and ions 1 and 2); whereas a liquid ion exchanger has four components (solvent, exchanger, and ions 1 and 2) and its chemical potential is therefore not a priori defined.

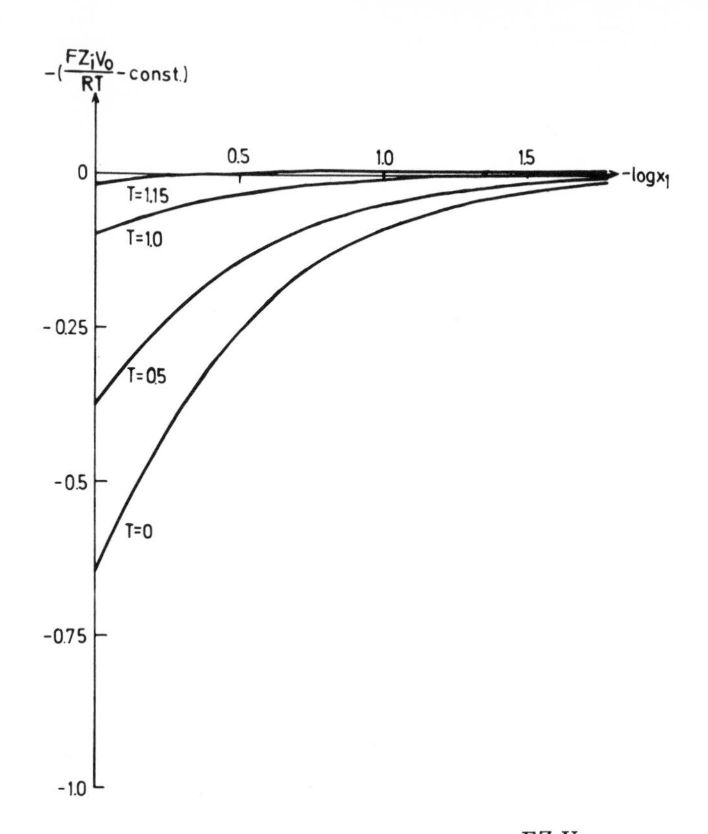

FIG. 7. Equation (16) is represented by plotting $\frac{FZ_iV_0}{RT}$ - const as a function of $-\log x_1$ where x_1 is the mole fraction of species 1 in the external solution. The parameters are the same as those used in Fig. 5. Note the maximum in the uppermost curve ($\tau = 1.15$).

In addition, the equilibria at the boundaries of liquid membranes involve two processes, namely the transfer of ions across the boundaries and the formation of ion pairs. This will lead to an entirely different set of boundary potentials than those derived for solid ion exchangers and consequently there seems to be little theoretical justification for applying Eq. (20) to liquid ion-exchange membranes.

Huston and Butler (46) give a slightly different equation for the potential of liquid membranes in the presence of monovalent and divalent ions:

$$V_0 = \text{const} + \frac{RT}{2F} \ln \left[m_2 (m_1 + 2m_2)^2 \gamma_{21}^3 + Km_1^2 (m_1 + 2m_2)^2 \gamma_{12}^4 \right] \quad (21)$$

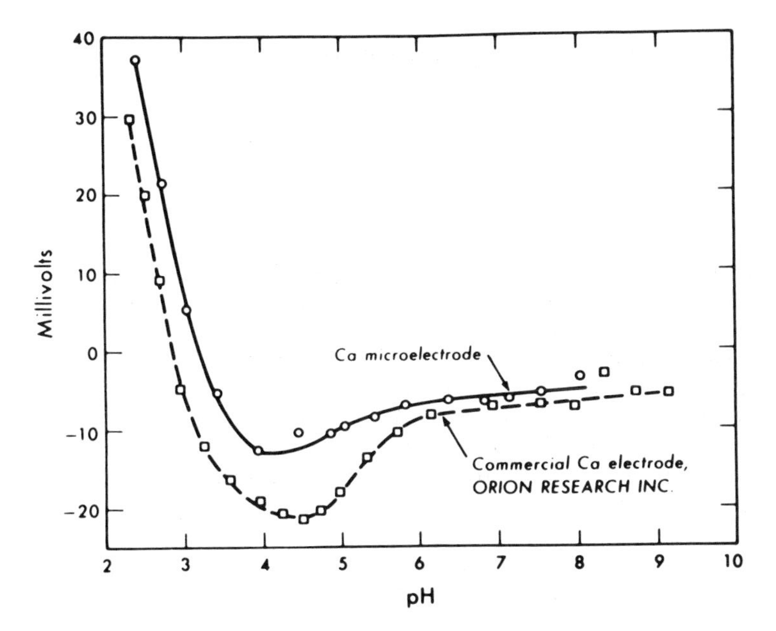

FIG. 8. Calcium electrode potential versus solution pH. The calcium concentration was held constant at 1 mM [Reprinted from Orme (82).]

where m_1 and m_2 are concentrations of monovalent and divalent ions, respectively, and γ_{12}, γ_{21} the corresponding activity coefficients.

Although these equations for monovalent-divalent mixtures have been published, no experimental verification of their validity seems to have been made and a rigorous theoretical derivation also remains to be done.

E. Fluxes and Current-Voltage Relationship

Some interesting properties also characterize liquid ion-exchange membranes with respect to ion fluxes and the current-voltage relationships. Explicit expressions for these can be obtained by integrating Eq. (11), although this presents considerable mathematical difficulties in the general case. For the limiting case of strong association, however, solutions have been presented (57) from which some properties of fluxes and resistance were deduced. The total flux is given by

$$J_i^* = \frac{I \cdot t_i}{z_i F} + J_{is} \tag{22}$$

where t_i is the transference number defined as

$$t_i = \frac{u_i c_i}{\sum_j u_j c_j + u_s c_s}$$

At zero current, the flux is seen from Eq. (22) to be entirely determined by the properties of the ion pairs. A comparison with the equivalent circuit (Fig. 2) then reveals that the permeability of an ion determined potentiometrically is entirely different from that obtained by a flux measurement. For example, if the conductance L_{ss} is small (see Fig. 2), the potential response will be determined by the properties of the left branch of the circuit whereas the fluxes will be dependent on the right branch, and the two branches need not at all have the same set of selectivities. It is important to realize this possibility in biological membranes where conclusions regarding fluxes are often inferred from potential measurements.

F. Resistance

As we have seen, the fluxes and the potential at zero current are solely determined by three parameters, namely the selectivity constant, the ion conductances, and the distribution coefficients. The resistance is also a function of these three parameters, but in addition it is explicitly dependent on the dissociation constants, defined as

$$K_i = \frac{C_i \cdot C_s}{C_{is}} \tag{23}$$

To derive expressions for the membrane resistance, therefore, requires additional assumptions such as constant K_i-values. Using this approximation, however, equations for the membrane resistance could be derived (57) and it was found to be strongly dependent on the degree of association of the components.

If all components are strongly associated, the current-voltage relationship shows only a slight nonlinearity apart from the presence of limiting currents (see Fig. 17). In the presence of one or more dissociated ions, however, the possibility for rectification arises (see Fig. 18), and in this respect a liquid ion-exchange membrane will resemble a solid ion-exchange membrane.

It appears that co-ions have a profound effect on the resistance of these membranes. In a theoretical analysis, Ciani and Gliozzi (33a) showed that the membrane resistance decreased asymptotically to zero with large fields in the presence of co-ions, and this seems to be consistant with experimental findings.

Although the expressions for the membrane resistance obviously represent rather crude approximations, they are nevertheless useful guidelines as to what may happen inside a liquid ion-exchange membrane, which the comparison between theory and experiment has demonstrated (12).

V. PHYSICAL PARAMETERS OF LIQUID MEMBRANES

If we turn our attention once more to Eq. (11) we can describe the properties of liquid ion-exchange membranes in terms of the physical parameters entering the equation. As one Faraday of current is passed reversibly across a thin section of the membrane, it is seen from Eq. (11) that part of the free energy change is associated entirely with the passage of charge through the solvent and part of it via the exchanger; and since the distribution of free energy between associated and dissociated forms is given by the quantity t we can write

$$d\Psi = (1-t)\, dG_{solvent} + t\, dG_{exchanger} \tag{24}$$

where G is the Gibbs free energy. This equation is of general validity regardless of the valence of the ions and of the particular chemical reactions in the membrane phase.

In evaluating the total free energy change, we must therefore consider three factors, namely ion solvation, ion conductance, and ion extraction. This is further confirmed by the integrated form of Eq. (11), namely Eq. (16), where these three factors appear through the distribution coefficients, the transference numbers, and the ion-exchange selectivity constant. It is implicitly assumed in this statement that the ion pairs have equal mobilities, which does not seem unreasonable, however, due to the lack of charge of the ion pairs and because the complexes have equal sizes as long as the exchanging ions are small.

All the parameters mentioned above can be determined by classical physicochemical methods and some experimental results obtained this way will be described in Sec. VI. In the original treatment of Sandblom et al. (35), the dissociation constants were introduced for theoretical reasons and appear in their expressions for V_0 and the current-voltage relationship. The dissociation constants have also been determined experimentally for a number of liquid ion-exchange systems (12,55,58), but in most weakly dissociated systems they are difficult to obtain by ordinary conductometric methods, and it is, therefore, preferable to use other measurable quantities, namely those appearing in Eq. (16).

A. Ion Solvation

The free energy change required to move an ion from an aqueous solution to a nonaqueous solvent can be decomposed into two terms, namely the free energy of hydration and the free energy of solvation

$$\Delta G_{total} = \Delta G_{hydration} + \Delta G_{solvation} \tag{25}$$

There have been many attempts to calculate ΔG theoretically and the simplest model assumes the validity of the Born equation, which describes the charging of the dielectric medium and neglects any specific interactions between ions and solvent molecules.

$$\Delta G_{el} = \frac{Ne^2 z_i^2}{2r_i} \left(\frac{1}{D_{org}} - \frac{1}{D_w}\right) \tag{26}$$

where D is the dielectric constant and r_i the ionic radius. Data for the free energy of hydration of the alkali halides in water (59) are fitted quite well by this equation but it fails to describe similar data for nonaqueous solvents (60). Nevertheless, it is expected to be valid for large ions where the charge is buried at the center. Grunwald et al. (61) have calculated a radius of 4.0 Å as being the limit below which the Born equation is inadequate. On the other hand, to assume the validity of this equation above 4.0 Å is an extrathermodynamic assumption, which has nevertheless been successfully used in evaluating single ion activity coefficients and conductances.

Below the 4-Å limit it is necessary to take into account the interactions between the charge of the ion and the complex distribution of charge on the solvent molecules. In general there are four such kinds of strong solvent-solute interactions, namely ion-dipole, dipole-dipole, π-complex-forming, and hydrogen bonding (62).

A useful classification from the point of view of ion solvation is to divide solvents into protic or hydrogen donors, and aprotic, which do not readily form hydrogen bonds (62). Examples of the former are water, formamide, methanol, etc., and examples of aprotic solvents are acetone, acetonitrile, nitrobenzene, etc. Protic and aprotic solvents differ with respect to their solvation of cations and anions.

Solvation of anions in nonaqueous liquids is well illustrated by a figure taken from Parker (Fig. 9) (63). In protic solvents such as methanol, the solvation of anions decreases strongly in the series F > Cl > Br > I, due to hydrogen bonding. Dipolar aprotic solvents (such as DMF), on the other hand, solvate by dipole interactions which increase with the size of the ions. For large anions in dipolar aprotic solvents the order is therefore reversed (62, 63).

For cations the situation is quite different and generalizations are more difficult to make, due to specific donor-acceptor interactions between cations and solvent molecules. In general, cations are strongly solvated in polar solvents with a negative charge located at oxygens, as found experimentally in methanol-water mixtures (see Fig. 10) (64), and in water-dioxane mixtures (61). In both cases the organic solvent prefers cations to anions.

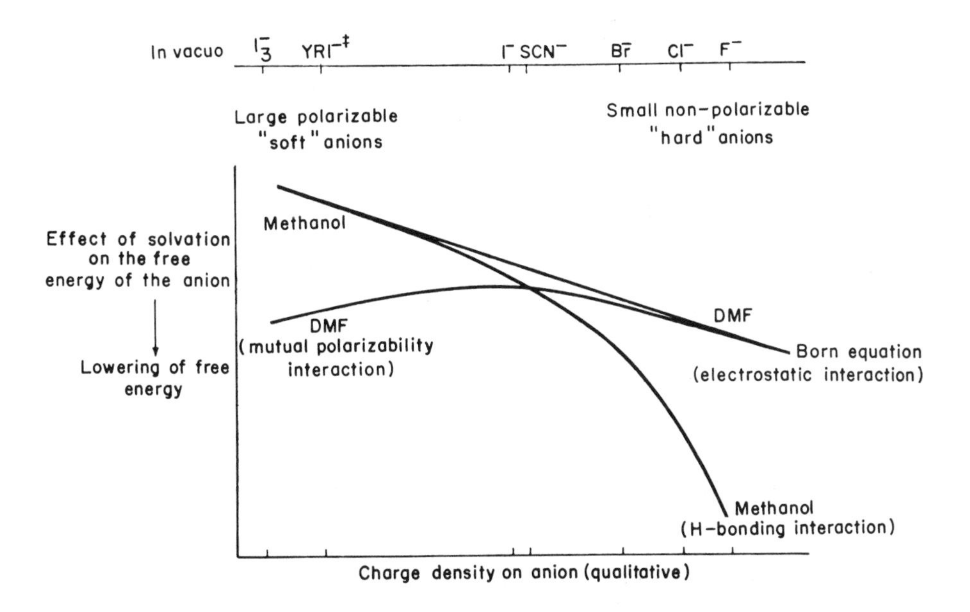

FIG. 9. Qualitative representation of the effect of electrostatic, hydrogen-bonding, and mutual polarizability interactions on the free energy of solvation of anions in methanol and in dimethylformamide (DMF). [Reprinted from Parker (63).]

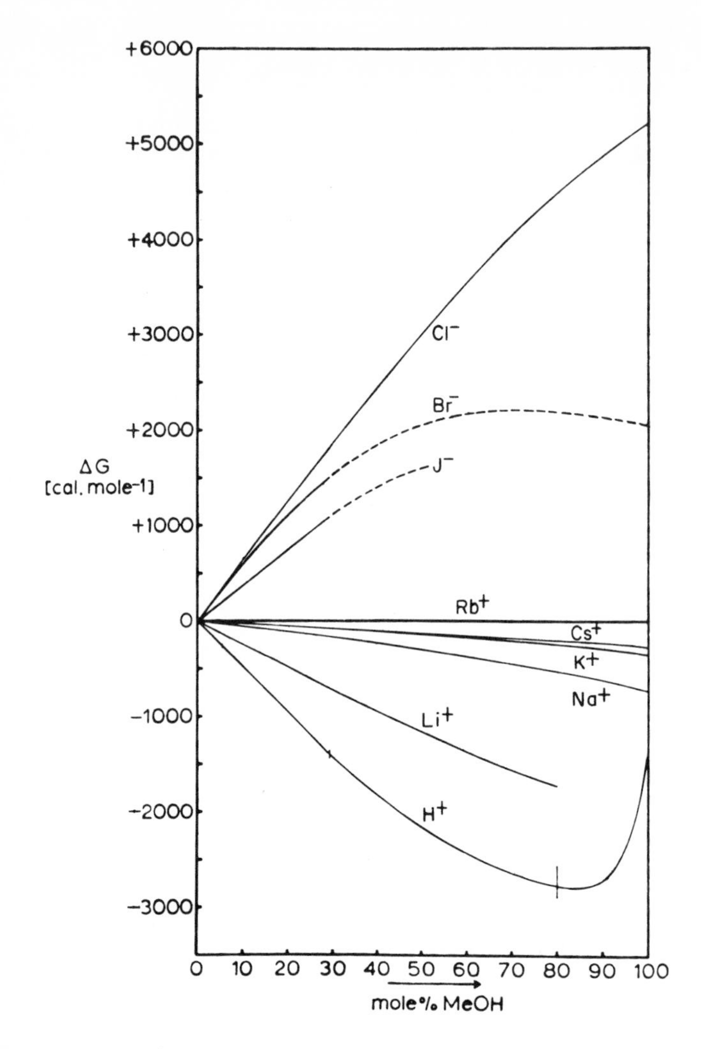

FIG. 10. The free enthalpies of transfer from water to water-methanol mixtures for some univalent ions. [Reprinted from de Ligny and Alfenaar (64).]

An equation which takes the ion-dipole interactions into account is the Born-Bernal-Fowler approximation (65):

$$\Delta G = \frac{Ne^2 z_i^2}{r_i}\left(\frac{1}{D_{org}} - \frac{1}{D_w}\right) + \frac{Ne_i z_i \cdot n}{r_i^2}(m_w - m_{org}) - T\Delta S \qquad (27)$$

where m is the dipole moment of the solvent molecules and n the coordination number of the ion. Although these general rules can be applied to the partition of ions between aqueous and nonaqueous solvents, the experimental data show large variations from solvent to solvent and often show deviations from theoretical behavior. Alexander and Parker (66) have listed activity coefficients for various anions and cations in different organic solvents obtained from solubility data. To obtain activity coefficients from solubility data, the extrathermodynamic assumption was made that the activity coefficients of the large ions tetraphenylborate and tetraphenylboride are equal.

Although a considerable amount of data on ion solvations and aqueous solvents exist in the literature, no systematic study on the properties of liquid membranes as a function of the organic solvent seems to have been undertaken.

B. Ion Conductance

The ion conductances or mobilities enter the potentiometric selectivity of liquid membranes through the transference numbers [see Eq. (16)], but in most liquid ion-exchange systems the limiting equivalent conductances are difficult to measure due to the very strong association, which decreases the conductivity. On the other hand, it is possible to measure transference numbers even in nonaqueous media with very low dielectric constant according to a potentiometric method described by Gemant (67); and since only the transference numbers enter the equation for the potential response [Eq. (16)], there is no need to measure ion conductances in order to make a theoretical prediction of V_0.

The ionic mobilities will in general depend on the crystal radii, the size of solvation sheaths, and specific interactions with the solvent molecules. These factors were taken into account in a paper by Fuoss (68), where he suggested that the electrostatic coupling between the ion and the solvent acts effectively as an increase in viscosity, namely through the extra work that must be done in orienting the dipoles as the ion passes among them. Using this argument, Fuoss derived the following expression for the mobility or the limiting equivalent conductance of an ion

$$\lambda_0 = \frac{F^2}{\eta \cdot N \cdot 1800 \cdot R_\infty \left(1 + \frac{A}{R_\infty^2}\right)} \tag{28}$$

where η is the viscosity coefficient, N Avogadro's number, D the dielectric constant, and R_∞ the hydrodynamic radius of the ion in a hypothetical solvent of infinite dielectric constant where all electrostatic forces would be reduced to zero. Other expressions for the mobilities have also been derived which take the ionic radii and the dielectric properties of the solvent into account. An excellent survey of these phenomena has been given by Conway (68a).

In general it is found that the transference numbers of univalent ions do not vary greatly from one solvent to another with the exception of the hydrogen ion (69, 70). On the other hand, the mobilities are expected to vary with concentration, partly because of electrostatic interactions taken into account in the Onsager theory and also because of changes in the size of the solvation sheath. Bonhoeffer et al. (30) found a strong concentration dependence of the transference numbers of LiCl in a chinolin membrane, which they interpreted as being due to a successive replacement of water molecules with solvent molecules in the solvation sheath as the electrolyte concentration increased. Hydration may also explain the decreased equivalent conductance of HCl in alcohols as water is added (71).

C. Ion Pair Extraction

The process by which an ion forms a complex with another agent and dissolves in the organic phase is called solvent extraction. This process has become a very important separation method in analytical chemistry and an enormous variety of complexing agents have therefore been described. The literature on this subject is quite extensive but is well covered in the books by Morrison and Freiser (72) and by Dyrssen et al. (73).

There are essentially two different ways in which an ion can form an extractable complex with the complexing agent, namely by coordination, in particular chelation, and by the formation of an ion pair. We shall mostly be concerned with the latter process since a large number of organic ions can be extracted by this mechanism (74), which therefore seems to have a wide range of applicability in this context.

An ion pair extraction can be written in the following way (74)

$$Q_{aq}^{+} + X_{aq}^{-} \rightleftarrows QX_{org} \tag{29}$$

and the extraction constant is given by

$$E_{QX} = \frac{QX_{org}}{Q_{aq}^{+} \cdot X_{aq}^{-}} \tag{30}$$

By comparing this equation with Eq. (19) it is seen that the ratio of extraction constants is related to the ion-exchange selectivity constant by

$$K_{21} = \frac{E_{Q_2X}}{E_{Q_1X}} \cdot \frac{\gamma_1}{\gamma_2} \tag{31}$$

and since data on extraction constants are available for a large number of systems, they can be used to correlate ion-exchange properties with the potential response of liquid membranes.

The magnitude of the extraction constant depends on the nature of the organic solvent, on the electron donor properties of the ion, and on the structure of the ion pair (74).

A class of exchangers commonly used for cations are the dialkylphosphoric acids (75):

```
R - O \       // O
        P
R - O /       \ OH
```

With divalent ions this compound can form a coordinated complex with ionic bonds between the oxygens and the ion

```
R - O         O           O          O - R
      \     /   \      .'   \\     /
        P          M'          P
      /   \\    .'   \      /   \
R - O         O           O          O - R
```

which is the reason why this exchanger is highly selective to divalent ions.

The type of alkyl group attached to the acid-in-ester linkage plays a very important role in the extent of extraction, which increases as the electronegativity of the R-group increases (76).

Anions are usually extracted with alkyl amines or quarternary ammonium ions (77), and the extraction is increased with increasing length of the hydrocarbon chain (74). In general, the order of preference of the weakly basic amines for various anions follows the sequence (3):

$$Cl0_4 > I > Br > NO_3 > Cl > HCO_4 > F$$

In calculating the free energy change for exchange reactions involving liquid systems, a series of different energy steps must be considered. The usual procedure in analyzing solvent extraction data is to split the total reaction into the following steps (72).

1. Entry of the exchanger into the aqueous phase

2. Dissociation of the exchanger.

3. The reverse steps of 2 and 1 for the second ion

The free energy change of each of these steps can be determined experimentally or by physicochemical calculations.

If, however, the exchanger is negligibly soluble in the aqueous phase, it seems more advantageous to use the following steps (65).

1. Solvation of the ion

2. Formation of ion pairs

3. The reverse steps of 2 and 1 for the second ion

Reaction 1 in the latter scheme has already been considered in Sec. V.A and reaction 2 can be analyzed in terms of theories of association (68a). This second series of steps was used in the original theoretical treatment of Sandblom et al. (35).

Finally, one may consider reaction steps similar to those used by Eisenman in his successful theory of specificity developed for solid ion-exchanger membranes (77a). The corresponding steps for liquid membranes would be:

1. Transfer of the ion from water to vacuum

2. Chemical binding to the exchanger in vacuum

3. Transfer of the ion pair from vacuum to the solvent

4. The reverse steps of 3 - 1 for the second ion

The first step in this third scheme is simply given by the free energy of hydration. The second step may be calculated from electrostatic interactions or by thermochemical methods as suggested by Eisenman (77b). However, the field strength or the electronegativity of the site will be altered by polar or inductive effects. Attachment of side groups to a molecule will shift electrons toward or away from an exchange site and thereby change its strength. A discussion of these effects as well as pertinent data have been given by Rozen (65).

The last step in the third scheme will depend on dipole moments of the solvent and of the ion pairs as well as on the size of the latter. In general the change of free energy in moving the ion pair from vacuum into the solvent will involve the same factors that were considered in the case of ion solvation, although coulombic forces will not be present.

Figure 11 shows a schematic representation of the various energy steps involved in the ion-exchange reaction according to the three different schemes.

Although no attempts have been made to calculate the various steps according to the third scheme for liquid membranes, it is probably the one best suited for ion-exchange reactions, and with the proper atomic models it should yield theoretical results for comparison with potentiometric data which are rapidly being accumulated in the literature.

D. Side Reactions

The ion-exchange process is not only dependent on interactions with the exchanger but is also greatly influenced by different kinds of side reactions. The most important of these can be listed as follows:

1. Polymerization of ion pairs
2. Adduct formation
3. Decomposition of the membrane solvent

1. Polymerization of Ion Pairs

Alkyl phosphoric acids form hydrogen bonded dimers and higher polymers in many solutions (76):

```
R - O \     // O . . . . . . . OH \     / O - R
       P                           P
R - O /     \ OH . . . . . . O  //     \ O - R
```

The effect is greatest in monoalkylphosphoric acid but decreases as the solvent becomes more polar and starts to compete for the hydrogen bonding. Higher aggregates are found in systems with very polar molecules, for example the di-ethylhexyl phosphoric acid system (HDEHPA),

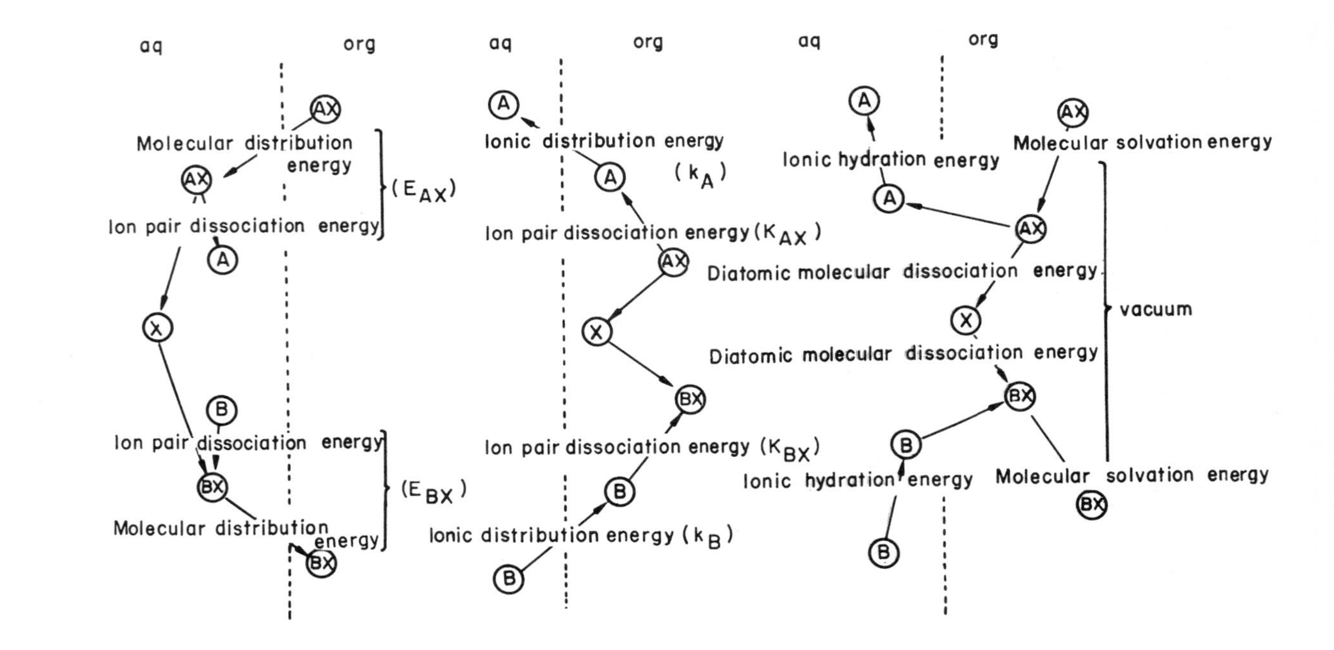

FIG. 11. Energy steps involved in a liquid ion exchange according to three different schemes. Note that the energy state of the exchanger X is constant during the exchange process and may therefore be given an arbitrary location. The equilibrium constants corresponding to the various energy steps have been defined in the text.

which in the presence of sodium will contain aggregates of the form HDEHPA · 3 Na HDEHPA (78,79) in wet benzene solutions.

This may affect the selectivity constant as well as the transport processes, since it introduces additional coupling between the transported components.

A particular type of aggregation involving co-ions was proposed in order to account for the departure from Nernst behavior of a calcium electrode consisting of an alkylphosphoric acid in di-n-octylphenylphosphonate (46). If, namely, species such as CaClR are transported through the organic phase instead of CaR_2, this would have the effect of making the membrane permeable to co-ions and destroy the Nernst behavior.

2. Adduct Formation

A more interesting phenomenon, which should be highly relevant to membrane transport, is that of adduct formation between ion pairs and other components. The phenomenon is also referred to as synergism, the enhancement of extraction by the presence of neutral molecules (65,72). Such molecules can consist either of particular substances added to the system or of the solvent molecules themselves. Depending on the coordination number of the exchanging ion, these adduct forming molecules will pack around the ions and thereby lower the free energy of the complex. The adduct forming molecules replace water molecules in the coordinating position and in principle the phenomenon is analogous to the solvation of cations by the neutral cyclic compounds.

Therefore, it is possible not only to alter the extraction constants but also the selectivity of a system by adding adduct forming compounds, a fact which can be used to explain biological specificity. It also has an immediate practical application, namely to increase the selectivity of electrodes.

3. Decomposition of the Membrane Solvent

A third type of reaction which can alter the membrane properties is a decomposition of the membrane. This involves reactions between the membrane solvent and the exchanger and/or the exchanging ion. As an example, one can mention the formation of alkylhalides between alcohols and halide ions (69):

$$HA + ROH \rightleftarrows RA + H_2O$$

This is a very slow reaction but it can nevertheless be quite extensive in acid media.

E. Water Content

Some very important properties of liquid membranes are associated with the partial water solubility in the membrane phase, which in general will depend on the concentration of electrolytes as well as of the exchanger. The presence of water, even in very small quantities, will greatly alter the properties of the membrane and its physical parameters. The selectivity of the membrane, for example, will be greatly reduced, and the conductivity increased in the presence of water, due to its effect on the dissociation constants. Also, the transference numbers change in the presence of water, which was pointed out in the previous section. Consequently the solubility of water in the membrane phase makes it very difficult to predict the properties of liquid membranes from the physical chemistry of dry systems.

Changes in water content with concentration could account for some of concentration dependence of the selectivity constants observed in liquid electrode systems (49), and it has also been suggested that water transport may be responsible for the departure from Nernst behavior in some electrode systems (47). The sulfonic acids, for example, are known to form micelles in contact with aqueous solutions. This could be the reason why the observed water transport is considerable in membranes containing dinonylnaphtalenesulfonic acid and which in turn could explain their departure from Nernst behavior

To conclude this section, it should be emphasized that only a very brief outline of the physical chemistry of liquid membranes has been made. A more thorough description does not seem justified here, considering that very few systematic investigations of how these affect the properties of liquid membranes have been undertaken. It is clear, however, that the physical parameters of liquid systems offer intriguing possibilities to accomplish specificity in the design of electrodes; and as models for biological membranes, liquid membranes clearly possess sufficient variability to be able to account for most biological transport phenomena.

VI. EXPERIMENTAL SYSTEMS

The various experimental systems used to study liquid membranes have differed greatly from each other but essentially they fall into three categories, namely: (a) thick nonconvection-free membranes, to which most commercial electrodes belong; (b) thin convection-free membranes; and, (c) systems designed to study surface diffusion.

A. Thick Systems

Most of the thick systems have been studied potentiometrically, although Shean and Sollner (9) have also performed flux measurements and compared them with potential measurements.

In spite of the fact that the selectivity properties of thick membranes have been the subject of numerous investigations in order to develop ion specific electrodes, relatively little has been done in the way of systematic studies which are suitable for a theoretical analysis. It is naturally of interest to know how the selectivity data obtained from potential measurements correlate with solvent extraction data. Shean and Sollner (9), as well as Coetzee and Freiser (7), claim at least qualitative agreement between potential and extraction data, although they have not presented any evidence to support their statements.

A crude illustration of this correlation can be obtained, however, by comparing the potential data of Coetzee and Freiser on organic and inorganic salts of methyltricaprylylammonium ion (Aliquat 336 S) dissolved in n-decanol, with the extraction data on a similar system, namely tetrabutylammonium ion in chloroform (74). Figure 12 shows the potential data plotted versus the extraction data, and the correlation is reasonable considering that we are dealing with two different systems.

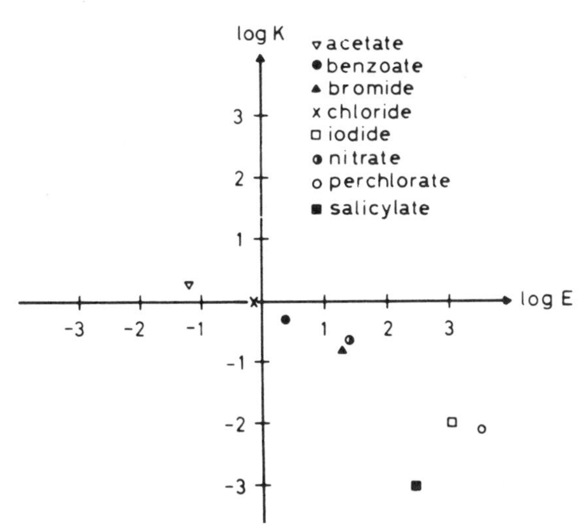

FIG. 12. The logarithm of the selectivity coefficients measured potentiometrically with an ion selective electrode (7) and plotted against the logarithm of the corresponding extraction coefficients (74). See text for further explanation.

In order to compare theoretical predictions with experimental results, Eisenman (55) undertook the ambitious task of measuring the physical parameters appearing in Eq. (16) by conductometric measurements. He chose a system consisting of di-2-ethylhexyl phosphoric acid dissolved in water saturated n-pentanol. The dissociation constant and limiting conductances were evaluated according to the method of Fuoss and Krauss (80), which requires that the conductance values are extrapolated to infinite dilution. This method, however, has two disadvantages, one being that the parameters are concentration dependent and need not have the same value at infinite dilution as at the actual membrane concentration. Secondly, it is difficult to extrapolate conductance values in solvents of low dielectric constants, although the system with HDEHP in wet n-pentanol chosen by Eisenman was sufficiently dissociated to permit reasonable extrapolations ($pK_{NaX} = 6.2$, $pK_{HX} = 8.15$).

The biionic potential for N-Na calculated on the basis of the conductance data turned out to be 240 mV as opposed to the 140 mV actually measured.

A somewhat different approach was used by the authors. They measured the exchange isotherm directly, at the same concentration of the exchanger as that used in the potential measurements. By solvent extraction procedure they obtained the curve shown in Fig. 13, from which an ion-exchange selectivity constant of $K_{21} = 2000$ was obtained.

The ratio of partition coefficients was measured by a method similar to that used by Shedlovsky and Uhlig (26). The value for k_H/k_{Na} was calculated from the distribution coefficients of HCl and NaCl between water and pentanol. These in turn were measured by equilibrating aqueous solutions of HCl and NaCl with n-pentanol, after which samples of the n-pentanol were redissolved in water and analyzed for chloride by electrometric titration. The distribution coefficients S between alcohol and water obtained this way were extrapolated to zero concentration and the limiting distribution coefficients (see Fig. 14) were found to be 0.037 and 0.0031, respectively, for HCl and NaCl. Squaring the ratio of the distribution coefficients gives the value of 100 for the partition coefficient ratio.

The values for the transference numbers were calculated from the transference number of NaCl measured potentiometrically in wet n-pentanol. To obtain the values with respect to HDEHP which are needed in Eq. (16), the conductometric data of Eisenman (55) were used to supply the missing information. A value of τ equal to 0.88 was given by Eisenman, which yields a calculated value for the potential of 190 mV, using the values for the selectivity constant and the partition coefficient ratio given above. This is in better agreement with the observed 140 mV, but a value of τ equal to 0.3 is required to obtain exact agreement between calculated and measured

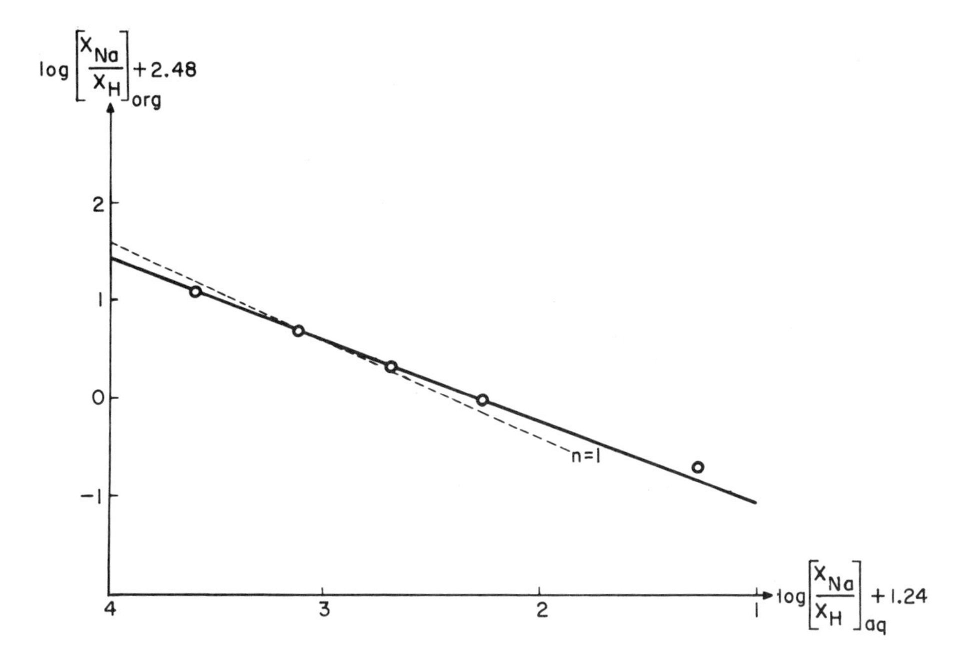

FIG. 13. Ion exchange isotherm for sodium-hydrogen exchange in 10% HDEHPA in n-pentanol. Note the slight departure from ideal behavior.

values for the potential. It seems unlikely that the conductometric data contain errors of this magnitude and the reason for the discrepancy is not quite clear.

It was also pointed out on theoretical grounds that the potential should be independent of J_s^* for external solution changes. Consequently the potential at zero current should be independent of the geometry as long as only external solution changes are made. To test this theoretical prediction, the potential was measured using the same chemical system but different geometrical conditions. Figure 15 shows the potential measured on two different systems, namely: (a) a thick membrane consisting of a 0.5-cm thick droplet located at the tip of a calomel electrode; and, (b) a thin (0.5-mm) convection-free membrane described by Sandblom (12), in which a stationary state was reached. The potentials are virtually the same in both systems and Eq. (16) is therefore expected to apply to thick systems as well as thin stationary state systems.

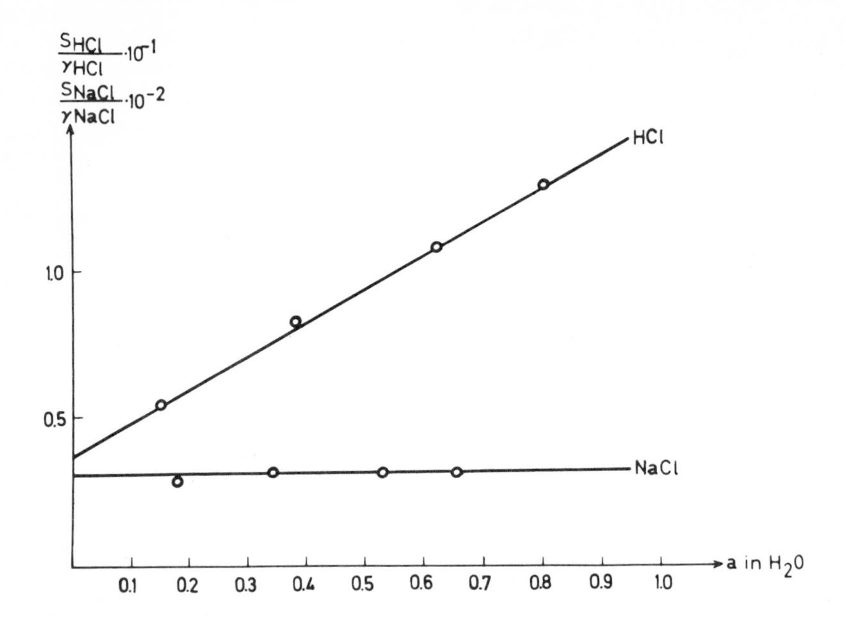

FIG. 14. Distribution coefficients of HCl and NaCl between water and wet n-pentanol plotted as a function of the activities in H_2O. Such plots yield straight lines according to the equations of Shedlovsky and Uhlig (26) and the extrapolations to zero concentrations are used to calculate the partition coefficient ratio (see text).

B. Thin Systems

As a corollary of the last statement it must be concluded that the instantaneous potential must be equal to the steady state potential. Theoretically this was predicated by Sandblom et al. (35), and it was tested on a thin 0.5-mm thick membrane consisting of HDEHP dissolved in wet n-pentanol (12). The membrane was maintained in a 0.1-mm diameter hole in a lucite disk where the surface tension kept the droplet intact and where the dimensions of the hole ensured a convection-free membrane. The e.m.f. and resistance were measured simultaneously following step changes in external solution conditions, and the time response of the two parameters are shown in Fig. 16. It is seen from the recording that, whereas the e.m.f. shows a practically instantaneous response, the resistance attains its stationary state value with a time constant of approximately 15 min, which corresponds to the relaxation of concentration profiles in the membrane.

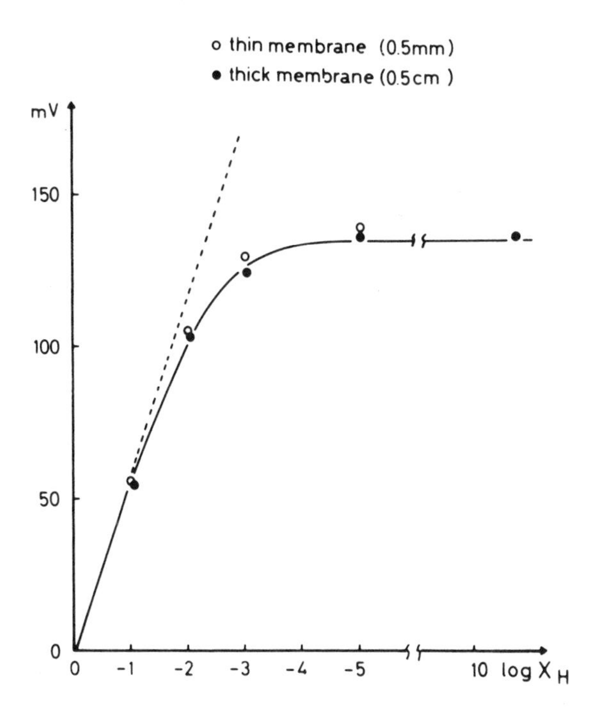

FIG. 15. The membrane potential of 10% HDEHPA in n-pentanol for mixtures of NaCl and HCl plotted against the mole fraction of hydrogen ion concentration.

The thin convection-free system has the advantage that the profiles in the membrane reach a stationary state. The pertubation of profiles by an electric field can therefore be studied by means of the current-voltage relationship, for which theoretical expressions have been derived. Figure 17 shows the current-voltage relationship for such a thin system in mixtures of HCl and NaCl (12). The chloride ions are effectively excluded and both cations form strongly associated ion pairs with the exchanger. The resistance is therefore nearly linear, exhibiting only a small nonlinearity in the biionic case, which is predicted by the theory (dotted line represents calculated values).

On the other hand, if dissociated ions are present, the membrane can show rectification, as demonstrated in Fig. 18. This figure shows the same membrane in solutions of HCl under conditions when co-ions (in this case Cl ions) are not excluded (12). Since the chloride ions are essentially

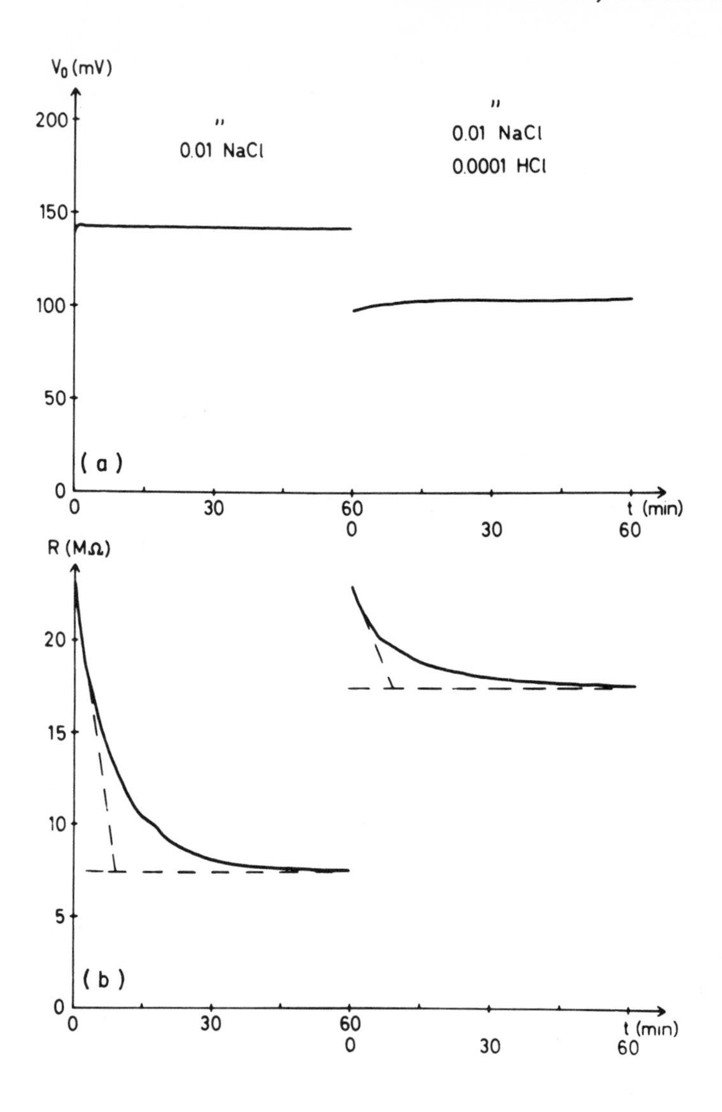

FIG. 16(a). The time course of the membrane potential at zero current following step changes in the external solution conditions. The left compartment contained a solution of 10 mM HCl and the right compartment was filled with solutions of compositions indicated in the figure. (b) The time course of the membrane resistance for the corresponding solution compositions of Fig. 16(a). The resistance changes were measured by applying a constant current of 5×10^{-9} A and recording the electric potential [Reprinted from Sandblom (12).]

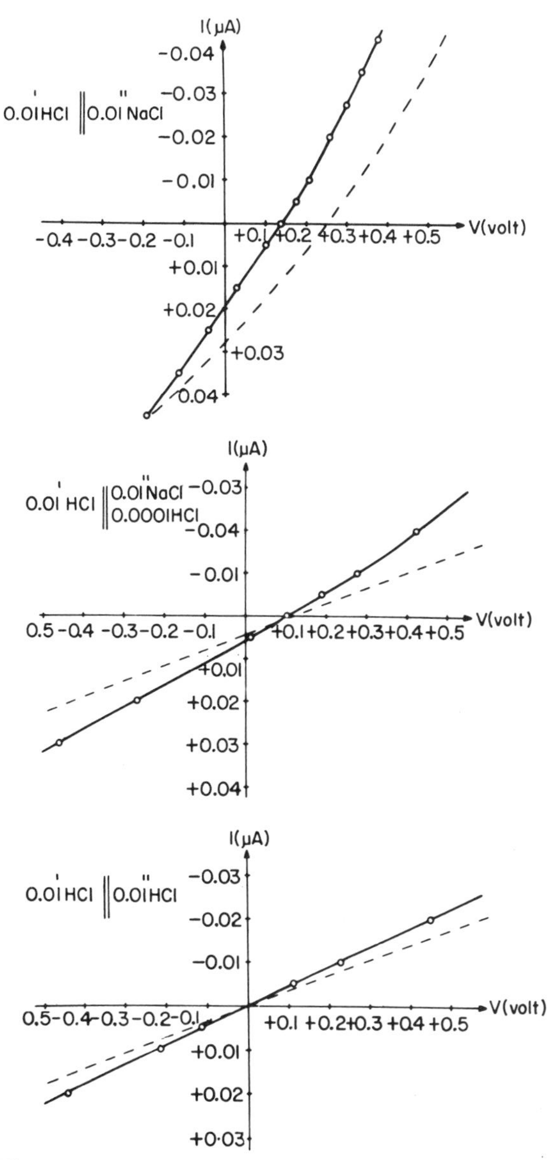

FIG. 17. The current-voltage characteristics for various external solution conditions. The left side of the membrane was exposed to 10 mM HCl and the right side was exposed to solutions indicated in the figure. Positive current flows from right to left and the left compartment is considered to have ground potential. The circles represent experimentally measured values and the dotted lines are theoretically calculated. [Reprinted from Sandblom (12).]

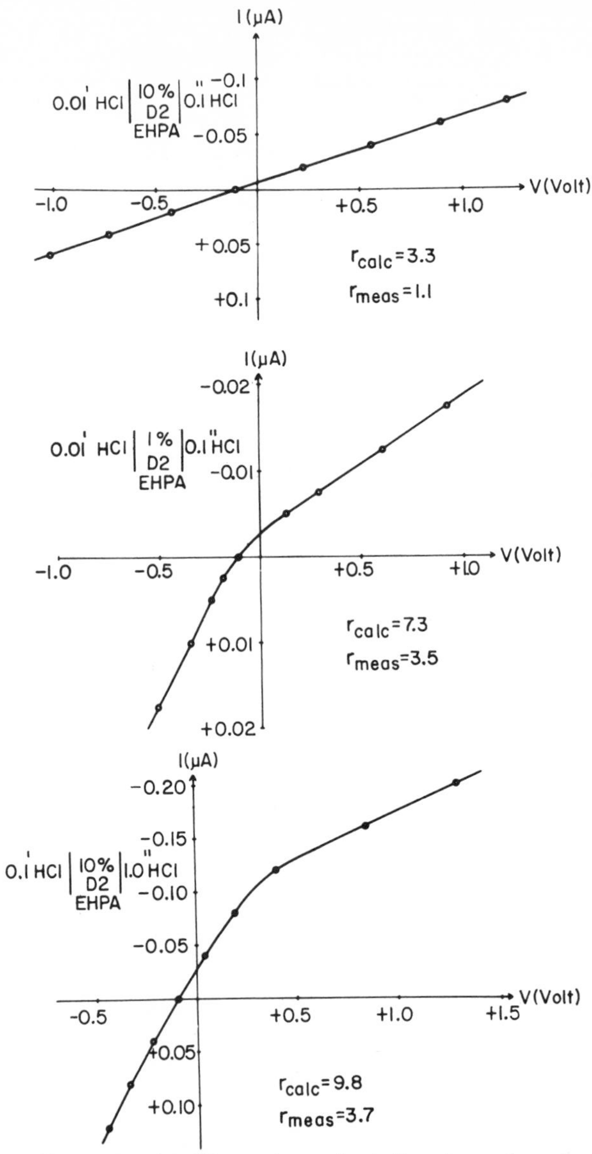

FIG. 18. The current-voltage characteristics for external solutions containing various concentrations of HCl. The solution conditions and the concentrations of the exchanger are indicated in the figure. A comparison between measured and calculated values of the rectification are also shown in the figure. Positive current flows from left to right and the left compartment is considered to have ground potential. [Reprinted from Sandblom (12).]

dissociated, they can contribute to a rectification of the electric current when the concentration is different on the two sides of the membrane. This rectification is at least qualitatively predicted by the theory.

C. Surface Diffusion

In order to study the surface diffusion in liquid membranes, the membrane phase as well as the external solutions must be vigorously stirred. The unstirred layers of the interfaces are estimated to be approximately 30μ thick and from such estimates, it was concluded by Rosano et al. (10) that the diffusion barrier of the boundaries are much higher than either of the liquid phases. The magnitude of activation energies also suggested that ions are transferred across the boundaries by being stripped of one water molecule.

An interesting finding, however, is that in the presence of an amphoteric phospholipid carrier in the nonaqueous membrane interface, the transport is markedly increased when the salts are opposed. This phenomenon is exactly similar to that described in the theoretical section for bulk transport of ions. It would, therefore, seem reasonable that the presence of surface active agents can lower the activation energy of the surface barrier, resulting in a bulk rate limiting transport of ions in the membrane phase. The analogy with the neutral carrier molecules is again apparent since the transport inducing effects of the neutral carriers in bimolecular lipid membranes is well described by a bulk phase rate limiting liquid model. Since bimolecular lipid membranes presumably have very high surface resistances for uncomplexed ions, it suggests that the neutral carriers have the effect of lowering the surface resistance, in addition to providing mediators for cation transport.

VII. APPLICATIONS

A. Potentiometric Measurements of Ion Activities

We have already described how liquid membranes can be used as electrodes to measure ion activities in aqueous solutions. A number of such systems have been studied and a variety of interesting applications have been suggested.

Coetzee and Freiser (7), in working with an anion exchanger, showed that it was possible to adapt a wide variety of solvent extraction systems for potentiometetric determinations of ion activities. In this way a new tool for the quantitative measurement of many organic ions can be developed.

A similar use was proposed by Bonner and Lunney (17), who measured the activity of p-toluene sulfonic acid solutions, a bulky electrolyte for which cells with solid ion-exchange membranes would be unsuitable.

Other ions for which solid ion exchangers are also generally unsuitable as electrodes are the polyvalent ions. Although polyvalent ions are preferred over monovalent ions in solid membranes, their affinity for the exchanger is usually accompanied by a very poor mobility, which also reduces their contribution to the electrode potential. In liquid membranes, on the other hand, the mobility differences probably play a relatively unimportant role (see Sec. V.B on ion conductance) and the electrode potential is largely determined by the selectivity parameters.

A liquid membrane sensitive to divalent ions was described by Ross (38) and consisted of an alkyl phosphoric acid dissolved in di-n-octophenylphosphate. This system was found to be very specific towards calcium ions and has been developed into a commercial electrode. Since then a considerable amount of work has been published involving the use of this electrode system, particularly with respect to complexometric titrations of calcium with EDTA (42,45).

Another liquid membrane system sensitive to polyvalent cations has been described, namely dinonylnaphtalene sulfonic acid in naphta (47), although this exchanger is highly water soluble and showed a rather large deviation from a Nernst slope for calcium ions. It was suggested by Harrell et al. (47) that this may be due to water transport through the membrane in the form of micelles.

It is customary in designing electrodes to convert the exchanger into the form corresponding to the ion whose activity is to be measured. From a theoretical point of view, however, this would appear to be an unnecessary precaution since the electrode potential is independent of the concentration profiles in the membrane, as explained in the theoretical section. Instead, it has a practical advantage, namely to avoid contamination of the solution with other ions which could influence the potential response.

A form of liquid membrane electrodes which show great promise are those containing neutral carriers instead of liquid ion exchangers. Very recently a potassium electrode containing valinomycin in different phospholipids has been described (81), having a potassium-to-sodium selectivity (K_{KNa}) of about 10^5 as opposed to the order of unity for a similar electrode containing an alkyl phosphoric acid.

Liquid membranes have also been used as microelectrodes (82). These are usually made of glass capillaries--at the tip of which is placed a droplet of the liquid ion exchanger, and the remaining part of the capillary is filled with the reference solution. Good potential responses with these electrodes have been obtained with tip diameters down to 1-5μ (83a, 118).

Many other types of electrodes have been designed and used for various purposes (84a, 84b), and no complete review will be presented here. The best current source on the applications of liquid ion-exchanger electrodes is perhaps the book edited by R.A. Durst (84c).

B. Carrier Transport of Ions

Although the cell membrane cannot be considered as a homogeneous liquid, the behavior of bimolecular lipid structures is similar to that which can be derived from a liquid model (8). Liquid membranes are therefore excellent models for carrier transport, and several characteristic properties of biological membranes which have been attributed to carrier transport have been demonstrated with liquid ion-exchange membranes (84). Notable examples are saturation kinetics, counter transport, discrepancies between resistance and flux measurements, etc.

It is also believed that drug absorption in many cases involves complex formation and it has been specifically suggested that quarternary ammonium ions can be absorbed as ion pairs in a manner similar to the ion-exchange processes described in this chapter (85). The distribution and rate of transport of these ions across biological membranes can, therefore, be studied by means of the same theoretical formalism presented in this chapter.

Since liquid membranes obviously provide excellent models for various biological transport phenomena, a further study of their physical chemistry--particularly with respect to water transport, adduct forming compounds, and ionic specificity--is certain to increase our understanding of biological membranes. Such studies will also help to design new and better membranes for quantitative measurements of many organic compounds of biological importance.

APPENDIX: RECENT DEVELOPMENTS

Since this chapter was originally prepared, a number of developments in the field of liquid ion-exchanger electrodes have taken place. Of particular interest are studies of natural and synthetic macrocyclic compounds as "carrier" molecules. Eisenman and co-workers (86-88) have published a detailed theory of the action of such compounds in bilayer membranes. Ashton and Steinrauf (89) have studied the transport properties of a number of ring antibiotics in a simple system. Spectroscopic studies by Shemyakin and co-workers (90) have shown correlations between structure and ion selectivity. In a discussion of organic compounds which may act as cation "carriers," Eigen and Winkler (91) have emphasized kinetic aspects of the problem. Finally, Mueller and Rudin (92) have published an extensive review of organic compounds which induce permeability changes in cell and bilayer membranes ("translocators"). They discuss a class of compounds, "gated translocators," which undergo conformational changes in certain states. Such compounds can be used to construct models of action potentials, generator potentials, and other bioelectric phenomena.

Several recent works discuss the oil/water partition of ions from a theoretical as well as a practical point of view. The most important of these is the textbook by Marcus and Kertes (93) on solvent extraction, which is the most complete review of the subject available. Recent advances have been covered in a review by Freiser (94), which has a short section on liquid membrane electrodes. Much useful information is also contained in recent reviews on the hydration (95) and solvation (96, 97) of ions. Partition of ions between water and organic solvents is also considered by Parker (98) in his review of the rates of bimolecular reactions.

Reviews of electrochemistry in dimethyl sulfoxide (99) and acetonitrile (100) are useful to those interested in the effect of the solvent on transport as well as equilibrium properties.

New practical developments include work on several new types of electrodes for inorganic (101, 102) and organic ions (103-105). The biologist will be interested in the use of liquid ion-exchanger electrodes to measure activity coefficients in salt mixtures (106-108), and in binding constants to compounds of biological interest (109-112). Liquid membranes, in the form of microelectrodes, have been used for the first time to estimate intracellular ion activities (113-117). Finally an excellent review summarizing the electrochemical properties of liquid membranes has recently been presented by Sollner (118).

REFERENCES

1. W. Nernst and E. H. Riesenfeld, Ann. Physik 8, 600 (1902); E. H. Riesenfeld, ibid, 8, 609 (1902); E. H. Riesenfeld, ibid, 8, 616 (1902).
2. M. Cremer, Z. Biol., 47, 562 (1906).
3. R. Kunin and A. G. Winger, Angew. Chem. Intern. Ed., Engl., 1, 149 (1962).
4. J. W. Ross, Science, 156, 1378 (1967).
5. Orion Research Inc., Bull. No 92-20 B (1966).
6. C. J. Coetzee and H. Freiser, J. Anal. Chem., 40, 2071 (1968).
7. C. J. Coetzee and H. Freiser, J. Anal. Chem., 41, 1128 (1969).

7a. M. Calvin, H. H. Wang, G. Entine, D. Gill, P. Ferruti, M. A. Harpold, and M. P. Klein, Proc. Natl. Acad. Sci., 63, 1 (1969).
7b. W. L. Hubbel and H. M. Mc Connel, Proc. Natl. Acad. Sci., 63, 16 (1969).
7c. W. Stoeckenius and D. M. Engleman, J. Cell Biol., 42, 613 (1969).
8. G. Eisenman, S. M. Ciani, and G. Szabo, Federation Proc., 27, 1289 (1968).
9. G. M. Shean and K. Sollner, Ann. N.Y. Acad. Sci., 137, 759 (1966).
10. H. L. Rosano, P. Duby, and J. H. Schulman, J. Phys. Chem., 65, 1704 (1964).
11. H. P. Ting, G. L. Bertrand, and D. F. Sears, Biophys. J. 6, 813 (1966).
12. J. Sandblom, J. Phys. Chem., 73, 257 (1969).
13. M. Dupeyrat, J. Chim. Phys. 61, 301 (1964); M. Dupeyrat, ibid. 61, 323 (1964).
14. M. Dupeyrat and M-C Ménétrier, C. R. Acad. Sci. Paris, 258, 4734 (1964).
15. G. Colacicco, Nature, 207, 936 (1965).
16. G. Colacicco, Nature, 207, 1045 (1965).
17. O. D. Bonner and D. C. Lunney, J. Phys. Chem., 70, 1140 (1966).
18. R. Beutner, Physical Chemistry of Living Tissues and Life Processes, Williams and Wilkins, Baltimore, 1933.
19. M. Kahlweit, Arch. Ges. Physiol., 271, 139 (1960).
20. E. Högfeldt, in Ion Exchange, (J. A. Marinsky, ed.), Vol. 1, Dekker, New York, 1967.
21. E. Baur and S. Kronmann, Z. Physik. Chem., 92, 81 (1917).
22. G. Ehrensvärd and L. G. Sillén, Nature, 141, 396 (1938); G. Ehrensvärd and L. G. Sillén, ibid, 142, 788 (1938).
23. W. J. V. Osterhout and W. M. Stanley, J. Gen. Physiol., 15, 667 (1931).
24. W. J. V. Osterhout, Cold Spring Harbor Symp. Quant. Biol., 8, 51 (1940).
25. L. G. Longsworth, J. Gen. Physiol., 17, 211 (1933); L. G. Longsworth, ibid., 17, 627 (1933).
26. T. Shedlovsky and H. H. Uhlig, J. Gen. Physiol., 17, 549 (1933); T. Shedlovsky and H. H. Uhlig, J. Gen. Physiol., 17, 563 (1933).
27. F. M. Karpfen and J. E. B. Randles, Trans. Faraday Soc., 49, 823 (1953).
28. K. H. Meyer, H. Hauptmann, and J. F. Sievers, Helv. Chim. Acta, 19, 946 (1936).
29. E. Lange and J. Schücker, S. Elektrochemie, 57, 22 (1953).
30. K. F. Bonhoeffer, M. Kahlweit, and H. Strehlow, Z. Physik. Chem. (Frankfurt), 1, 21 (1954).
31. M. Kahlweit, H. Strehlow, and C. S. Hocking, Z. Physik. Chem. (Frankfurt), 4, 212 (1955).
32. M. Kahlweit and H. Strehlow, Z. Elektrochem., 58, 658 (1954).
33. F. Conti and G. Eisenman, Biophys., 6, 227 (1966).

33a. S. Ciani and A. Gliozzi, Biophysik, 5, 145 (1968).
34. J. L. Walker, Jr. and G. Eisenman, Biophys. J., 6, 513 (1966).
35. J. Sandblom, G. Eisenman, and J. L. Walker, Jr., J. Phys. Chem., 71, 3862 (1967).
36. J. Sandblom, G. Eisenman, and J. L. Walker, Jr., J. Phys. Chem., 71, 3971 (1967).
37. K. Sollner and G. M. Shean, J. Am. Chem. Soc., 86, 1901 (1964).
38. M. E. Thompson and J. W. Ross, Science, 154, 1643 (1966).
39. J. A. King and K. Mukerji, Naturwissenschaften, 53, 702 (1966).
40. S. C. Glauser, E. Ifkovitz, E. M. Glauser, and B. W. Sevy, Proc. Soc. Exp. Biol. Med., 124, 131 (1967).
41. F. S. Nakayama and B. A. Rasnick, J. Anal. Chem., 39, 1022 (1967).
42. A. Shatkay, J. Anal. Chem., 39, 1056 (1967).
43. A. Shatkay, J. Phys. Chem., 71, 3858 (1967).
44. A. Shatkay, Biophys. J., 8, 912 (1968).
45. G. A. Rechnitz and Z. F. Lin, J. Anal. Chem., 40, 696 (1968).
46. R. Huston and J. N. Butler, J. Anal. Chem., 41, 200 (1969).
47. J. B. Harrell, A. D. Jones, and G. R. Choppin, J. Anal. Chem., 41, 1459 (1969).
48. Z. Stefanac and W. Simon, Microchem. J., 12, 125 (1967).
49. K. Srinivasan and G. A. Rechnitz, J. Anal. Chem., 41, 1203 (1969).
50. M. J. D. Brand and G. A. Rechnitz, J. Anal. Chem., 41, 1185 (1969).
51. J. Sandblom, Arkiv Fysik, 35, 329 (1967).
52. A. Katchalsky and R. Spangler, Quart. Rev. Biophys. 1, 127 (1968).
53. W. Wilbrandt and T. Rosenberg, Pharmacol. Rev., 13, 109 (1961).
54. L. W. Holm, Arkiv Kemi, 5, 151 (1956).
55. G. Eisenman, Anal. Chem., 40, 310 (1968).
56. A. H. Truesdell and C. L. Christ, in Glass Electrodes for Hydrogen and Other Cations: Principles and Practice (G. Eisenman ed.), Dekker, New York (1967).
56a. G. Eisenman, "On the Elementary Atomic Origin of Equilibrium Ionic Specificity," in Symposium on Membrane Transport and Metabolism, (A. Kleinzeller and A. Kotyk, eds.), Academic Press London, 1961, p. 163.
56b. G. Eisenman, "The Electrochemistry of Cation Sensitive Glass Electrodes," in Advances in Analytical Chemistry and Instrumentation (C.N. Reilley, ed.), IV, Interscience, New York, 1965, p. 215-369.
57. J. Sandblom, J. Phys. Chem., 73, 249 (1969).
58. J. L. Walker, Jr., G. Eisenman, and J. Sandblom, J. Phys. Chem., 72, 978 (1968).
59. W. M. Latimer, K. S. Pitzer, and C. M. Slansky, J. Chem. Phys., 7, 108 (1939).

60. D. Feakins and P. Watson, J. Chem. Soc. (London), 4, 4734 (1963).
61. E. Grunwald, G. Baughman, and G. Kohnstamm, J. Am. Chem. Soc., 82, 5801 (1960).
62. A. J. Parker, Quart. Rev. (London), 16, 163 (1962).
63. A. J. Parker, in Advances in Physical Organic Chemistry (V. Gold, ed.), Academic Press, London, 1967.
64. C. L. de Ligny and M. Alfenaar, Rec. Trav. Chim., 84, 81 (1965).
65. A. M. Rozen, in Solvent Extraction Chemistry (D. Dyrssen, J. O. Liljenzin, and J. Rydberg eds.), North-Holland Publishing Co. 1967, p. 195.
66. R. Alexander and A. J. Parker, J. Am. Chem. Soc., 89, 5549 (1967).
67. A. Gemant, J. Chem. Phys., 12, 79 (1944).
68. R. M. Fuoss, Proc. Natl. Acad. Sci., 45, 807 (1959).
68a. B. E. Conway, in Physical Chemistry, an Advanced Treatment (H. Eyring, ed.) Academic, New York, 1970.
69. G. J. Janz and S. S. Danyluk, Chem. Rev., 60, 209 (1960).
70. C. A. Kraus, Ann. N.Y. Acad. Sci., 51, 789 (1949).
71. C. A. Kraus, The Properties of Electrically Conducting Systems, N.Y. Chemical Catalog Co., 1922.
72. G. H. Morrison and H. Freiser, Solvent Extraction in Analytical Chemistry, Wiley, New York, 1957.
73. Solvent Extraction Chemistry (D. Dyrssen, J.O. Liljenzin, and J. Rydberg, eds.) North-Holland Publishing Co. 1967.
74. G. Schill, Svensk Kem. Tidskr., 80, 10 (1968).
75. K. Kimura, Bull. Chem. Soc. Japan, 33, 1038 (1960).
76. J. R. Ferraro and D. F. Peppard, Nucl. Sci. Eng., 16, 389 (1963).
77. C. F. Coleman, C. A. Blake, Jr., and K. B. Brown, Talanta, 9, 297 (1962).
77a. G. Eisenman, in Membrane Transport and Metabolism (A. Kleinzeller and A. Kotyk, eds.) Academic, London, 1961.
77b. G. Eisenman, Biophys. J., 2, Vol. 2, 259 (1962).
78. A. L. Myers, W. J. McDowell, and C. F. Coleman, J. Inorg. Nucl. Chem., 26, 2005 (1964).
79. W. J. McDowell and C. F. Coleman, J. Inorg. Nucl. Chem., 28, 1083 (1966).
80. R. A. Fuoss and C. A. Krauss, J. Am. Chem. Soc., 55, 476 (1933); R. A. Fuoss, ibid, 57, 488 (1935).
81. M. S. Frank and J. W. Ross, Jr., Science, 167, 987 (1970).
82. F. Orme, in Intracellular Glass Microelectrodes (M. Lavallee, O. Shanne and N. C. Hebert), Wiley, New York, 1968.
83. J. Walker, Liquid Ion-Exchanger Microelectrodes for Ca^{++}, Cl^- and K^+ (Unpublished manuscript).
83a. F. Orme, Development of a Magnesium Sensitive Electrode (Unpublished manuscript).

84. J. Sandblom, Acta Univ. Upsaliensis, Abstr. Uppsala Diss. in Med., 44 (1967).
84a. A. K. Covington, Chem. Brit. 5, 388 (1969).
84b. E. J. Zinser and J. A. Page, Chem. Can., 12, 31 (1969).
84c. Ion Selective Electrodes R. A. Durst, ed., Natl. Bureau of Standards, Washington, D.C., 1969.
85. G. Schill, 29th Intern. Congr. Pharmacol. Sci., (London), 1969.
86. S. Ciani, G. Eisenman, and G. Szabo, J. Membrane Biol. 1, 1 (1969).
87. G. Eisenman, S. Ciani, and G. Szabo, J. Membrane Biol., 1, 294 (1969).
88. G. Szabo, G. Eisenman, and S. Ciani, J. Membrane Biol. 1, 345 (1969).
89. R. Ashton and L. K. Steinrauf, J. Mol. Biol., 49, 547 (1970).
90. M. M. Shemyakin, Yu. A. Ovchinnikov, V. T. Ivanov, V. K. Antonov, E. I. Vinogradova, A. M. Shkrob, G. G. Malenkov, A. V. Evstratov, I. A. Laine, E. I. Melnik, and I. D. Ryabova, J. Membrane Biol. 1, 402 (1969).
91. M. Eigen and R. Winkler, in The Neurosciences: Second Study Program (F. O. Schmitt, ed.), Rockefeller Univ. Press, New York, 1970.
92. P. Mueller and D. O. Rudin, Curr. Top. Bioenerg. 3, 157 (1969).
93. Y. Marcus and A. S. Kertes, Ion Exchange and Solvent Extraction, Wiley, New York, 1969.
94. H. Freiser, Critical Rev. Anal. Chem., 1, 47 (1970).
95. J. E. Desnoyers and C. Jolicoeur, Mod. Aspects Electro-chem., 5, 1 (1969).
96. O. Popovych, Critical Rev. Anal. Chem., 1, 73 (1970).
97. H. G. Hertz, Angew. Chem. Intern. Ed., 9, 124 (1970).
98. A. J. Parker, Chem. Rev., 69, 1 (1969).
99. J. N. Butler, J. Electroanal. Chem., 14, 89 (1967).
100. J. F. Coetzee, Progr. Phys. Org. Chem., 4, 45 (1967).
101. I. Nagelberg, L. I. Braddock, and G. J. Barbero, Science, 166, 1403 (1969).
102. J. Ruzicka and K. Rald, Anal. Chim. Acta, 53, 1 (1971)
103. M. Matsui and H. Freiser, Anal. Lett., 3, 161 (1970).
104. G. Baum, Anal. Biochem., 39, 65 (1971).
105. T. Higuchi, C. R. Illian, and J. L. Tossounian, Anal. Chem., 42, 1674 (1970).
106. J. N. Butler and R. Huston, Anal. Chem., 42, 676 (1970).
107. J. V. Leyendekkers and M. Whitfield, Anal. Chem., 43, 322 (1971).
108. J. Bagg, Australian J. Chem., 22, 2467 (1969).
109. G. A. Rechnitz and M. S. Mohan, Science, 168, 1460 (1970).
110. J. de Moura, D. Le Tourneau, and A. C. Wiese, Arch. Biochem. Biophys., 258 (1969).

111. L. M. Crawford and J. M. Bowen, Am. J. Vet. Res., 32, 357 (1971).
112. J. S. Jacobs, R. S. Hattner, and D. S. Bernstein, Clin. Chim. Acta, 31, 467 (1971).
113. J. L. Walker and A. M. Brown, Science, 167, 1502 (1970).
114. S. Y. Chow, D. Kunze, A. M. Brown, and D. M. Woodbury, Proc. Nat. Acad. Sci., 67, 998 (1970).
115. M. C. Cornwall, D. F. Peterson, D. L. Kunze, J. L. Walker, and A. M. Brown, Brain Res., 23, 433 (1970).
116. J. L. Walker, Anal. Chem., 43, 89A (1971).
117. F. Orme, Dissertation, Univ. of California, 1971.
118. K. Sollner, The Basic Electrochemistry of Liquid Membranes, in Diffusion Processes (J. Sherwood and A. Chadwick, eds.), Proc. Thomas Graham Mem. Symp., Univ. of Strathclyde, Vol. 2, Gordon and Breach, London, 1971.

Chapter 4

CATION-EXCHANGE PROPERTIES OF DRY SILICATE MEMBRANES

Harmon Garfinkel

Corning Glass Works
Technical Staffs
Sullivan Park
Corning, New York 14830

I. INTRODUCTION

The basis for the origin of the potential of the glass electrode is the idea that the glass itself, upon exposure to aqueous solution, is a cation-exchange membrane, whose electrode potential represents a sum of contributions from phase-boundary and diffusion processes. It may not be the membrane properties of the glass itself, but rather of a hydrated layer, that are significant. The use of fused salts as the liquid phase allows study of the cation-exchange properties of the dry glass, thus removing the uncertainty of the role of water in the exchange process.

Although ion exchange between glass and molten salts has been known for some time (1), it has only recently received considerable attention. These studies have both scientific and technological relevance, and their impact is felt on diverse fields ranging from glass structure to membrane biophysics.

Much of this interest has been generated because the phenomenon can be used to alter the mechanical, optical, and electrical properties of glass (2). The most promising commercial application so far is in the area of chemical tempering (3-6), which is defined as a method of increasing the strength of glass by changing the surface composition of the sample. Ion exchange has been used also to prepare thin-layer photochromic glasses (7) and new silicate minerals (8). Recently, there has been increased interest in the cation-selective properties of glass membranes in molten salts (9-15), because this system permits study of the "dry" state. Thus, a comparison can be made between the properties of the hydrated and unhydrated membrane, hopefully giving clues as to the role of water in membrane transport.

This review is not comprehensive in nature, but rather focuses on work done at our laboratory on the ion-exchange properties of glass and glass ceramics. The reader is referred to other reviews on this subject for additional references (1, 16-19). We are concerned mainly with the properties of the exchanger and not the fused salt. Again, the reader is referred to several excellent reviews on molten salts (20-22).

In the theory of the origin of the potential of a fixed-site ion-exchange membrane, the cation selectivity of the membrane is a function of equilibrium and kinetic parameters. Data are presented on ion-exchange equilibria and the kinetics of exchange, as well as electrochemical measurements in several model systems. These results are discussed in terms of the Nernst-Planck flux equations and will be compared, wherever possible, to aqueous systems.

II. THE EXCHANGER

A. Structure

Both glass and glass ceramics have been studied as ion exchangers in molten salts. The unit of structure in the glass considered here is the silica tetrahedron in which each silicon is surrounded by four oxygens and each oxygen is shared, in turn, between two silicons. These tetrahedral groups are approximately fixed in position with respect to each other. Silica, when melted and subsequently cooled, forms a three-dimensional solid, which has short-range structure like β-cristobalite with similar density, but is disordered (amorphous) over longer distances. Only one or two diffuse rings can be seen in the X-ray diffraction pattern of glasses, and, in this sense, they are much like normal liquids. "Network modifying oxides" (e.g., Na_2O, CaO) added to fused silica disrupt the network, and, as a result, the modifying cations associate with the oxygen anions. Zirconium, boron, phosphorus, and germanium can substitute for silicon in the network, while aluminum, depending upon its concentration, can occur in the network or act as a network modifier.

Controlled crystallization of glass can yield a nonporous, fine-grained material called a glass ceramic with a high degree of crystallinity (24-27). In some cases the glass is self-nucleating; in others it requires a nucleating agent such as titania or zirconia. The polycrystalline material is held together by residual glass high in silica. Some examples are as follows: nephelines; solid solution of β-quartz and β-eucryptite; keatite (a polymorph of silica and β-spodumene); and cordierite, to mention a few.

It is the oxygen anions tied firmly to the silicate network that give glass and glass ceramics their cation-exchange properties. It is well known that monovalent ions move readily through the glass structure (23). These mobile monovalent cations give silicates their ion-exchange character by exchanging with ions in solution at the glass or glass ceramic surface. There is no evidence, morever, for anions entering the glass structure by ion exchange of conventional silicate glasses or glass ceramics.

B. Composition

A wide range of compositions is known for glasses, and, to a lesser extent for glass ceramics. The compositions of some of the glasses studied are shown in Table 1. The membrane properties of Corning Code 7740 borosilicate glass were also studied quite extensively. The exchange properties of several other glasses including Corning Code 0080, a soda-lime-silica glass, were investigated, but not to the same extent as those mentioned above. The ion-exchange capacities of the silicates studied ranged from 3×10^{-3} to 1×10^{-2} moles/cm^3. Corning Code 7740 borosilicate glass, containing relatively low alkali oxide for good chemical durability, has a calculated capacity of about 3×10^{-3} moles/cm^3--equal to that of a standard polysulfonate resin.

TABLE 1

Glass Composition in Mole % Oxide

	A	B	C	D	E
Li_2O	11.4	5.86	--	--	--
Na_2O	--	--	16.3	11.4	7.00
Al_2O_3	16.5	11.5	13.2	5.82	--
B_2O_3	--	--	--	18.7	23.00
TiO_2	--	3.78	3.81	--	--
MgO	--	4.09	--	--	--
SiO_2	71.5	74.7	66.7	64.1	70.0

C. Stress

During exchange above the transformation range, the glass can actually flow to accommodate the exchanged species. With a sodium-containing glass, for example, Li^+-Na^+ exchange below the "strain point" (i.e., the temperature at which the viscosity of the glass is 10^{14} P will result in microcracking of the glass surface because the surface is placed in tension; however, the surface of the glass remains intact when exchanged above the transformation range, because of this accommodation.

When large-ion-for-small-ion exchange is effected in the dry state below the strain point, there is little tendency for the glass to swell or contract. This rigidity, which is due to the stiffness of the silicon-oxygen network, results in the introduction of substantial stress with large-ion-for-small ion-exchange (3). This stress can affect the properties of the glass or glass ceramic including those important in ion exchange, although this effect has been little studied. It has been shown, however, that the stress and concentration profiles can be related by means of a thermoelastic model (28, 29), the results of which indicated very little stress relaxation well below the transformation range. At temperatures approximately 100° or less below the transformation range there is some relaxation (5); alumina and/or zirconia-containing silicate glasses experience less relaxation than other glass systems (6).

III. ION-EXCHANGE EQUILIBRIA

A. Introduction

For the generalized ion-exchange reaction

$$\bar{A}\text{ (glass)} + B\text{ (salt)} = \bar{B}\text{ (glass)} + A\text{ (salt)}$$

where $\bar{A}$ and $\bar{B}$ are the counterions in the exchanger phase and A and B are the counterions in the liquid phase, an equilibrium constant K_{AB} is defined so that

$$K_{AB} = \frac{\bar{a}_B \, a_A}{\bar{a}_A \, a_B} \tag{1}$$

The $\bar{a}_i$'s in Eq. (1) are the respective activities in the exchanger phase and the a_i's are the respective activities in the liquid phase. The value of K_{AB} depends upon the reference state chosen to define activities. For the salt, it is convenient to use the pure material as reference state, so that

$$\lim_{N_A \to 1} \gamma_A = 1$$

for component A, and similarly for component B. The reference state for the solid exchanger is that in which all of the exchangeable cations are of the ion in question.

Ion-exchange equilibrium is characterized by the ion-exchange isotherm, which is a graphical representation covering all experimental conditions at constant temperature. Of prime significance is the selectivity of the exchanger, i.e., the selection of one counterion in preference to the other by the ion exchanger. The equilibrium constant K_{AB} is an integral measure of selectivity.

B. Methods of Measurement

Since the experimental techniques are described elsewhere (30), they are only outlined briefly here. This discussion is concerned with two direct measurements as opposed to indirect measurements involving membrane potentials.

In the first method, powdered exchanger, which had passed a -270 (53μ) or -325 (44μ) mesh screen, were equilibrated for 200-500 hr, depending upon temperature, in a molten salt bath with continuous stirring. The glass powders were analyzed either by flame photometry, or by gamma counting for those instances in which the bath was tagged with radiotracer.

In the second method, the concentration-distance profile was determined on an ion-exchanged sample in the form of a rod. The sample was etched away with HF-H_2SO_4, and the etchates were analyzed either by gamma counting or by flame photometry, depending upon whether the bath was tagged with radiotracer or not. Then the concentration-distance plot was extrapolated to the surface x = 0 to obtain the surface concentration.

C. Results

1. Considerations

Certain assumptions must be used to get at the experimental quantities in Eq. (1). By following the suggestion of Rothmund and Kornfeld (31), the ratio of the activities of the ions in the exchanger is given by

$$\frac{\bar{a}_B}{\bar{a}_A} = \left(\frac{\bar{N}_B}{\bar{N}_A}\right)^n \tag{2}$$

where $\bar{N}$ is the cation fraction. This has been referred to as n-type behavior in aqueous systems (32).

Most of the molten nitrate mixtures used can be characterized as regular solutions (21). Although the heat-of-mixing of these salts is a slight function of composition, for our purposes here, the approximation is quite satisfactory. Therefore, we may write

$$RT \ln\left(\frac{\gamma_A}{\gamma_B}\right) = A(1-2N_A) \tag{3}$$

where A is a constant independent of temperature. Values of the parameter A given in the literature (21) for systems of interest here are shown in Table 2. Substitution of Eqs. (2) and (3) into Eq. (1) yields

$$\log\left(\frac{N_B}{N_A}\right) - \frac{A}{2.303\ RT}(1-2N_A) = n \log\left(\frac{\bar{N}_B}{\bar{N}_A}\right) - \log K_{AB} \tag{4}$$

which is similar to the semiempirical relation proposed by Kielland for aqueous exchangers (33). Thus, a plot of log (a_B/a_A) for the salt, against log $(\bar{N}_B/\bar{N}_A)$ for the exchanger is characterized by a slope of n if K_{AB} is constant with concentration.

2. Homogeneous Exchangers

In this section, glass exchangers consisting of a single phase are discussed. Figure 1 shows the ion-exchange isotherms for several different exchange reactions for those glass compositions given in Table 1. The cation fraction of the foreign counterion in the exchanger phase is shown

TABLE 2

Values of the Parameter A in Eq. (3)

System	A (Cal/mole)
$NaNO_3$ - $LiNO_3$	-470
$NaNO_3$ - KNO_3	-442
$NaNO_3$ - $AgNO_3$	590
$LiNO_3$ - $AgNO_3$	650

as a function of the cation fraction of the foreign counterion in the liquid phase. The two features to note in Figs. 1(a), 1(b), and 1(c) are the sigmoid character of the isotherms and the failure to reach greater than 90% conversion in the examples. This sigmoid isotherm is usually characterized as S type (34); i.e., the selectivity is reversed in the course of conversion.

The most probable explanation for this irregular behavior is that proposed by Barrer and Falconer (35). In the graph shown in Fig. 1(a), for example, irregular behavior occurs because occupancy of an exchange site by Na^+ influences the relative affinities of the adjacent sites for sodium and lithium. Since accommodation of Na^+ becomes increasingly more difficult as conversion proceeds, occupancy of two neighboring sites by two Na^+ ions is energetically less favorable than occupancy by one Na^+ and one Li^+ or by two Li^+ ions as in basic cancrinite (35). Conversion may even remain incomplete, because of space requirements. Although the structure of glass is less regular than the structure of zeolites, it seems plausible that a similar explanation may hold for glass.

Conversion of composition A has been observed to be as low as 40% in molten potassium nitrate ($N_K = 1$) at 400°C and 90% at 500°C, while conversion of Corning Code 0088 soda-lime-silica glass in potassium nitrate ($N_K = 1$) at 450°C is about 78%. Although conversion is incomplete in the examples of Na^+-Li^+ exchange in Figs. 1(a) and 1(b), and in the Ag^+-Li^+ exchange in Fig. 1(c), it is almost 100% in the examples of Ag^+-Na^+ exchange in Fig. 1(d), K^+-Na^+ exchange in Fig. 1(e), and Li^+-Na^+ exchange in Fig. 1(f). At 266°C, conversion is 96% complete for composition D from tracer diffusion studies in tagged $AgNO_3$. From self-diffusion studies of

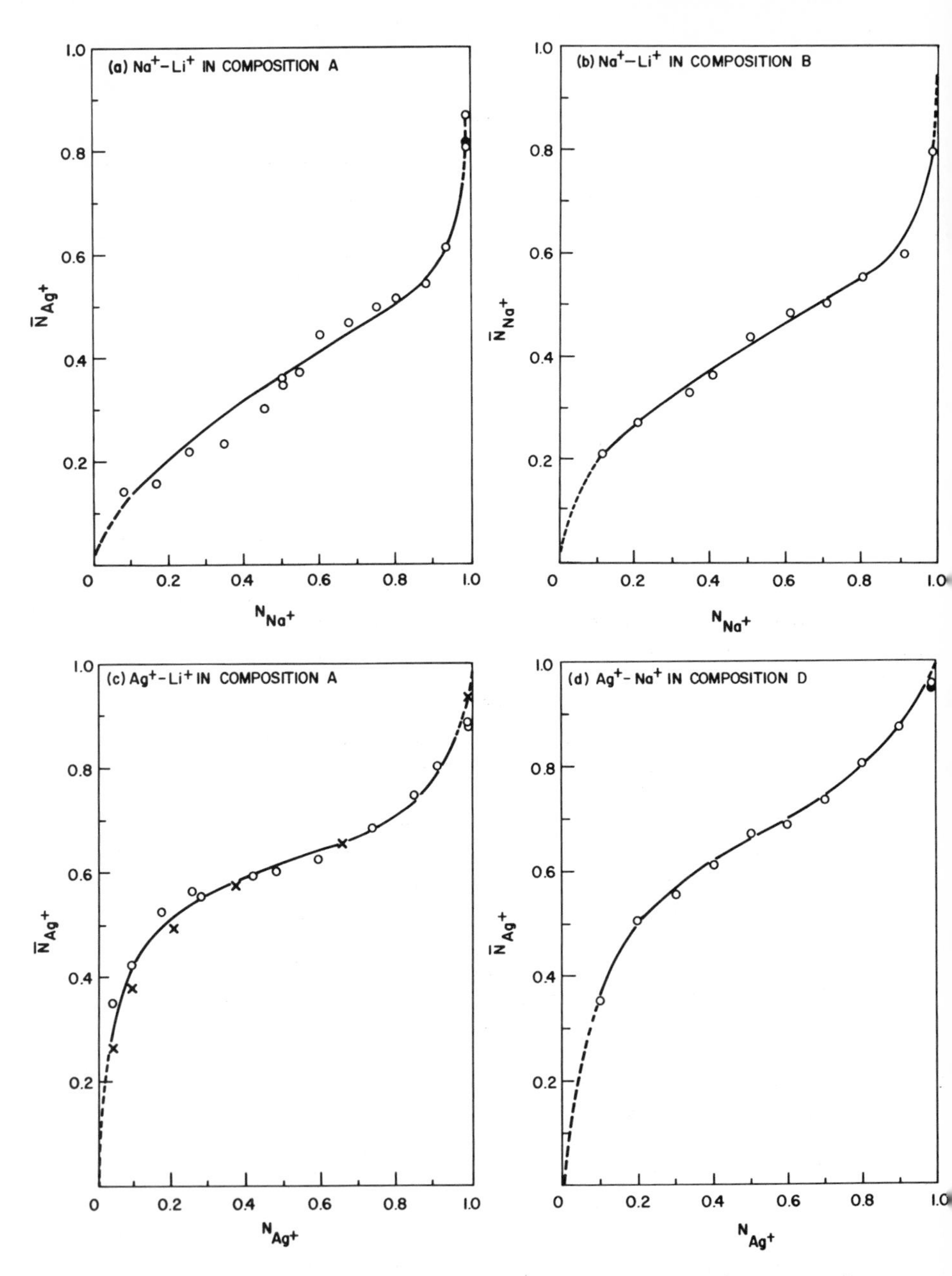

FIG. 1. Ion-exchange isotherms: (a) o, 400°C; •, 451°C; (b) 400°C; (c) o, 300°C; x, 451°C; (d) 300°C; o, powder, • tracer.

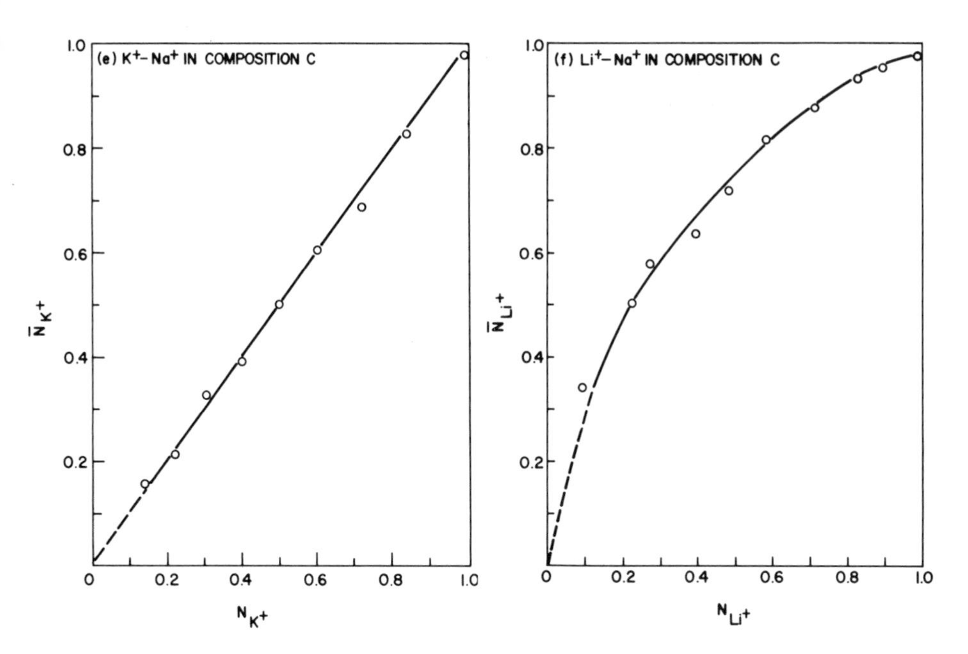

FIG. 1 (continued). (e) 500°C; (f) 400°C.

silver in composition D with sodium nitrate dilute in radioactive silver, a partition factor of 500 was obtained, which is similar to results reported for silver exchange in soda-lime-silica glass (36).

Moreover, there is no selectivity reversal in Figs. 1(d), 1(e), and 1(f). In addition, no evidence of sieving action has been observed even in the lithium-containing glasses for rubidium and cesium exchange. Some peculiarities were observed in the kinetics of Rb^+-Li^+ and Cs-Li^+ exchange, which is discussed later.

Figure 2 shows the data in Fig. 1 plotted according to Eq. (4). For Na^+-Li^+ exchange in composition A, least-squares analysis of the results in Fig. 2(a) gave n = 1.9 and $K_{LiNa} = 0.28$. Since $K_{LiNa} < 1$, the glass prefers lithium. Analysis of the data in Fig. 2(b) gave n = 2.2 and $K_{LiAg} = 3.2$ at 300°C for Ag^+-Li^+ exchange in composition A. From these results it is estimated that the equilibrium constant for Ag^+-Na^+ exchange in composition A is at least $K_{NaAg} = 11$ at 300°C. Preliminary results with potassium show that it is even less preferred in glass composition A than sodium. Therefore, the integral selectivity order with composition A is $Ag^+ > Li^+ > Na^+ > K^+$.

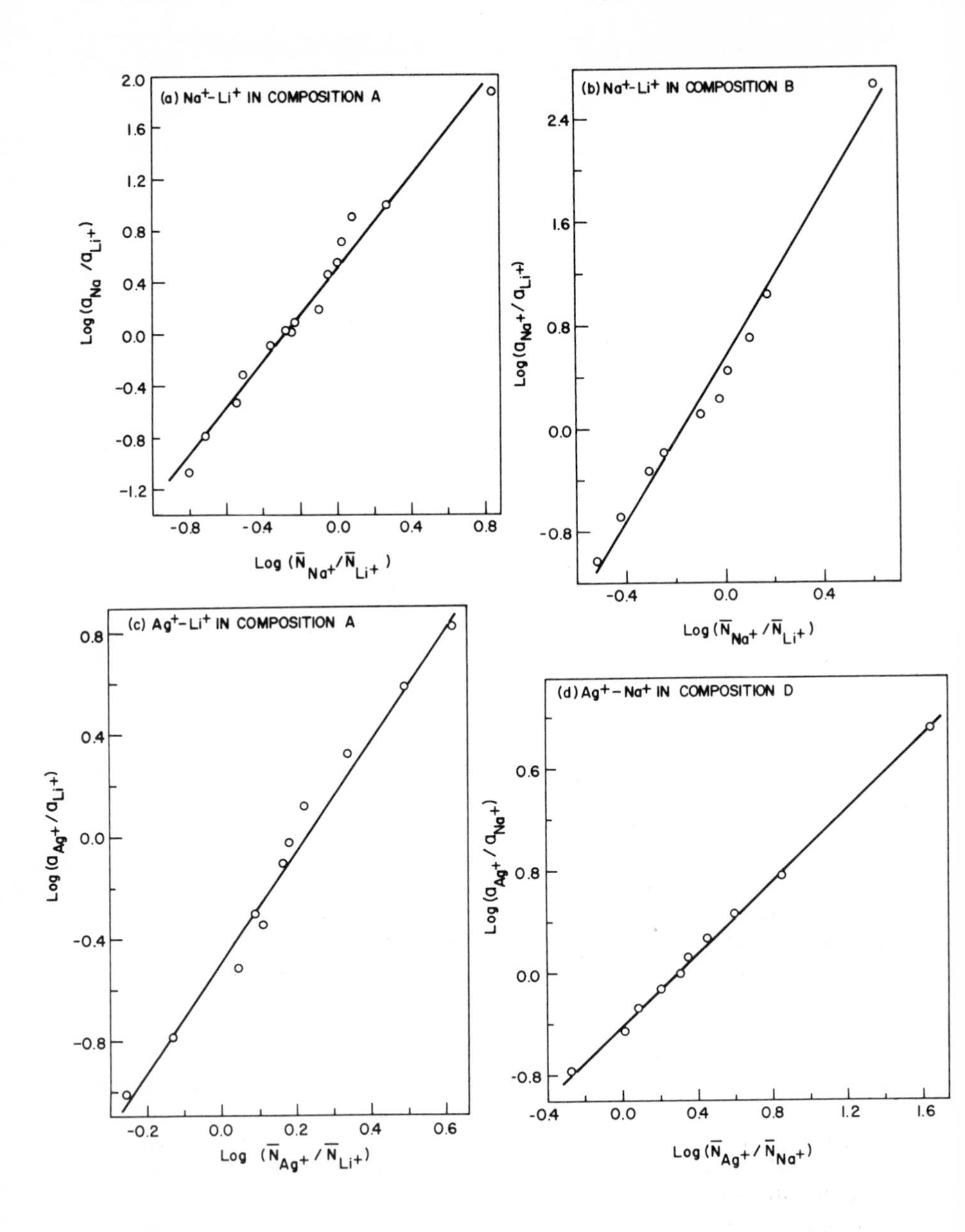

FIG. 2. Test of n-type behavior for the exchange systems shown in Fig. 1: (a) 400°C; (b) 400°C; (c) 300°C; (d) 300°C

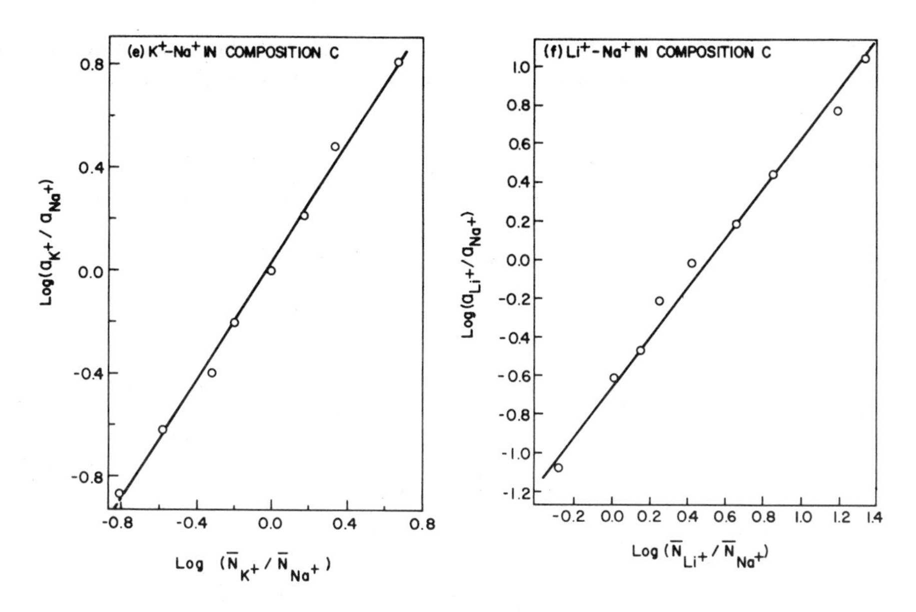

FIG. 2 (continued). (e) 500°C; (f) 400°C.

The exchange isotherm for K^+-Na^+ exchange in composition C shown in Fig. 1(e) is almost ideal. Analysis of the data in Fig. 2(e) gave K_{NaK} = 0.94 with n = 1.2 at 500°C. Thus, the glass exhibits almost equal preference for the two ions at this temperature, with sodium being slightly favored.

In the first five examples shown in Fig. 1, the ion-exchange reaction introduces compressive stresses to different degrees as diffusion proceeds. Li^+-Na^+ exchange was investigated in composition C to determine whether Eq. (4) held for systems in which tensile stresses accompanied the exchange. The results shown in Fig. 2(f) indicate that Eq. (4) does in fact describe both situations. In this case K_{NaLi} = 4.4 ± 0.4 with n = 1.27 ± 0.05 at 400°C. Thus, at temperatures less than 500°C, the integral selectivity of composition C is $Li^+ > Na^+ > K^+$.

The results seen so far indicate that these exchange reactions are relatively temperature insensitive, which is similar to behavior reported for aqueous systems (34, 37). The enthalpy change of two of the exchange reactions studied was determined by a solution-colorimetry technique (30, 38). The heat of solution in hydrofluoric acid was determined for the products and reactants at 26.9°C. The heat of reaction was calculated at 26.9°C from the difference of the sum of the heats of solution of the products and reactants. The actual heat effect at temperature was determined by calculating the difference in heat content necessary to bring the reactants and products to that temperature. This procedure yielded -1 ±1 kcal/g-ion Na^+ exchanged as the heat of reaction for Na^+-Li^+ exchange in glass composition A at 400°C. A value of 3.8 ± 0.6 kcal/g-ion of K^+ exchanged was determined for K^+-Na^+ exchange in composition C at 500°C. Since this value of the heat of reaction should approximate ΔH°, the integral selectivity is a slight function of temperature; above 520°C, the glass would prefer potassium over sodium. These results are summarized in Table 6 on page 196.

3. Two-Phase Exchangers

This section focuses on the exchange properties of crystallized glasses and phase-separated glasses. Glass composition B can be crystallized by the proper heat treatment to yield a nonporous, fine-grained polycrystalline material called a glass ceramic (24). This glass ceramic is characterized as a solid solution of β-spodumene and the metastable polymorph of silica called keatite.

The isotherm for Na^+-Li^+ exchange in the glass ceramic is shown in Fig. 3 at 400°C. Comparison of the isotherms in Fig. 1(b) for the glass and Fig. 3 for the glass ceramic, both ostensibly of the same composition, shows quite different behavior. The crystalline material very much prefers to remain in the lithium form over the entire range in composition. It is evident that the large selectivity for lithium reflects the preferred crystal structure. The data in Fig. 2 for the glass ceramic could not be represented adequately by Eq. (4). Instead, two straight-line segments were obtained.

The glass ceramic is about 95% crystalline and 5% amorphous (glass). Chemical analysis showed that 95% of the total cation content is lithium and 5% is sodium, and it is most likely that sodium exists in the glass phase. The results for the glass ceramic can be described by assuming that there are two types of exchange sites in the material.

With this assumption, we find that $K'_{LiNa} = 2 \times 10^{-2}$ with n = 0.6 for the original lithium sites, and $K''_{LiNa} = 2 \times 10^{-6}$ and n = 4 for the original sodium sites from the two straight-line segments. Thus, the isotherm is a superimposition of nonideal isotherms for each of the types of sites. It is probably somewhat fortuitous that the value of n for the sodium sites is approximately equal to the value of n for the base glass. These results show that the sodium sites in the glassy phase prefers Li^+ about 10^4 times more than do the Li^+ sites in the crystalline phase.

Membrane potential measurements with Corning Code 7740 borosilicate glass membranes in molten potassium nitrate indicate n = 1 (13).

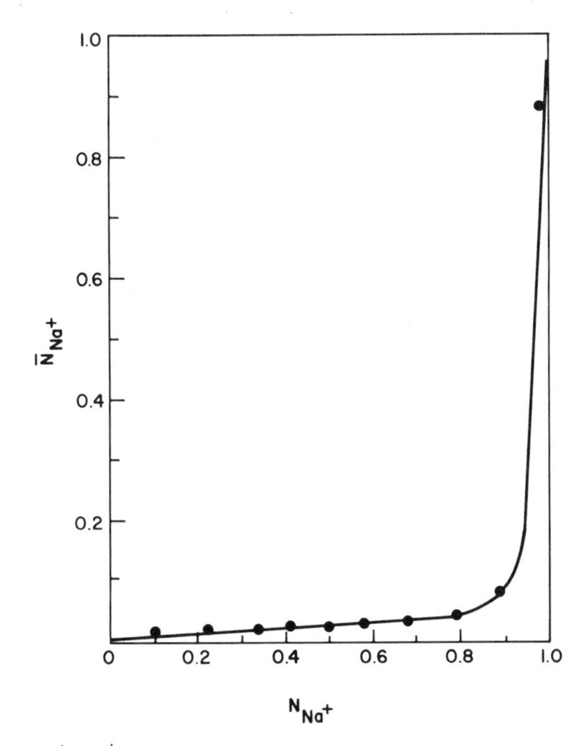

FIG. 3. Na^+-Li^+ exchange isotherm for crystallized composition B at 400°C.

Equilibrium exchange measurements were made with the borosilicate glass to determine K_{NaK} and to obtain an independent check of the finding from the emf data that n = 1. Results were obtained at 50, 80, and 100 mole % KNO_3 at three different temperatures. K_{NaK} was calculated with these data with n = 1 in Eq. (4). The results are shown in Table 3. It is apparent that the coefficient K_{NaK} is not a constant, although the membrane potential measurements indicated that n was unity.

Charles (39) concluded that the borosilicate glass is a two-phase system by comparing its dc electrical properties with those of a borosilicate glass known to be phase separated. Doremus and Turkalo (40) reported they could see phase separation in the borosilicate glass with the electron microscope on a 30-Å scale after heightening the contrast by silver exchange.

The results in Table 3 can be understood in terms of the work reported by Doremus (41), who showed that the results of silver exchange in the borosilicate glass were consistent with the assumption that the glass

TABLE 3

Results of Equilibrium Measurements on the Borosilicate Glass

Temperature, °C	N_K [a]	N_{Na}/N_K	$\bar{N}_K$	K_{NaK}
380	0.5	1.00	0.453	0.827
	0.8	0.206	0.727	0.550
	1.0	0.000952	0.855	0.00501
435	0.5	1.00	0.542	1.18
	0.8	0.207	0.804	0.851
	1.0	0.00108	0.946	0.0190
465	0.5	1.00	0.570	1.32
	0.8	0.209	0.847	1.15
	1.0	0.00117	0.993	0.167

[a] Nominal bath composition.

consisted of two exchangeable and thermodynamically ideal phases each with its appropriate exchange constant with n = 1.

Therefore, $K_I = \alpha m_{K,I}/m_{Na,I}$ and $K_{II} = \alpha m_{K,II}/m_{Na,II}$, in which α is the ratio of sodium to potassium activities in the melt, and the m's refer to the number of gram-ions of each species in the two phases per unit volume of the two-phase glass. Since the total number of exchanging ions in each phase is assumed to be constant, $m_I = m_{K,I} + m_{Na,I}$ and $m_{II} = m_{K,II} + m_{Na,II}$. The mole fraction of exchange sites in each phase (presumably a constant for a specific glass composition of given thermal history) is $N_I = m_I/(m_I + m_{II})$ and $N_{II} = m_{II}/(m_I + m_{II})$. With n = 1, the measured coefficient K_{NaK} given in Table 3 is

$$K_{NaK} = \frac{\alpha(m_{K,I} + m_{K,II})}{m_{Na,I} + m_{Na,II}}$$

Combining the above equations yields

$$K_{NaK} = \frac{\alpha N_I (K_I - K_{II}) + K_{II} (\alpha + K_I)}{\alpha + K_I - N_I (K_I - K_{II})} \quad (5)$$

An alternate approach is to consider the exchanger as consisting of a single, nonideal phase. A new exchange constant K'_{NaK} is defined as

$$K'_{NaK} = \frac{\alpha \bar{a}_K}{\bar{a}_{Na}} \tag{6}$$

With Eqs. (5) and (6) and an integration of the relation

$$\bar{N}_K \frac{\partial \ln \bar{a}_K}{\partial \bar{N}_K} - \bar{N}_{Na} \frac{\partial \ln \bar{a}_{Na}}{\partial \bar{N}_{Na}} = 0$$

expressions for the activities of the ions in the glass are found in terms of N_i, K_I, K_{II}, and α, from which K'_{NaK} is found to be

$$K'_{NaK} = \sqrt{K_I K_{II}} \left(\frac{K_{II}}{K_1}\right)^{(N_{II}-N_I)/2} \tag{7}$$

The results of the two-phase model from the data in Table 3 are shown in Table 4. The average value of $N_I = 0.08 \pm 0.05$ brackets the value of 0.05 reported for Ag^+-Na^+ exchange in the borosilicate glass (41). The temperature dependence of K'_{NaK} estimated from the data at 435 and 465°C is quite large, being about 16 kcal/mole. From the values of K'_{NaK}, one can conclude that below 450°C the two-phase exchanger prefers sodium, while above 450°C it prefers potassium.

TABLE 4

Summary of Results of the Two-Phase Model of Ion Exchange in the Borosilicate Glass

Temperature, °C	K_I	K_{II}	N_I	K'_{NaK}
380	(7×10^{-7})[a]	1.1	0.13	(0.2)[a]
435	5.4×10^{-4}	1.4	0.080	0.76
465	4.9×10^{-3}	1.4	0.030	1.2

[a] Numbers in parentheses are uncertain.

The results of the two-phase model suggest that one phase is primarily sodium borosilicate and contains about 92% of the sodium ions in the unexchanged glass, while the second phase is high in silica and contains the remaining sodium. The high-silica phase strongly prefers sodium, and the borosilicate phase prefers potassium. It is the strong attraction of the silica phase for sodium that is responsible for the large variation in K_{NaK} at lower sodium concentrations. Table 5, showing some of the properties of the two phases in the borosilicate glass, was constructed from these results. The alumina present in the borosilicate glass was arbitrarily put into phase I; the densities listed in Table 5 are not too different from the bulk density of 2.23 gm/cm^3.

The exchange properties of a phase-separable borosilicate glass containing 7 Na_2O, 23 B_2O_3, and 70 mole % SiO_2 (or 6.9 Na_2O, 25.7 B_2O_3, and 67.4 wt % SiO_2), were also studied. This glass composition was chosen because its phase-separation properties are well known (42). Before the exchange isotherm was determined, the glass powder was heat treated at 580°C for 16 hr to develop the two phases. Elmer et al. (43) showed the effect of heat treatment on microstructure, using electron microscopy. Their results indicate that the relative amount of the two phases was essentially independent of heat treatment. They reported that the amount of silica-rich phase was about 60% by volume.

TABLE 5

Calculated Properties of the Two Phases in the Borosilicate Glass[a]

	Phase I	Phase II
M % Na_2O	0.39	28.3
M % B_2O_3	3.33	67.9
M % SiO_2	94.9	3.80
M % Al_2O_3	1.35	--
Density (gm/cm^3)	2.19	2.30
Volume fraction	0.86	0.14

[a]Based on results reported in Refs. 13 and 39.

Figure 4 shows the K^+-Na^+ exchange isotherm at 450°C for the phase-separated glass. The data in Fig. 4 could not be represented by Eq. (4). Again two straight-line segments were obtained, and the data were treated in the same way as the glass-ceramic data. Thus, $K_I = 0.66$ with $n = 0.7$, and $K_{II} = 2.2$ with $n = 1.4$. By comparison with the Code 7740 borosilicate results, phase I is probably the high-silica phase and phase II is the borosilicate phase.

4. Regular-Solution Model

The results obtained for the various exchange reactions are summarized in Table 6. The single and two-phase systems are included for comparison. The relationship given by Eq. (4) provides an adequate description of the exchange equilibria. Garrels and Christ (44) demonstrated that the description represented by Eq. (4) is related to the regular-solution theory of binary mixtures, used so successfully to describe ion-exchange equilibria in aqueous systems (37).

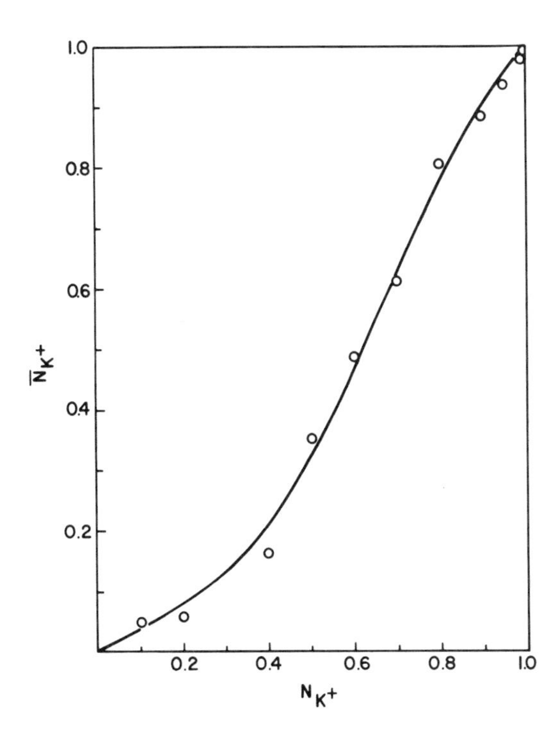

FIG. 4. K^+ - Na^+ exchange isotherm at 450°C for phase separated glass E.

TABLE 6

Summary of Equilibrium Results

Exchange Pair B - A	Glass	Temperature, °C	K_{AB}	n	W_{A-B} kcal/mole	ΔH° kcal/mole	ΔS° e.u.	Preference
$Na^+ - Li^+$	A	400	0.28	1.9	-2.4	-1 ± 1	-	Lithium
$Ag^+ - Li^+$	A	300	3.2	2.2	-2.7	--	-	Silver
$K^+ - Na^+$	C	500	0.94	1.2	-0.6	3.6 ± 0.6	4.5	Sodium
$Li^+ - Na^+$	C	400	4.4	1.3	-0.8	--	--	Lithium
$K^+ - Na^+$	Code 7740[a]	465	1.2	--	--	(16)[b]	--	Potassium
$Na^+ - Li^+$	B	400	0.27	3.2	-5.9	--	--	Lithium
$Na^+ - Li^+$	B(cryst.)[a]	400	(0.02)[b]	(0.6)[b]	(1)[b]	--	--	Lithium
			(2×10^{-6})[b]	(4)[b]	(-7)[b]	--	--	Lithium
$K^+ - Na^+$	E[a]	450	0.66	0.7	0.9	--	--	Sodium
			2.2	1.4	-1.1	--	--	Potassium
$Ag^+ - Na^+$	D	300	2.5	1.4	-0.9	--	--	Silver

[a] Based on a two-phase model.

[b] Numbers in parentheses indicate uncertain values.

The assumption that the exchanger is a regular solution gives the ratio of the activity coefficients of species A and B in the binary mixture as

$$\ln\left(\frac{\bar{\gamma}_A}{\bar{\gamma}_B}\right) = -\frac{W_{A-B}}{RT}(1 - 2\bar{N}_B) \tag{8}$$

Assuming n-type behavior and introducing the definition of the rational activity coefficient, we find that

$$\ln\left(\frac{\bar{\gamma}_A}{\bar{\gamma}_B}\right) = (n-1)\ln\left(\frac{\bar{N}_A}{\bar{N}_B}\right) \tag{9}$$

Expansion of $\ln(\bar{N}_A/\bar{N}_B)$ gives

$$\ln\left(\frac{\bar{N}_A}{\bar{N}_B}\right) = 2\left[(1-2\bar{N}_B) + 1/3\,(1-2\bar{N}_B)^3 + 1/5\,(1-2\bar{N}_B)^5 + \ldots\right] \tag{10}$$

Since only the first term of Eq. (10) need be considered for the concentrations used here, we find that

$$n = 1 - \frac{W_{A-B}}{2RT} \tag{11}$$

Calculated values of W_{A-B}, the excess interaction energy of neighboring A and B ions, are shown in Table 6. This n-type description for an exchanger with identical sites is unrealistic, because it requires the activity coefficient of the dilute component to be a function of concentration even at greater dilution. Under these conditions one should use the regular-solution model.

Since a negative value of W_{A-B} indicates repulsion between the A and B ions, inspection of Table 6 shows that the deviations from ideality of the exchanger are caused by repulsions of the counterions. The strain around the exchanged site also affects this behavior.

There are no direct heat-of-mixing data available for these systems. Tischer (45) investigated the heats-of-mixing in the binary $Li_2Si_3O_7$-$K_2Si_3O_7$ system. Although he found that the heats-of-mixing depended upon composition, the sodium-potassium silicate system showed smaller (negative) deviations from ideality than the sodium-lithium silicate system--in qualitative agreement with the results in Table 3.

IV. KINETICS OF EXCHANGE

A. Introduction

The type of binary ion-exchange reaction under consideration here can be characterized as an equimolar, countercurrent diffusion process. Thus, the fluxes of the two interdiffusing ions are equal in magnitude, even though the counterion mobilities may be different. This is a consequence of the electroneutrality requirement and the constancy of the number of exchange sites for a given exchanger. The faster ion, of course, tends to diffuse at the higher rate. Any excess flux of an ion is equivalent to a net transfer of electric charge, which produces an electric field that slows down the faster ion and accelerates the slower ion. In this way the fluxes become equal. These electric forces, among the most important factors in ion-exchange kinetics, will be discussed in greater detail later.

Since ionic diffusion in the molten salts is much more rapid than in the glass or glass-ceramic exchanger, diffusion of the counterions in the exchanger is rate controlling. Univalent-for-univalent exchange occurs most rapidly. The rate of the interdiffusion process decreases and the temperature coefficient of the rate increases with increasing size and charge of exchanging species. The temperature dependence of the interdiffusion and self-diffusion coefficients follows the Arrhenius equation

$$D = D_0 \exp(-E^{\ddagger}/RT)$$

in which $E^{\ddagger}$ is the activation energy.

In this section the exchange kinetics of some of the systems discussed in Sec. III are examined. Results of self-diffusion and interdiffusion are presented, and the relation between the two is discussed in terms of the Nernst-Planck equations.

B. Methods of Measurement

Since most of the techniques used to study exchange kinetics have been described in some detail in the literature, they are only discussed briefly here. In all the methods described in the following sections, the conditions are experimentally fixed so that it can be assumed that: (1) net diffusion is unidirectional and (2) the sample extends semi-infinitely in the direction parallel to the gradients in concentration. Weight change and sectioning were used most extensively, while resistance change and electron microprobe were used to a lesser extent. The methods are described individually below.

1. Weight Change

By measuring the weight of the exchanger before and after the exchange, the total amount in moles of A taken up per unit surface area of glass exposed to the molten salt is

$$Q_A = \frac{\Delta w}{S(M_A - M_B)}$$

where M is the gram-atomic weight, Δw is the change in weight of the sample following ion exchange, and S is the superficial area.

Integration of Fick's first law with constant interdiffusion coefficient $\tilde{D}_{AB}$ gives

$$Q_A = 2C_{o,A}\left(\frac{\tilde{D}_{AB}t}{\pi}\right)^{\frac{1}{2}} \quad (12)$$

where $C_{o,A}$ is the surface concentration per unit volume. Thus, the amount of material taken up by the exchanger is proportional to the square root of time. If the interdiffusion coefficient is a function of concentration, then $\tilde{D}_{AB}$, determined from Eq. (12), will be some mean value over the initial and final concentration in the sample. In this case, the uptake is written

$$Q_A = \int_o^t J_A^o \, dt = -2(\tilde{D}_{AB}^o t)^{\frac{1}{2}} \left(\frac{\partial c_A}{\partial w}\right)_{w=o} \quad (13)$$

where $w = x/(\tilde{D}_{AB}^o t)^{\frac{1}{2}}$; J_A^o is the flux of A and $\tilde{D}_{AB}^o$ is the interdiffusion coefficient, respectively, at $x = 0$. Thus, Q_A is still proportion to the square root of the time of exchange, because $(\partial c_A/\partial w)_{w=0}$ is constant for a constant profile shape irrespective of the depth of exchange (36, 46). The square-root-of-time law for uptake is well established for ion exchange of glass in molten salts (1, 4, 18, 36, 46) and aqueous solutions (47).

2. Changes in Electrical Resistance

The change in resistance was measured in two ways. In the first method, it was followed at temperature as diffusion progressed; by measuring the resistance of the unexchanged glass at temperature, the change in resistance is known as a function of time of exchange (36, 46). In the second method, the resistance was measured on exchanged disks as a function of temperature to 100°C below the exchange temperature to insure that the concentration distribution would remain fixed. The specific resistivity of the unexchanged sample was also determined. With these data the resistance before and after ion exchange was extrapolated to the actual temperature of exchange to give the change in resistance (46).

The electrical resistance of the exchanger after exchange for time t is given by

$$R_t = \frac{1}{S} \int_0^L \frac{dx}{\sigma_t} \tag{14}$$

where S is the superficial area of the exchanger exposed to the molten salt, L is the thickness of the exchanger, and σ_t is the specific conductivity of the exchanger at a point x from the glass surface. Therefore, the resistance of the exchanger at any time t of exchange is given by (36)

$$R_t - R_o = \frac{(D_A t)^{\frac{1}{2}} R_o}{L} \int_0^{L/(D_A t)^{\frac{1}{2}}} \frac{(\sigma_o - \sigma_t)}{\sigma_t} \, dy \tag{15}$$

where $y = x/(D_A t)^{\frac{1}{2}}$, D_A is the self-diffusion coefficient of the ion originally present in the exchanger, R_o is the resistance, and σ_o is the specific conductivity of the exchanger in its original form. Since the integral in Eq. (15) is constant as long as the depth of exchange is short compared to the thickness of the exchanger, i.e., $L \gg (D_A t)^{\frac{1}{2}}$, the change in resistance is proportional to the square root of the time of ion exchange.

3. Sectioning

The concentration distribution can be determined by sectioning the exchanger following ion exchange. Thin, successive portions of the exchanged sample are removed with an HF-H_2SO_4 solution. The depth removed is calculated from the density of the glass, the mass of the section removed, and the total surface area exposed to the etchant, i.e.

$$d = \frac{\Delta w - m_B(M_B - M_A)}{\rho_0 S}$$

where m_B is the total number of moles of B in the layer, M is the molecular weight, and ρ_0 is the density of the unexchanged sample. Analysis of the etchates by flame photometry or titrimetry gives the ionic composition of the removed layer. If the bath is tagged with radiotracer, then counting the etchates or remaining sample will yield information about the diffusion profile.

For the case in which the etched solution is analyzed either by flame photometry or counting methods, the diffusion coefficient is determined from (48)

$$A_j/A_0 = 1 - \text{erf}\, \xi_j \tag{16}$$

where A_j/A_0 is the ratio of the activity of the jth layer to that at the surface, erf ξ_j is the error function or probability integral, and $\xi_j = x_j/2(Dt)^{\frac{1}{2}}$. For the residual counting technique, i.e., the method in which the entire sample was counted before and after etching, the diffusion coefficient is obtained from (48)

$$A_j/A_0 = \pi^{\frac{1}{2}}\, \text{ierfc}\, \xi_j \tag{17}$$

where A_j/A_0 is the ratio of activity left in the sample after removal of the jth layer to the total activity in the sample, $\xi_j = x_j/2(Dt)^{\frac{1}{2}}$, and ierfc ξ_j is the first integral of the error function. Equations (16) and (17) are applicable only for the case in which the diffusivity is independent of concentration; these equations are especially applicable to self-diffusion studies. If the interdiffusion coefficient is a function of concentration in the sample, then numerical methods like Fujita's solution (49) must be used.

4. Electron Microprobe

The sample, in the form of a rod, is exchanged in the molten salt. A section is removed from the center of the rod, mounted in a plastic support, and polished. The profile is then determined with the electron microprobe (50).

C. Results

1. Self-Diffusion

Small alkali metal ions move most rapidly through glass and glass-ceramic exchangers. Johnson et al. (51) showed that the self-diffusion coefficient of sodium D_{Na}, in binary sodium silicate glasses, varied exponentially with the sodium concentration. Substitution of CaO for SiO_2 at the same Na_2O level decreased D_{Na}; calcium and the other divalent ions exhibit a "blocking" effect on the diffusion of univalent ions. The self-diffusion coefficient of sodium varied between 10^{-10} and 10^{-8} cm^2 sec^{-1} over the temperature range 350-500°C. The activation energy for self-diffusion was 14-22 kcal/mole over the same temperature range for the compositions studied. In a study of the self-diffusion of sodium in a nepheline glass ceramic (52) of approximate composition $Na_2O \cdot Al_2O_3 \cdot 4SiO_2$, D_{Na} was found to be 1.8×10^{-8} cm^2 sec^{-1} at 741°C with an activation energy of about 31 kcal/mole over the temperature range 721-798°C (53).

Johnson et al. (51) found the self-diffusion coefficient of calcium, D_{Ca}, in soda-lime-silica glass was 10^{-10} - 10^{-8} cm^2 sec^{-1} over the temperature range 600-900°C with an activation energy of about 30 kcal/mole. The value of D_{Ca} at 654°C was more than two orders of magnitude smaller than D_{Na} in a similar glass at the same temperature. The self-diffusion coefficient of calcium in 36.3 CaO, 26.8 Al_2O_3, 36.9 mole % SiO_2 is approximately 4×10^{-13} cm^2 sec^{-1} at 900°C; the activation energy over the temperature range 900-1000°C is 36 kcal/mole (53). In contrast, the self-diffusion coefficient of calcium in stoichiometric calcium metaphosphate glass is 2×10^{-11} - 3×10^{-10} from 350 to 500°C with an activation energy of 21 kcal/mole (53).

If the electrical current in an exchanger is carried by a single ionic species, then the electrical conductivity of the exchanger can be related to the diffusion coefficient of this species by the Nernst-Einstein equation

$$D = \frac{\sigma RT}{c(zF)^2} \tag{18}$$

where σ is the specific conductivity in Ω^{-1} cm^{-1}, R is 8.314 J/deg mole, c is the concentration of the diffusing species in moles/cm^3, z is the ionic value, and F is the Faraday or 96,500 C/eq.

It has been found that the measured electrical conductivity of glass is less than that calculated from diffusion data (1, 18). These results have been interpreted in terms of a model which accounts for the difference in terms of the two kinds of processes. In this model a correlation factor, f, which depends on the structure of the exchanger and on the transport mechanism, is used to modify the Nernst-Einstein equation so that

$$f = D_A/D_\sigma \tag{19}$$

where D_A is the measured self-diffusion coefficient, and D_σ is the diffusion coefficient calculated from the conductivity with Eq. (18). Thus, f is a measure of the efficiency of the Nernst-Einstein equation. If diffusional motion is completely random, f = 1. However, f is less than unity if the next jump depends upon the direction of the previous jumps, i.e., the jump direction is not entirely random. Since Compaan and Haven (54) showed that such correlations do not arise in electrical conduction, the factor f can be attributed to the correlations in diffusion. Haven and Stevels (55), and later Haven and Verkerk (56), discussed the possible mechanisms and the magnitude of the correlation effects associated with these mechanisms in glass that could account for the observed relation between diffusion and ionic conductivity.

A comparison of the tracer-diffusion coefficient and that calculated from the conductivity with Eq. (18) is shown in Fig. 5 for the Code 7740 borosilicate glass (46). It is seen from the data that f is unity; the activation energy for diffusion and conduction is 22 kcal/mole. It is important that the samples for both measurements have the same thermal history. For unannealed samples of the borosilicate glass, the activation energy for

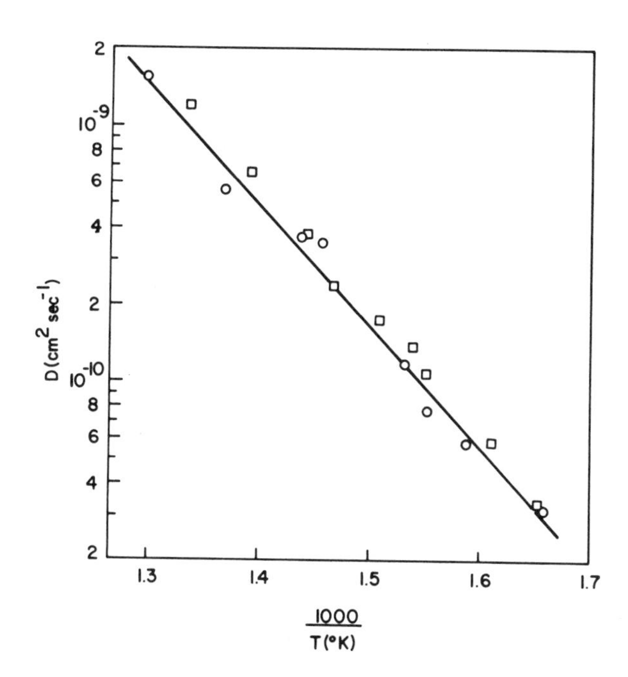

FIG. 5. Temperature dependence of the diffusion coefficient in Code 7740 borosilicate glass; o, D_{Na}; □, D_σ.

self-diffusion was found to be 15 kcal/mole. This lower activation energy results from the more open structure of the unannealed glass. The correlation factor f is approximately unity for composition C, while f is about 0.8 for composition D (53).

Doremus reported f equal to unity in a commercial soda-lime-silica glass (36), a 96% silica glass (16), and fused silica (57). He also reported a value of f of about 0.4 for the Code 7740 borosilicate glass and suggested that deviations from Eq. (18) arise because of phase separation and not because of any special mechanism of diffusion (16,58).

The diffusion coefficients determined by the tracer method (48), and by conductivity (53) are shown in Fig. 6 for two alumina containing glasses that are not phase separated; both are cation-selective electrode glasses described by Eisenman (47). The correlation factor, however, is less than unity in both cases; f is 0.5 for the glass in Fig. 6(a) and 0.4 for the glass in Fig. 6(b). Heckman et al. (59) reported correlation factors of 0.2-0.7 for a series of soda-alumina-silica and soda-lime-alumina-silica glasses and

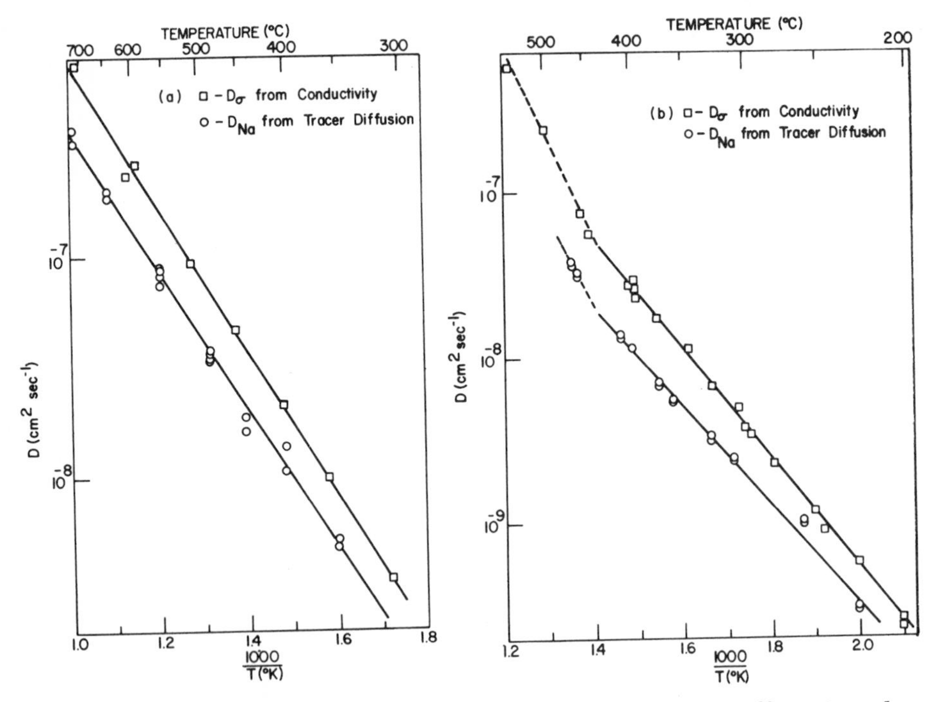

FIG. 6. Relationship between measured self-diffusion coefficients and those calculated from the conductivity with Eq. (18) in: (a) 11 Na_2O, 18, Al_2O_3, 71 mole % SiO_2; (b) 27 Na_2O, 3 Al_2O_3, 70 mole % SiO_2.

suggested the deviations from Eq. (18) are due to a defect mechanism. Terai (60) reported correlation factors of about 0.2-0.3 for a series of soda-alumina-silica glasses and also suggested that a defect mechanism was responsible for the observed deviations from Eq. (18).

In an effort to test this defect model, Doremus (57) quenched a silica glass containing trace sodium rapidly into water, from 900°C, to quench the defects in the glass. The resistance of the glass below 200°C showed no decrease from the value before quenching. This test was considered to be optimum for finding quenched-in defects, due to the large distances between ionic sites in fused silica and the low mobility at low temperatures. Thus, the defect mechanism for ionic transport in glass remains without direct support.

Nevertheless, the conductivity can be used with lithium containing glasses and glass ceramics to estimate the self-diffusion coefficient of lithium, for which there is no suitable tracer. Although Urnes (61) has suggested that diffusional and conduction migration processes are different in glass, there is no experimental evidence to support this hypothesis.

2. Interdiffusion

In binary ion exchange, the diffusing species A and B are charged, and tend, in general, to diffuse at different rates. Thus, there is a tendency for one ion to move faster than the other, leading to a build up of electrical charge. There is, however, a gradient in electrical potential along with this charge, which acts to slow down the faster ion and speed up the slower one. Despite the difference in the mobilities of the two ions, the gradient in electrical potential forces the fluxes of the two ions to be equal and opposite, thus preserving electrical neutrality.

The driving force for transport of species A or B is the negative of its gradient in total chemical potential μ. The total chemical or electrochemical potential is the sum of the gradient in activity and the gradient in electrical potential. This assertion is based upon the assumption that no other gradients or thermodynamic forces, such as gradients in pressure or gradients in temperature, are present in the system. With the additional assumption that the mobilities of the ions are the same in self-diffusion and interdiffusion, the flux of diffusing species per unit time in the x direction is (36)

$$J_A = -D_A \left[\frac{\partial \bar{c}_A}{\partial x} \frac{\partial \ln \bar{a}_A}{\partial \ln \bar{c}_A} + \bar{c}_A \frac{FE}{RT} \right] \qquad (20)$$

where D_A is the diffusion coefficient, $\bar{c}_A$ the ionic concentration, $\bar{a}_A$ the activity of species A, and E the potential gradient or electric field; a similar expression can be written for the flux J_B. Only univalent-for-univalent exchange is under consideration here. This set of flux equations is the Nernst-Planck equations.

The conditions of electrical neutrality require that $J_A = -J_B$. Since the number of negative exchange sites is constant and fixed, $\partial\bar{c}_A/\partial x = -\partial\bar{c}_B/\partial x$. Therefore, the potential gradient is

$$E = \frac{RT}{F}\left[\frac{D_B - D_A}{\bar{c}_A D_A + \bar{c}_B D_B}\right]\frac{\partial \bar{c}_A}{\partial x}\frac{\partial \ln \bar{a}_A}{\partial \ln \bar{c}_A} \tag{21}$$

Substitution of Eq. (21) into Eq. (20) gives

$$J_A = -\left[\frac{D_A D_B}{\bar{N}_A D_A + \bar{N}_B D_B}\right]\frac{\partial \bar{c}_A}{\partial x}\frac{\partial \ln \bar{a}_A}{\partial \ln \bar{c}_A} \tag{22}$$

where $\bar{N}$ is the cation fraction in the exchanger; a similar equation can be written for J_B. The term in the brackets is the same in the expressions for J_A and J_B. By comparison with Fick's first law for diffusion, one may write an interdiffusion coefficient

$$\tilde{D} = \frac{D_A D_B}{\bar{N}_A D_A + \bar{N}_B D_B}\frac{\partial \ln \bar{a}_A}{\partial \ln \bar{c}_A} \tag{23}$$

which depends upon cation concentration. If the results from the previous section on self-diffusion are applied, then Eq. (23) must be modified to give

$$\tilde{D} = \frac{D_A D_B (\bar{N}_A f_A + \bar{N}_B f_B)}{(\bar{N}_A D_A f_B + \bar{N}_B D_B f_A)}\frac{\partial \ln \bar{a}_A}{\partial \ln \bar{c}_A} \tag{24}$$

where f is the correlation factor. If f_A equals f_B, the results are unaffected by the value of f, but this is not the case for f_A not equal to f_B.

Helferrich and Plesset (62, 63), who first used the Nernst-Planck equations for diffusion in ion exchangers, found that these equations adequately described the interdiffusion of hydrogen and sodium ions in polysulfonate exchangers in aqueous solution.

By means of tracer and interdiffusion measurements of various kinds, the applicability of the Nernst-Planck equations was verified for the interdiffusion of silver and sodium ions in a soda-lime-silica glass from molten nitrates (36). In this system the factor $\partial \ln \bar{a}_A / \partial \ln \bar{c}_A$ was unity, and the mobility ratio $u_A/u_B = D_A/D_B$ was independent of concentration.

Composition A was studied most extensively. Figure 7 shows concentration-distance profiles for sodium, potassium, and silver exchange in the lithium glass (53). In each example the interdiffusion coefficient varies with ionic concentration, although to different degrees. The data in Fig. 7 were fitted from concentration profiles and residual profiles calculated with the computer for different values of D_A/D_B according to Fujita's solution (49) for a diffusion coefficient varying according to $\tilde{D} = D_0/(1-\alpha c)$. The concentration dependence is least pronounced with Na^+-Li exchange in Fig. 7(a). The ratio of the self-diffusion coefficients of lithium and sodium, D_{Li}/D_{Na}, used to fit the data in Fig. 7(a) was 1.3. The self-diffusion coefficient of sodium measured in composition A with Na (22) was found to be 2×10^{-9} cm^2 sec^{-1} at 400°C with an activation energy of about 16 kcal/mole. The self-diffusion coefficient of sodium calculated from Eq. (23) with the thermodynamic factor given in Table 6 was 3×10^{-9} cm^2 sec^{-1} at 400°C, in fair agreement with the measured value. Similarly, the self-diffusion coefficient of lithium was estimated from Eq. (23) to be 4×10^{-9} cm^2 sec^{-1}. The self-diffusion coefficient of lithium was estimated from the conductivity as 5×10^{-9} cm^2 sec^{-1}; a correlation factor of $f = 0.5$ was assumed because of the similarity in composition of glass A with the glass shown in Fig. 6(a). The activation energy for conduction was 17 kcal/mole.

The interdiffusion coefficients for K^+-Li^+ exchange [Fig. 7(b)] and Ag^+-Li^+ exchange [Fig. 7(c) and 7(d)] have a pronounced dependence on ionic concentration. For silver exchange, D_{Li}/D_{Ag} is 19 at 289°C and for potassium exchange, D_{Li}/D_K is 30. The electron microprobe results in Fig. 7(d) for Ag^+-Li^+ exchange are included for comparison. In each example given, the ion originally present in the exchanger has a greater mobility than the foreign ion. Conversely, the results for 8.3 Li_2O, 4.4 MgO, 7.8 B_2O_3, 9.2 Al_2O_3, 70.3 mole % SiO_2 shown in Fig. 8 gave $D_{Li}/D_{Na} = 0.21$, indicating that the foreign ion has greater mobility than the ion originally in the exchanger (53). This point was verified by the finding that the dc resistance of a glass disk at 398°C prior to exchange was twice the value measured following exchange at 398°C in sodium nitrate for 17.7 hr.

The uptake and the temperature dependence of the uptake for sodium, potassium, and silver exchange is shown in Fig. 9 (53). It is seen that the uptake follows the square root of time as required by Eq. (13), and the

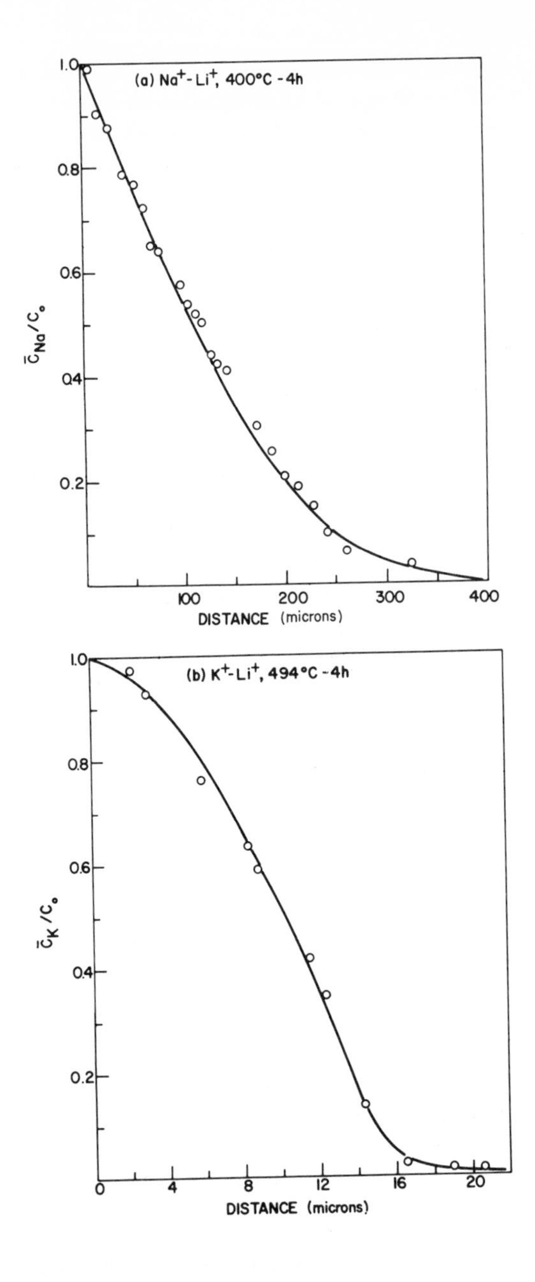

FIG. 7. Concentration profiles for several exchange reactions in exchanger A by various methods: (a) tracer sectioning; (b) chemical sectioning.

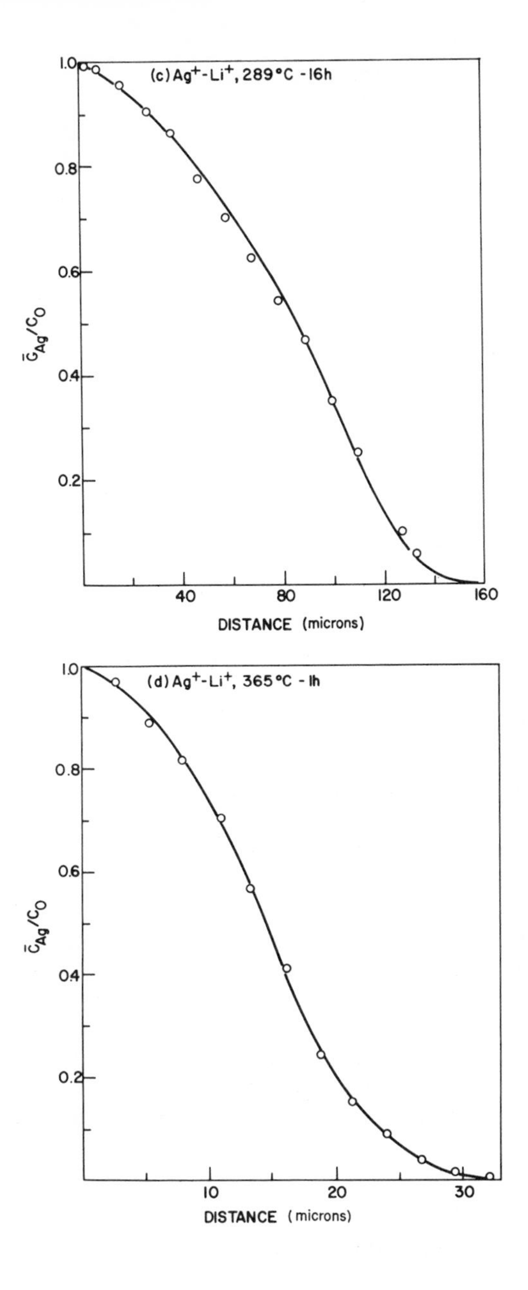

FIG. 7 (continued). (c) tracer sectioning; (d) electron microprobe.

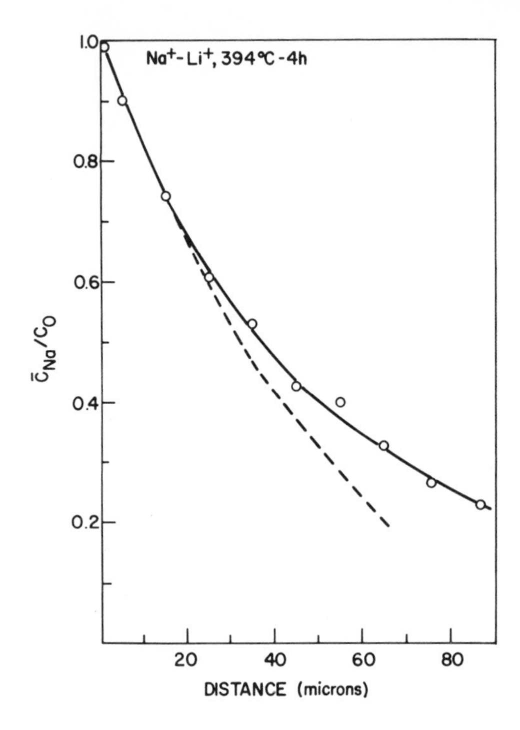

FIG. 8. Relative concentration of sodium as a function of distance in an exchanger of composition 8.3 Li_2O, 4.4 MgO, 7.8 B_2O_3, 9.2 Al_2O_3, 70.3 mole % SiO_2 with $D_{Li}/D_{Na} = 0.21$, $\bar{D} = 1.8 \times 10^{-9}$ cm^2 sec^{-1}.

temperature dependence of the uptake is fitted by an Arrhenius-type equation. However, some irregularities were noted with Rb^+-Li^+ and Cs^+-Li^+ exchange.

With Rb^+-Li^+ exchange the uptake was not linear with the square root of time as shown in Fig. 10(a) (53). It was assumed that in this case the surface is not saturated immediately. Thus, the surface concentration is given by

$$C(0,t) = C_0 + kt \tag{25}$$

It can be shown (49) that the uptake follows

$$Q = \left(\frac{Dt}{\pi}\right)^{\frac{1}{2}} (2C_0 + \frac{4}{3} kt) \tag{26}$$

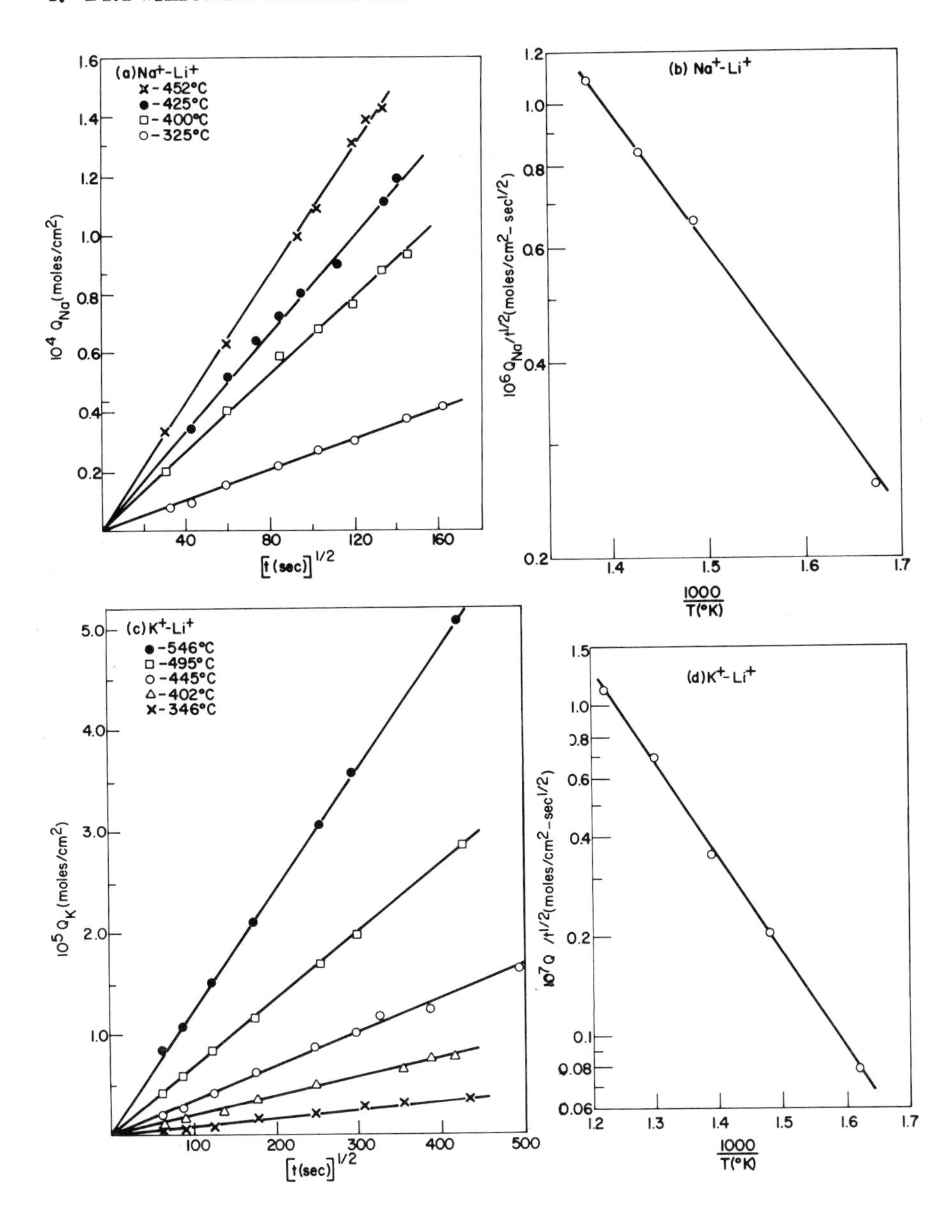

FIG. 9. Uptake of sodium, potassium, and silver in exchanger A: (a), (c), (e) as a function of $t^{\frac{1}{2}}$; (b), (d), (f) as a function of temperature.

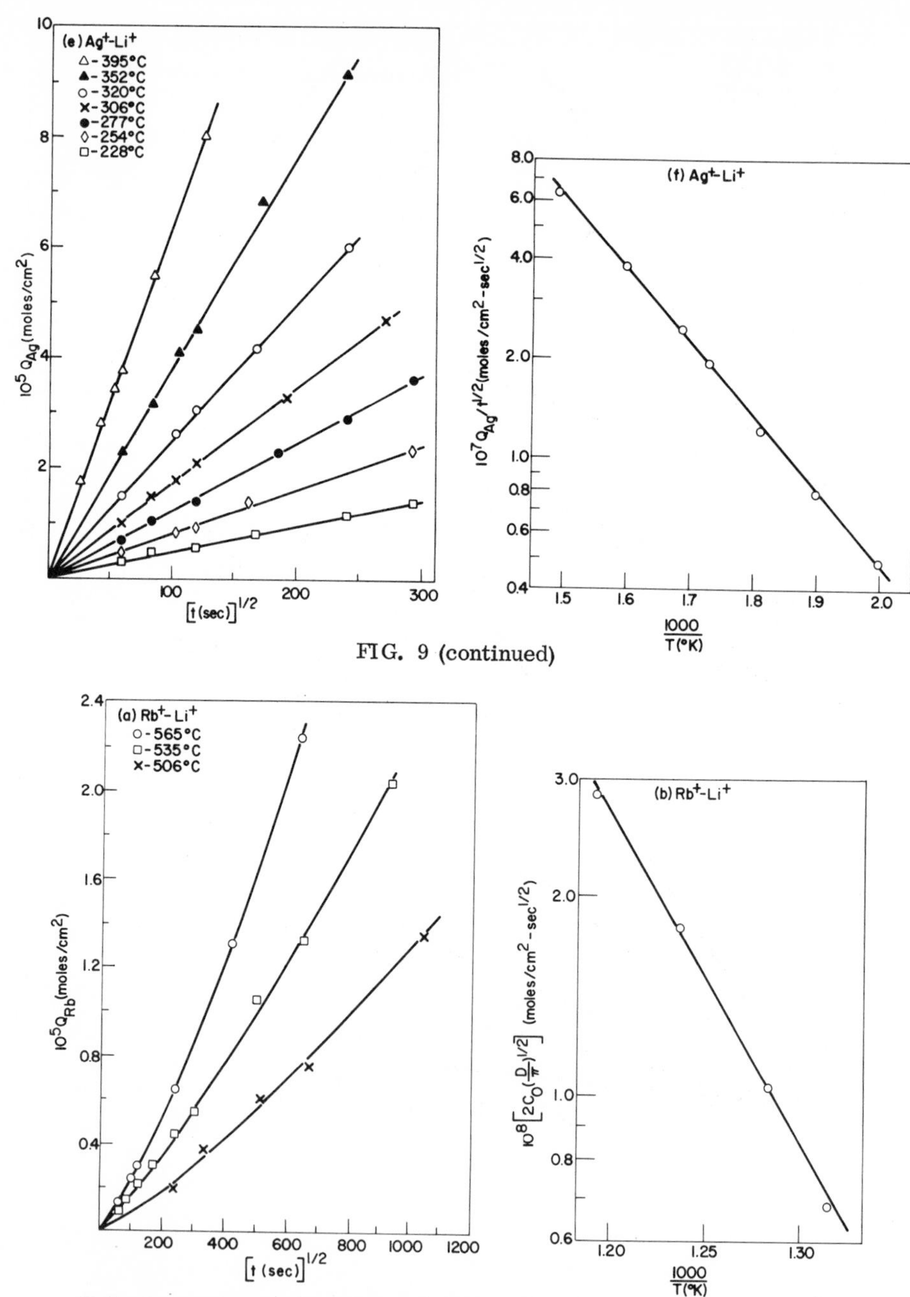

FIG. 9 (continued)

FIG. 10. Uptake of rubidium in exchanger A: (a) as a function of $t^{\frac{1}{2}}$; (b) temperature dependence of the mean interdiffusivity.

for the boundary condition given by Eq. (25). Equation (26) adequately fitted the data in Fig. 10(a). With the assumption that C_0 is essentially constant over the temperature range shown, the activation energy for interdiffusion was estimated to be 46 kcal/mole from the data in Fig. 10(b). Similarly, the temperature dependence of k was found to be about 58 kcal/mole. It was felt that this effect would be more pronounced in Cs^+-Li^+ exchange. On the contrary, the uptake varied linearly with the square root of time, and the rate of exchange had a relatively small temperature coefficient. The reasons for this behavior are not well understood at this time.

Concentration profiles for K^+-Na^+ exchange with four different compositions are shown in Fig. 11 for comparison (53). In every case the interdiffusion coefficient is independent of ionic concentration. The important feature to note is that the constancy of the diffusion coefficient is evident in a variety of compositions of different complexities.

The distribution of potassium in composition C following K^+-Na^+ exchange is shown in Fig. 12(a) at two different temperatures (53). The results at 500°C exhibited a slight concentration dependence and were fitted with the computer solution for $D_{Na}/D_K = 1.8$. An activation energy of 23 kcal/mole was estimated from the data in Fig. 12(b). The integral mean value

$$\bar{D} = \frac{1}{C_0} \int_0^{C_0} \tilde{D}(c)dc \tag{27}$$

where $\tilde{D} = D_0 = D_0/(1-\alpha c)$, was used for the value of the interdiffusion coefficient at 500°C.

The uptake of potassium and the temperature dependence of the uptake are shown in Fig. 13 (53). With the value of $C_{0,K}$ determined by the method described in Sec. III, and the slope of the line in Fig. 13(a) at 500°C, the mean value of the interdiffusion coefficient was calculated from Eq. (12) as 5.3×10^{-10} cm^2 sec^{-1}, in excellent agreement with the value derived from the concentration profile. The results in Fig. 14 show the uptake of potassium at 500°C as a function of bath composition. From these data the value of D_{Na}/D_K was found to be 1.8 with the method of moments of Fujita and Kishimoto (49), in agreement with the diffusion coefficient ratio obtained with the results in Fig. 12(a) at 500°C. The self-diffusion coefficient of sodium was calculated as 6.4×10^{-10} cm^2 sec^{-1} from Eq. (23) with $\tilde{D}^0$ determined from the data in Fig. 12(a) at 500°C and the value of $\partial \ln \bar{a}_K / \partial \ln \bar{c}_K$ given in Table 6 for exchanger C. The measured self-diffusion coefficient of sodium was 6.2×10^{-10} cm^2 sec^{-1}, in good agreement with the value

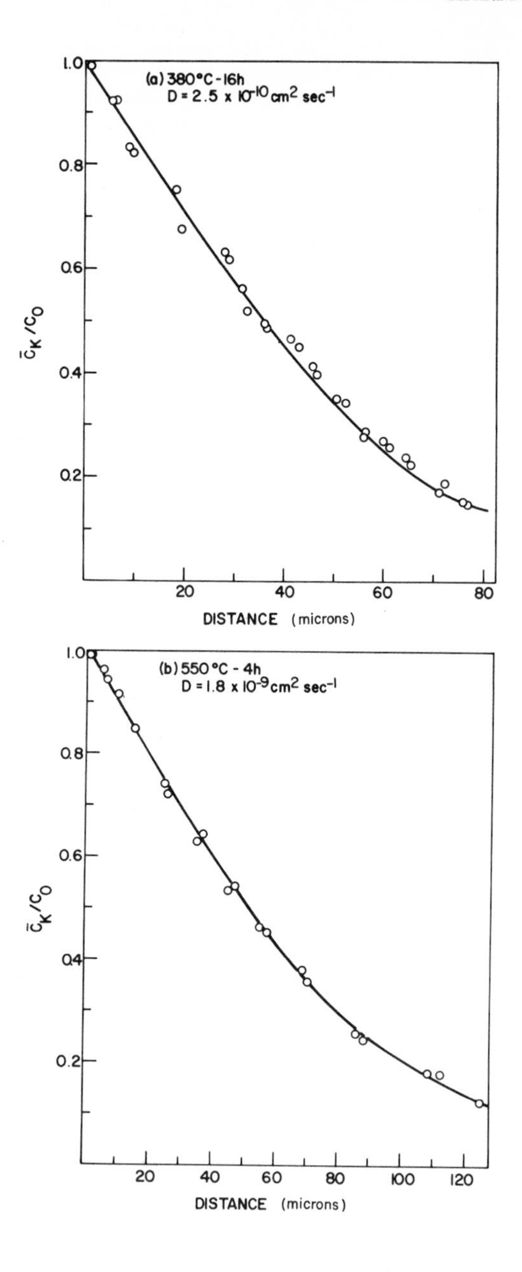

FIG. 11. Concentration profiles of potassium in several glass compositions: (a) 22.7 Na_2O, 23.5 Al_2O_3, 53.8 mole % SiO_2; (b) 17.7 Na_2O, 2.2 K_2O, 6.7 Al_2O_3, 7.2 ZrO_2, 66.2 mole % SiO_2.

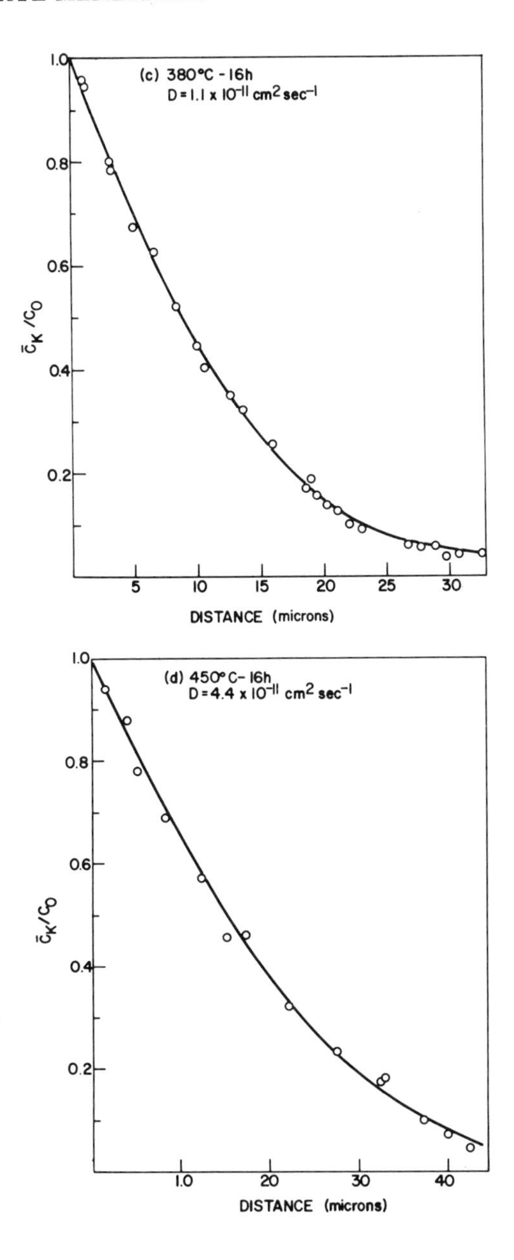

FIG. 11 (continued). (c) 17.5 Na_20, 1.5 K_20, 2.4 Ca0, 12.6 Al_20_3, 3.8 $Ti0_2$, 1.9 B_20_3, 60.3 mole % Si 0_2; (d) Corning Code 0088 soda-lime-silica.

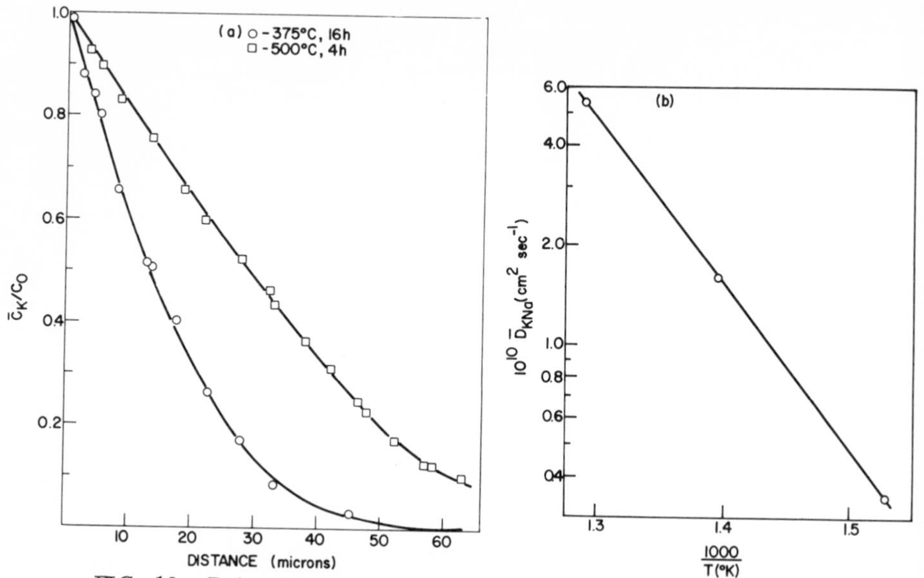

FIG. 12. Potassium interdiffusion in exchanger C: (a) concentration of potassium as a function of distance; (b) temperature dependence of the mean interdiffusion coefficient.

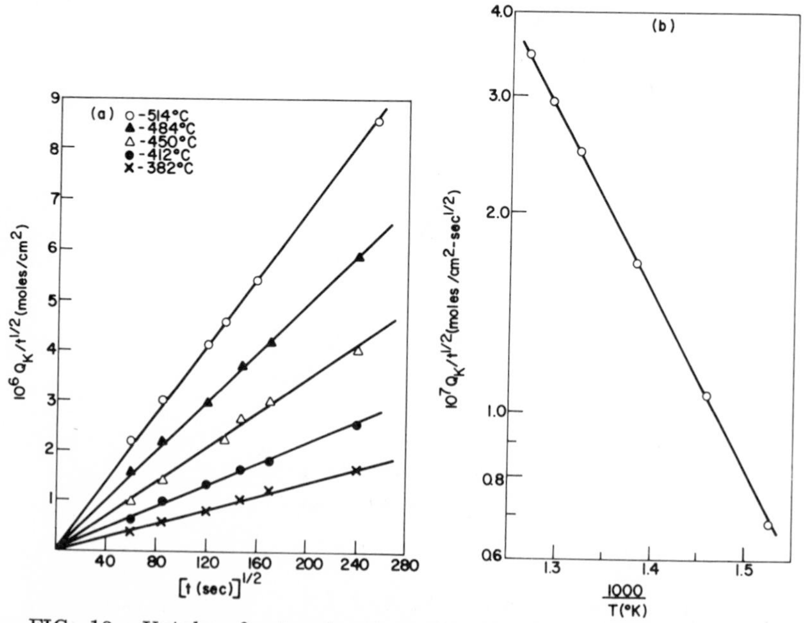

FIG. 13. Uptake of potassium in exchanger C as a function of: (a) square root of time; (b) temperature.

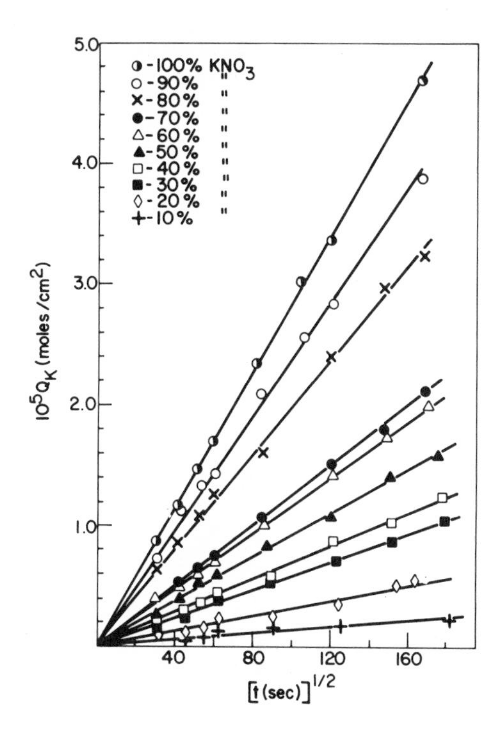

FIG. 14. Uptake of potassium in exchanger C as a function of $t^{\frac{1}{2}}$ for different bath concentrations at 500°C.

calculated from Eq. (23). With the value of $\tilde{D}^\circ$ and the uptake result in Fig. 13(a) at 500°C, a value of 0.499 was calculated for $-\frac{1}{C_o}(\partial\bar{c}_K/\partial w)_{w=o}$ according to Eq. (13). Inspection of Table 12 (page 241) shows that this value of $-\frac{1}{C_o}(\partial\bar{c}_K/\partial w)_{w=o}$ corresponds to $D_{Na}/D_K \approx 1.9$, in good agreement with the values reported by other methods.

The distribution of silver is shown in Fig. 15 for composition D following exchange in molten silver nitrate for 64 hr at 267°C and 7.5 hr at 354°C (53). The data at 267°C were fitted with $D_{Na}/D_{Ag} = 2.3$. The extent of silver exchange at the surface was about 96%. The self-diffusion coefficients of silver and sodium are 1.0×10^{-11} cm^2 sec^{-1} and 2.2×10^{-11} cm^2 sec^{-1}, respectively, from Eq. (23) and the data in Fig. 15 and Table 6. The measured self-diffusion coefficients of silver and sodium were found to be 1.1×10^{-11} cm^2 sec^{-1} and 1.9×10^{-11} cm^2 sec^{-1}, respectively, at 267°C.

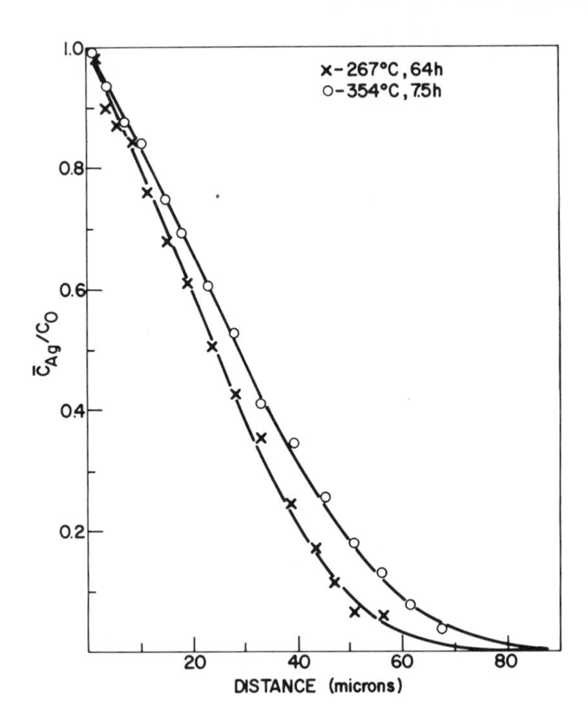

FIG. 15. Concentration profiles for silver exchange in composition D: o, 267°C, 64 hr; x, 354°C, 7.5 hr.

The data at 354°C were fitted with $D_{Na}/D_{Ag} = 2.2$. The fraction of exchange at the surface was identical to that at 267°C. The ratio D_{Na}/D_{Ag} is relatively insensitive to temperature. This behavior is similar to that reported for Ag^+-Na^+ exchange in Corning Code 7740 borosilicate glass (11).

Figure 16 shows the uptake and temperature dependence of the uptake for Ag^+-Na^+ exchange in exchanger D (53). Although silver chloride or silver nitrate were used as exchange media, no difference in uptake was noticed. The temperature dependence of the uptake is 10.6 kcal/mole. The mean value of the interdiffusivity, $\bar{D}$, is 2.1×10^{-11} cm^2 sec^{-1} and 2.5×10^{-11} cm^2 sec^{-1} from profile and uptake measurements, respectively. With the value of $\tilde{D}^o$ in Fig. 15, the value of $Q_{Ag}/t^{\frac{1}{2}}$ in Fig. 16(b) at 267°C, and the experimentally determined value of C_o, $-\frac{1}{C_o}(\partial \bar{c}_{Ag}/\partial w)_{w=o}$ was calculated from Eq. (13) as 0.491. Interpolation in Table 12 gave $D_{Na}/D_{Ag} = 2.0$, in good agreement with the value obtained by fitting the concentration profile.

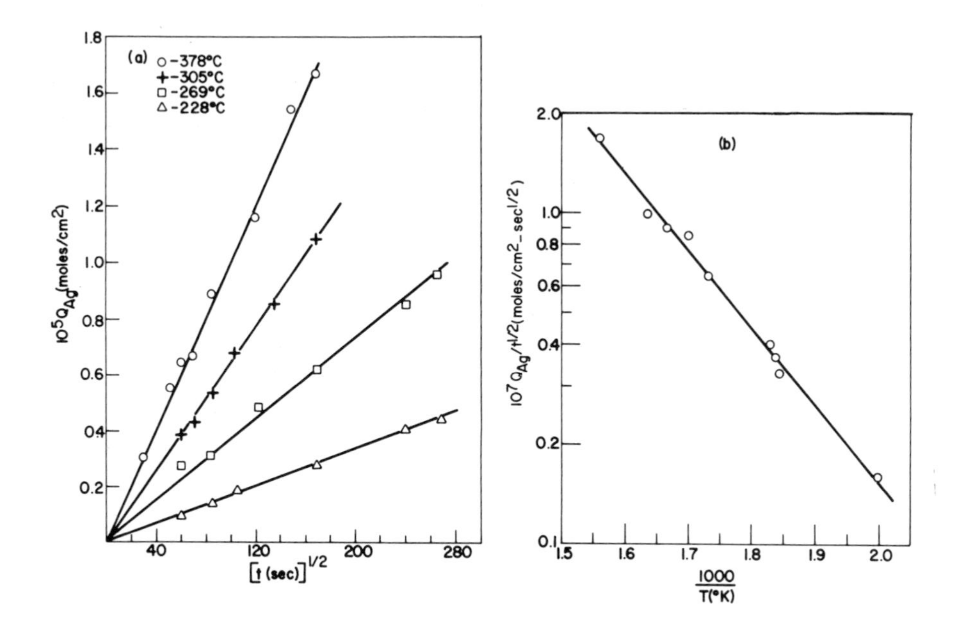

FIG. 16. Uptake of silver in exchanger D as a function of: (a) square root of time; (b) temperature.

Residual profiles described in Sec. IV.B.3. are shown in Fig. 17(a) for Na^+-Li^+ exchange at 349 and 394°C in the glass-ceramic exchanger B (66). The curves were calculated at both temperatures with constant interdiffusion coefficients. From the data in Fig. 17(b), the activation energy for interdiffusion is about 20 kcal/mole. The exchange rate is greater for the glass than the glass ceramic at all temperatures investigated, as indicated by the results in Fig. 18. Good agreement was obtained for the interdiffusivity determined from residual-concentration profiles with Eq. (17) and uptake results with Eq. (12), thus reinforcing the conclusion that D_{Li}/D_{Na} is unity. Furthermore, the resistance of the sample before and after Na^+-Li^+ exchange at 427°C for 4 hr was virtually unchanged, again indicating that lithium and sodium ions have identical mobilities.

The interdiffusion results are summarized in Table 7 for the exchangers in Table 1. One feature common to most of the exchange systems discussed in this section is the applicability of Eq. (23). This result would appear to confirm the assumption of equal mobilities in self-diffusion and interdiffusion used to derive Eq. (22) and the assumption of constant self-diffusion coefficient with ionic fraction. However, the latter assumption has been

TABLE 7

Summary of Results of Kinetic Studies

Exchanger	Ions A-B	Temperature, °C	D_B/D_A	$\bar{D}\left(\frac{cm^2}{sec}\right)$	$E^{\ddagger}_{A-B}$ (kcal/mole)
A	Na-Li	400	1.3	8.1×10^{-9} [c]	20 [u]
A	Ag-Li	289	19	6.7×10^{-9} [c]	
				6.0×10^{-9} [u]	20 [u]
A	K-Li	494	30	5×10^{-11} [c]	--
				7×10^{-11} [u]	29 [u]
A	Rb-Li	565	--	2×10^{-11} [u]	46 [b]
					58 [s]
A	Cs-Li	560	--	(10^{-9}) [u]	(10) [u]
C	K-Na	500	1.8	5.3×10^{-10} [c]	23
				5.3×10^{-10} [u]	24
		375	1.0	3.5×10^{-11} [c]	--
D	Ag-Na	267	2.3	2.4×10^{-11} [u]	21
				2.1×10^{-11} [c]	
		354	2.2	2.4×10^{-10} [c]	19
B(gl)	Na-Li	500	--	1.6×10^{-8} [u]	21
B(gl)	K-Li	500	--	2.6×10^{-11} [u]	26
B(cr)	Na-Li	393	1.0	8.6×10^{-10} [c]	20 [c]
				8.9×10^{-10} [u]	21 [u]
B(cr)	K-Li	500	--	5.4×10^{11}	27

[c] Integral-mean value obtained from c-x data with Eq. (27).

[u] Mean value obtained from uptake with Eq. (12).

[s] Surface; [b] Bulk.

gl = glass; cr = crystal.

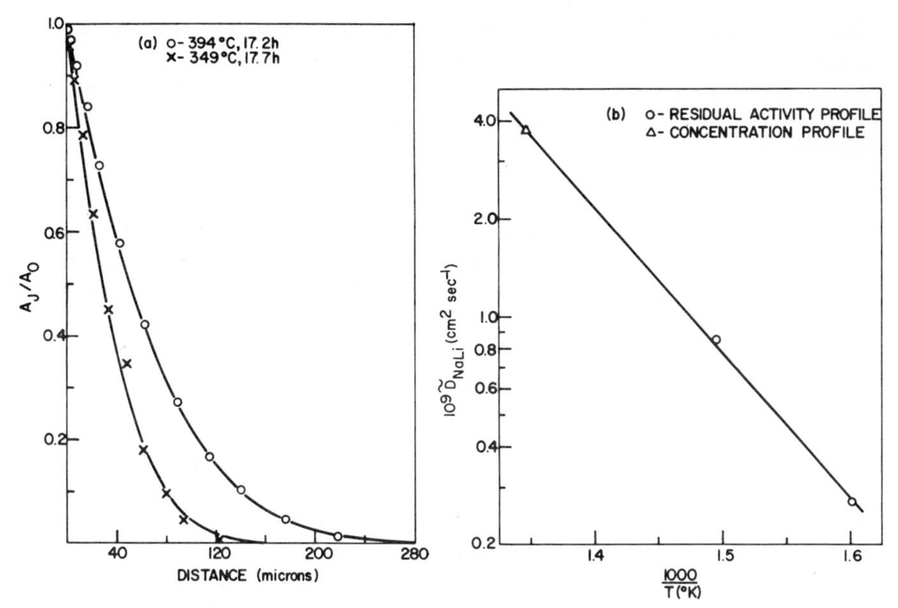

FIG. 17. Sodium exchange in the crystalline exchanger B: (a) residual-concentration profiles; (b) temperature dependence of the interdiffusion coefficient.

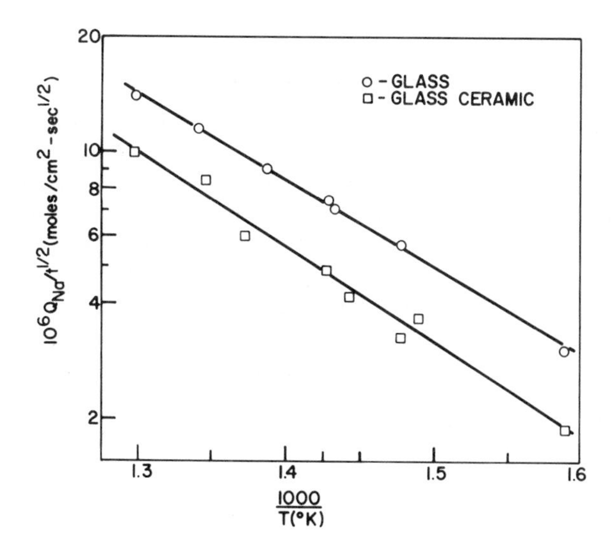

FIG. 18. Comparison of sodium uptake with temperature for the glass and crystalline forms of exchanger B.

shown to hold only at the two extremes of concentration. A more definitive experiment to test this assumption, which would involve self-diffusion measurements in samples equilibrated to different levels of the counterions, was tried but was unsuccessful, because of the difficulties associated with sample preparation.

The results for exchanger A in Table 7 show that the interdiffusion coefficients decrease in the order $Ag^+ > Na^+ > K^+ > Rb^+$, corresponding to an increasing ionic radius except for silver. However, the results suggest that the ionic mobililities relative to lithium decrease with increasing ionic radius even for silver. Cesium appears to be quite anomolous. The activation energies listed in Table 7 for the various exchange reactions are close to those reported for ion exchange on different zeolites (37). The diffusion-coefficient or mobility ratios in Table 7 can be compared with other reported values (16). In general, the mobility ratios are smaller for these exchangers shown in Table 7 than those given in Ref. 16

The results presented in this section indicate that a potential gradient, given by Eq. (21), is set up in the exchanger when a foreign ion of mobility different from the ion originally present diffuses into the exchanger. This point, which is pertinent to the selectivity of fixed-site membrane electrodes, is discussed in some detail in the next section.

V. ION-EXCHANGE MEMBRANE POTENTIALS

A. Introduction

When an ion exchanger is placed in contact with a solution with which it can exchange ions, an electrical potential is set up across the exchanger. This potential can be separated into two parts, viz., a boundary or Donnan potential and a diffusion potential. The Donnan potential results from a difference in the activities of the exchangeable ions between the solution and exchange phases, while the diffusion potential arises from differences of ionic mobilities in the exchanger (32, 64, 65).

If two different ions exchange across an exchanger membrane separating two solutions, the boundary potential for an ideal exchanger is (32)

$$V_s = \frac{RT}{F} \ln \frac{a_A + K_{AB} a_B}{a'_A + K_{AB} a'_B} \tag{28}$$

where a and a' are the ionic activities in the two solutions, and K_{AB} is the exchange constant given by Eq. (1).

There is no diffusion potential in the exchanger with only one kind of ion in the system. However, if two cations of different mobilities interdiffuse in the exchanger, a potential gradient arises in the exchanger as shown by Eq. (21). The diffusion potential V_D is (16,17)

$$V_D = \int_0^L E dx = \frac{RT}{F} \int_{\bar{N}_A(0)}^{\bar{N}_A(L)} \frac{(u_B/u_A - 1)}{\bar{N}_B(u_B/u_A) + \bar{N}_A} \frac{\partial \ln \bar{a}_A}{\partial \ln \bar{c}_A} d\bar{N}_A \tag{29}$$

This equation can be integrated, if the exchanger follows Eq. (2), i.e., $\partial \ln \bar{a}_A / \partial \ln \bar{c}_A = n$, and if the mobility ratio u_B/u_A is independent of concentration, to give

$$V_D = -\frac{nRT}{F} \ln \frac{\bar{N}_A(L) + \bar{N}_B(L)(u_B/u_A)}{\bar{N}_A(O) + \bar{N}_B(O)(u_B/u_A)} \tag{30}$$

where the $\bar{N}$'s are the ionic mole fractions in the glass at the surface $x = 0$ and $x = L$. By means of Eq. (1) for K_{AB} the diffusion potential is related to the ionic activities in the external solutions. Therefore, for $n = 1$ and the same solvent in the two solutions, the diffusion potential is (32)

$$V_D = -\frac{RT}{F} \left[\ln \frac{a'_A + K_{AB}(u_B/u_A)a'_B}{a_A + K_{AB}(u_B/u_A)a_B} - \ln \frac{a'_A + K_{AB}\, a'_B}{a_A + K_{AB}\, a_B} \right] \tag{31}$$

The total membrane potential is obtained from the sum of Eqs. (28) and (31) to give (32)

$$V = \frac{RT}{F} \ln \frac{a_A + K_{AB}(u_B/u_A)a_B}{a'_A + K_{AB}(u_B/u_A)a'_B} \tag{32}$$

Eisenman tested the validity of Eq. (32) by measuring K_{AB}, u_B/u_A, and the membrane potential for a 27 Na_2O, 4 Al_2O_3, 69 mole % SiO_2 glass selective for potassium (64). He found reasonable agreement with Eq. (32) for those properties measured directly and indirectly.

Molten salts provide an alternative environment to water for studies with ion-exchange membranes because the complicating effect of the hydration of the membrane is avoided. The membrane behavior of Corning Code 7740 borosilicate glass bulbs in molten sodium nitrate-potassium nitrate is discussed in this section in terms of the ion-exchange model for membrane potentials as represented by Eq. (32). These results are compared with the ion-exchange properties of the glass itself. Preliminary results with other glass compositions in fused sodium nitrate-potassium nitrate are presented and, whenever possible, compared with their behavior in aqueous solution.

B. Methods of Measurement

Potential measurements were conducted on the cell

$$Ag/AgBr(s)/\frac{KBr}{KNO_3}/\ frit/\frac{KNO_3\ \ (N_K)}{NaNO_3\ \ (N_{Na})}//\ glass/\frac{KBr}{KNO_3}/\ AgBr(s)/Ag$$

The details of the measurement have appeared in the literature (13). The emf was followed after each addition of sodium nitrate, and the steady value, but not necessarily the steady-state value, was recorded. This steady value was reached very quickly with the initial additions; however, it took longer to reach the steady value with larger additions because of the time required for the temperature to reestablish itself. The concentration range covered was up to 30 mole % sodium nitrate in potassium nitrate.

In addition to the potential measurements, the transference numbers of potassium and sodium into the borosilicate glass were determined as a function of temperature and composition. The technique used (67) was similar to that reported by Keenan and Duewer (68). The transport numbers were determined by measuring the change in composition of the borosilicate bulb filled with a silver nitrate-sodium nitrate melt and immersed in a sodium nitrate-potassium nitrate melt, after the passage of a known amount of current for a known amount of time.

C. Results

1. Membrane Potentials

Conditions were maintained for the cell so that $a'_K \approx 1$. The membrane potential of the cell, then, is

$$V = \frac{2.303\ RT}{F}\ \log\ (a_K + k_{KNa}a_{Na}) \tag{33}$$

where $k_{KNa} = (u_{Na}/u_K)K_{KNa}$. The quantity k_{KNa} is called the selectivity constant; its value is a quantitative measure of the electrode preference. The measured cell potential never exceeded 1 mV with just potassium nitrate in the cell; this value was subtracted from the potential measured when sodium nitrate was added.

A value of k_{KNa} was determined for each concentration according to

$$k_{KNa} = \frac{\exp(F\Delta V/RT) - a_K}{a_{Na}} \tag{34}$$

where ΔV is the measured cell potential minus the small initial potential recorded with no sodium nitrate present. If $n = 1$, k_{KNa} should be constant. Figure 19 shows some typical results for the borosilicate glass electrodes with the average values of the selectivity constant $\bar{k}_{KNa} = 17.7 \pm 0.9$ and 23.2 ± 1.4 at 360° and 374°C, respectively. The rational activities were calculated from the mole fraction and calculated activity coefficients. The fit verifies $n = 1$ over the concentration range, which extends to about 18 mole % sodium nitrate in potassium nitrate.

The effect of temperature on the selectivity constant of the borosilicate membrane is shown in Fig. 20. Each point was obtained with a different electrode. The cell used by Keenan et al. (10) was different from that used in Ref. 13. Therefore, their result at 350°C was normalized to a selectivity of unity for potassium and is included in Fig. 20 for comparison. The temperature coefficient of $\bar{k}_{KNa}$ was found to be 14.0 ± 0.7 kcal/mole from the data in Fig. 20. Preliminary results indicate the selectivity order of the borosilicate membrane at 360°C is Na ≈ Ag > Li > K.

The behavior of several other glass membranes was investigated in addition to the borosilicate glass, although not as extensively. The results obtained with composition C is shown in Fig. 21(a) at 361 and 367°C. The highest concentration used was about 10 mole % sodium nitrate in potassium nitrate. Again, increasing the temperature results in greater preference of the membrane for sodium. The results obtained with 11 Na_2O, 18 Al_2O_3, 71 mole % SiO_2 are shown in Fig. 21(b) at 359°C; the highest concentration is 5 mole % sodium nitrate in potassium nitrate. This glass, which is known to be highly selective towards sodium in aqueous solution (64), shows electrode preference to sodium in molten salts also. On the other hand, preliminary results show that 27 Na_2O, 3 Al_2O_3, 70 mole % SiO_2, a potassium selective electrode in aqueous solution (64), is sodium selective in molten potassium nitrate-sodium nitrate.

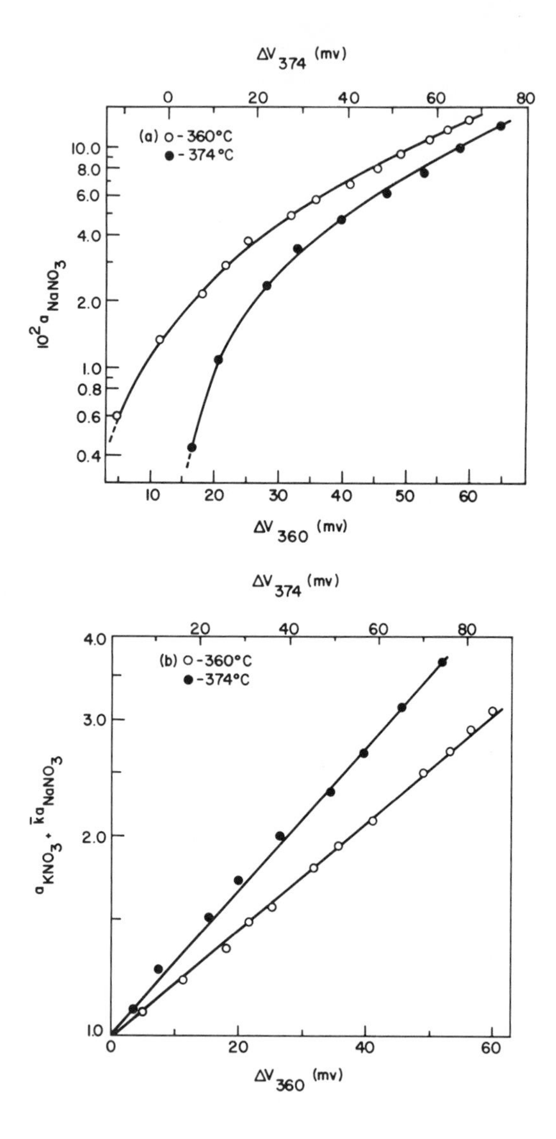

FIG. 19. Sodium response of the borosilicate membrane electrode: (a) plotted versus activity; (b) plotted according to Eq. (33).

The results presented for the borosilicate glass, glass C, and the 11 Na_2O, 18 Al_2O_3, 71 mole % SiO_2 glass indicate that n is unity; it has already been shown that n $\approx$1 for glass C at 500°C from the exchange isotherm (see Table 6). Figure 22 shows the response of the glass A with added sodium

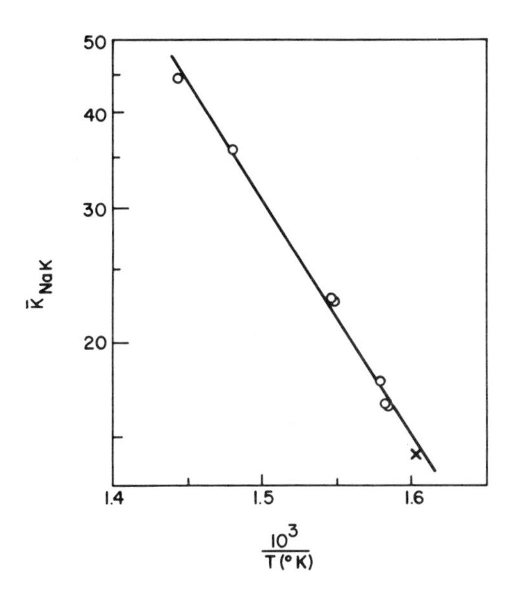

FIG. 20. Temperature dependence of the average selectivity constant $\bar{k}_{KNa}$ for the borosilicate membrane: o, Ref. 13; x, Ref. 10.

nitrate at 371°C; these data extend out to 25 mole % sodium nitrate in potassium nitrate. The apparent selectivity of this membrane is greater than 300. However, the selectivity is a function of concentration, as determined from Eq. (34), indicating a selectivity factor different from unity (see Table 6).

While little perceptible strain is introduced by K^+-Na^+ exchange in the borosilicate glass, the strain level of the 11 Na_2O, 18 Al_2O_3, 71 mole % SiO_2 glass and glass C are affected appreciably by the ion exchange (4). Yet the results for all three compositions indicate that any surface stresses present are steady with time, while the potential itself is relatively time independent. If there are pressure gradients in the glass, it can be shown under certain simplifying assumptions that a term of the form

$$V_p = \frac{\bar{V}}{F}(P' - P)$$

would have to be added to the total potential (17), where $\bar{V}$ is the partial molar volume of the two ions (assumed to be equal), P' is the pressure of the glass surface in contact with the solution in the reference compartment, and P is the pressure at the glass surface in contact with the external

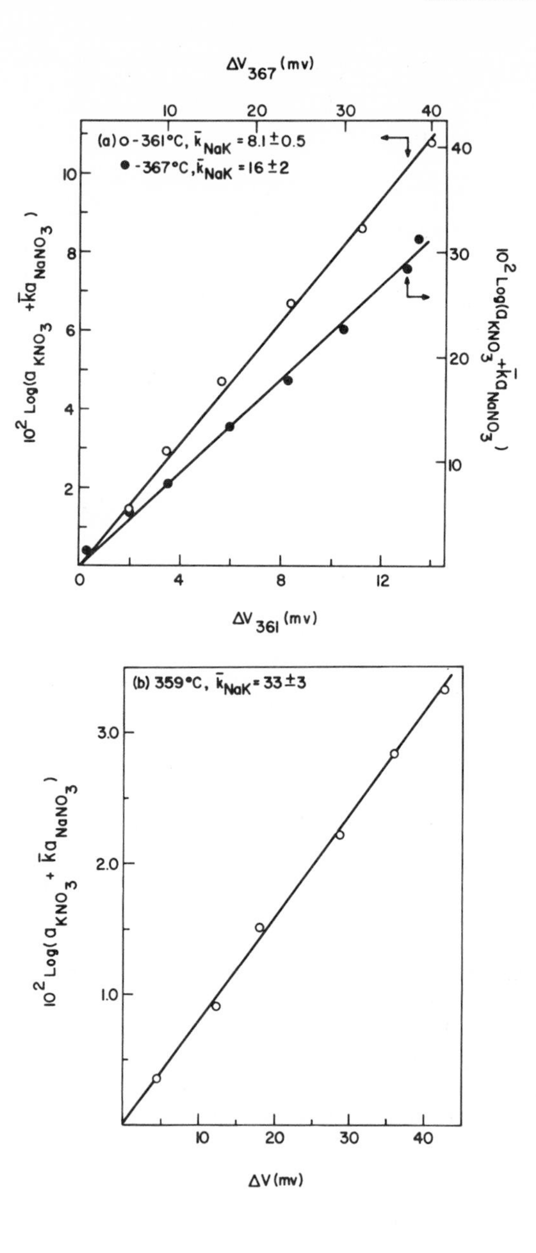

FIG. 21. Sodium response plotted according to Eq. (33) of: (a) exchanger C; (b) 11 Na_2O, 18 Al_2O_3, 71 mole % SiO_2.

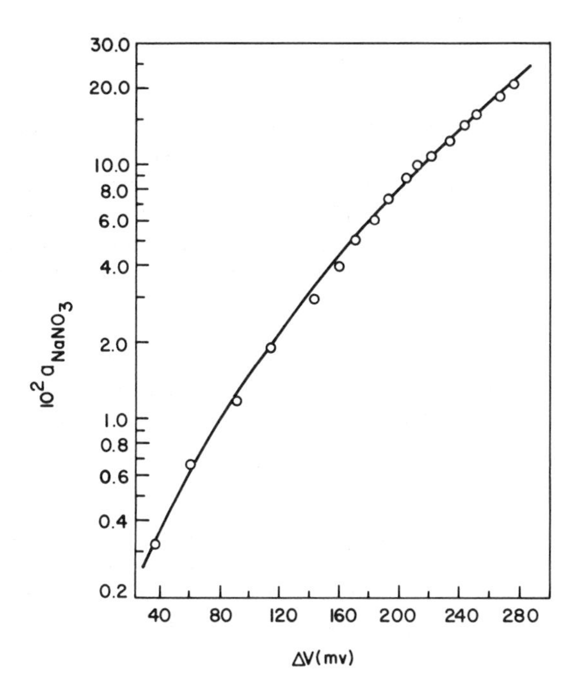

FIG. 22. Sodium response of exchanger A at 371°C.

solution. Surface stresses of the order of 6 x 10^9 dyn/cm are common in these exchange systems with alumina glasses (29). Although P is a function of the level of exchange at the surface, the concentration range of sodium nitrate in potassium nitrate investigated with glass C and the 11 Na_2O, 18 Al_2O_3, 71 mole % SiO_2 glass is probably too narrow to permit observation of any pressure effects. The effect of pressure, however, is most likely important for the results shown in Fig. 22 for glass A. The high-alumina glasses will eventually crack on long-time storage in molten potassium nitrate, indicating very high internal tension.

The relative electrode preference of the 11 Na_2O, 18 Al_2O_3, 71 mole % SiO_2 and glass C is in agreement with Eisenman's anionic field strength predictions (69). Thus, the selectivity of the glass membrane towards sodium increases with increasing Al/Na ratio in the glass. One would expect the Eisenman approach to be more applicable in this case, since we are dealing here with essentially unhydrated surfaces. The fact that 27 Na_2O, 3 Al_2O_3, 70 mole % SiO_2 is more selective towards sodium than potassium, a situation that is the reverse of the aqueous case, emphasizes the need for considering the effects of the solvation energies in the reference solutions as well as possible effects of hydration in the glass in predicting membrane behavior.

Although the exchange-isotherm results with the borosilicate glass are consistent with the concept of phase separation, the bulk glass seems to have homogeneous potential properties, which are probably averages for the two phases. Once the surface equilibria are established, the potential is independent of time. Moreover, these results point up the difficulty in trying to determine whether or not a glass membrane is homogeneous from potential measurements alone.

The large temperature dependence of the selectivity of the borosilicate membrane in molten sodium nitrate-potassium nitrate contrasts rather sharply with Doremus' results in fused sodium nitrate-silver nitrate (11), for which he found almost no variation of membrane selectivity with temperature.

The values of $K^{\prime}_{NaK}$ given in Table 4 can be used in the equations for the membrane potentials of the borosilicate glass in molten sodium nitrate-potassium nitrate (41). However, if the glass is considered as a single, nonideal phase, then the term $\partial \ln \bar{a}_K / \partial \ln \bar{c}_K$ in the integral for the diffusion potential is not a constant (17) so that an integral of the general form

$$V_D = \int_{\bar{N}_{A,1}}^{\bar{N}_{A,2}} \frac{(u_B/u_A - 1)\,\bar{N}_A}{\bar{N}_B(u_B/u_A) + \bar{N}_A} \, \frac{\partial \ln \bar{a}_A}{\partial \alpha} \, d\alpha \tag{35}$$

would have to be evaluated. On the other hand, if it is assumed that only the sodium-rich phase in the two-phase borosilicate glass is conducting, i.e., potential determining, the mobility ratio u_{Na}/u_K can be estimated from the data in Fig. 20 and Table 4. The values of the mobility ratio shown in Table 8 were calculated on the assumption that $u_{Na}/u_K = \bar{k}_{KNa} K_{II}$.

Although there are only few systematic reports on the variation of the mobility ratio with temperature, the trend shown in Table 8 does agree with the results reported for u_{Na}/u_K in soda-silica and u_{Na}/u_{Ag} in soda-lime-silica glasses (17). Doremus' results for the borosilicate glass in molten sodium nitrate-silver nitrate (11) indicate that the mobility ratio u_{Na}/u_K and the equilibrium constant K' are relatively temperature insensitive.

From the value of 22 kcal/gm-ion for the activation energy of the self-diffusion of sodium in the borosilicate glass (see Fig. 5), we must

TABLE 8

The Apparent Mobility Ratio $u_{Na}/u_K = \bar{k}_{KNa} K_{II}$ as a Function of Temperature

Temperature, °C	u_{Na}/u_K
380	27
435	79
465	118

conclude from the results in Table 8 that the activation energy for the self-diffusion of potassium in the borosilicate glass is small, being about 8 kcal/gm-ion. Thus, the slower ion, which is also the larger ion, has a smaller activation energy than the faster ion. However, studies of the electrical conductivity in the end members prepared by melting would be expected to indicate that the ions have about the same activation energies for self diffusion, although the absolute rates would be different (16). This emphasizes the fact that the structure of the glass is determined by its composition during melting at high temperatures. Once the glass is cooled below the glass transition temperature, the structure remains the same, even if ion exchange occurs. Data on the density variation in an ion-exchanged glass indicates that some relaxation does take place during the course of ion exchange. De Waal (70) has suggested that there is some relaxation process that occurs in ion-exchanged glass upon reheating at temperatures well below the transformation range. This most likely related to the relaxation phenomenon discussed recently with respect to strengthening (6).

While it seems reasonable, there is no direct experimental justification for the assumption that the high-soda phase conducts most of the current, so that the above discussion is quite speculative. Since the electrode potential of the borosilicate glass might still be constant, even if the glass is not ideal and the mobility ratio is a function of concentration, the direct measurement of u_{Na}/u_K in the borosilicate glass is required.

2. Transference Numbers

Recently Keenan and Duewer (68) reported that the transference numbers of sodium and silver ions into the borosilicate glass at 318° are equal to their respective mole fractions in the binary nitrate melt. They explained this result in terms of a junction model without surface exchange, in contrast to their earlier membrane potential studies, which required an ion-exchange model (9,10). However, one does not necessarily need to use a junction model to explain their results. The transport of sodium and potassium in the borosilicate glass will be examined first.

The transference numbers of potassium and sodium were determined as described in Sec. V,B. In calculating the transference numbers it was assumed from the equilibrium studies reported in Sec. III,C.3. that potassium would exchange and carry current only through the borosilicate phase; sodium would exchange and carry current through both phases. Therefore, it was assumed that the fraction of sodium transported through phase I, the major phase (i.e., high in silica), is equal to N_I, the mole fraction of exchange sites in phase I.

The transference number of potassium $t_{K,II}$ in phase II is given in Table 9 as a function of the mole fraction of potassium in the molten salt at 354°C. Thus, the transference number is a strong function of bath composition, and any junction model for membrane potentials of the borosilicate glass would have to be consistent with this concentration dependence.

TABLE 9

Transference Data for the Borosilicate Glass at 354°C

N_{KNO_3}	Current, in Faradays	$t_{K,II}$
0.400	7.9432	0.0165
0.500	6.0456	0.0278
0.600	6.1485	0.0455
0.700	6.0541	0.0672
0.800	3.6965	0.166
0.800	1.4750	0.140
0.900	1.1727	0.303
0.942	1.2170	0.457
0.950	2.3970	0.514
0.990	0.5854	0.791
0.990	0.6356	0.812

The transport number of sodium in phase II is

$$t_{Na,II} = \frac{\bar{N}_{Na,II} u_{Na,II}}{\bar{N}_{Na,II} u_{Na,II} + \bar{N}_{K,II} u_{K,II}} \tag{36}$$

Since it was assumed that the borosilicate phase is thermodynamically ideal

$$K_{II} = \frac{\bar{N}_{K,II} a_{Na}}{\bar{N}_{Na,II} a_K} \tag{37}$$

Combining Eqs. (36) and (37) yields the ratio of the transference numbers in the borosilicate phase as

$$\frac{t_{K,II}}{t_{Na,II}} = \frac{a_K K_{II}}{a_{Na} \beta} \tag{38}$$

where $\beta = u_{Na}/u_K$, so that the electrode selectivity is given by

$$k_{KNa} = \frac{t_{Na,II} a_K}{t_{K,II} a_{Na}} \tag{39}$$

Because of the inconsistency in the definition of the electrode selectivity k_{KNa}, it was erroneously stated in Ref. 67 that the mobility ratio β could be determined from the measured transference numbers.

Thus, from Eq. (39) a plot of log $(t_{K,II}/t_{Na,II})$ against log (a_K/a_{Na}) should have a slope of unity if k_{KNa} is independent of composition. Such a graph is shown in Fig. 23. The slope of the line is 1.02, indicating that k_{KNa} is indeed constant. However, the value of $k_{KNa} = 32.4$ obtained from the data in Fig. 23 does not agree very well with the value of $k_{KNa} = 15.6$ from the results in Fig. 20.

The transference numbers of potassium and sodium into the borosilicate glass were determined over the temperature range 235-401°C in an equimolar potassium nitrate-sodium nitrate bath. The results are summarized in Table 10. While there is some scatter, the general trend is for $t_{K,II}$ to decrease with increasing temperature. Although there was a great deal of scatter, the electrode selectivity constant k_{KNa} was found from the data in Table 10 to increase with increasing temperature. The temperature

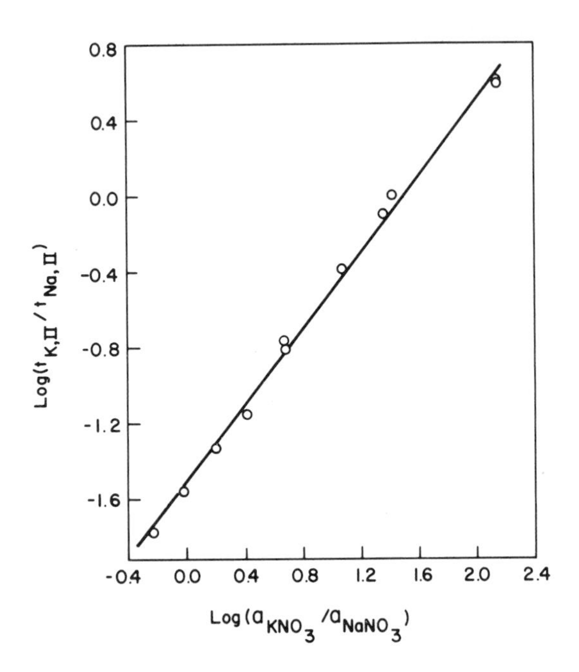

FIG. 23. Dependence of the transference numbers of potassium and sodium in the borosilicate phase on bath composition.

coefficient of the selectivity constant was estimated to be about 10 kcal/mole from these data as compared to 14 kcal/mole from the results in Fig. 20.

Now let us examine the results of Keenan and Duewar on the transference numbers of sodium and silver into the borosilicate glass. Doremus (41) showed that the ion-exchange selectivity coefficient for Ag^+-Na^+ exchange in the borosilicate glass at 335°C is essentially constant ($K_{NaAg} \approx 2$) from about 0.05-0.40 mole fraction silver nitrate in the melt; below 0.05 mole fraction silver nitrate the selectivity coefficient depended on concentration. With the assumption that the selectivity coefficient is constant over the range 0.1-0.9 mole fraction silver nitrate, Keenan's and Duewer's results are plotted according to the logarithmic form of Eq. (39) in Fig. 24 in the same way that the potassium nitrate-sodium nitrate results for the borosilicate glass were treated. The line drawn through the points was obtained by the least-squares method, which yielded a slope of 1.06 and a value of 1.01 ± 0.03 for k_{NaAg}, which is in good agreement with the result reported

TABLE 10

Transference Numbers of Potassium as a Function of Temperature in the Equimolar Potassium Nitrate-Sodium Nitrate Melt

Temperature, °C	$t_{K,II}$
235	0.187
237	0.175
237	0.109
248	0.142
250	0.0881
258	0.0823
269	0.0450
269	0.0375
286	0.0549
303	0.0436
304	0.0529
311	0.0375
345	0.0275
353	0.0278
378	0.0372
381	0.0259
381	0.0328
401	0.0208

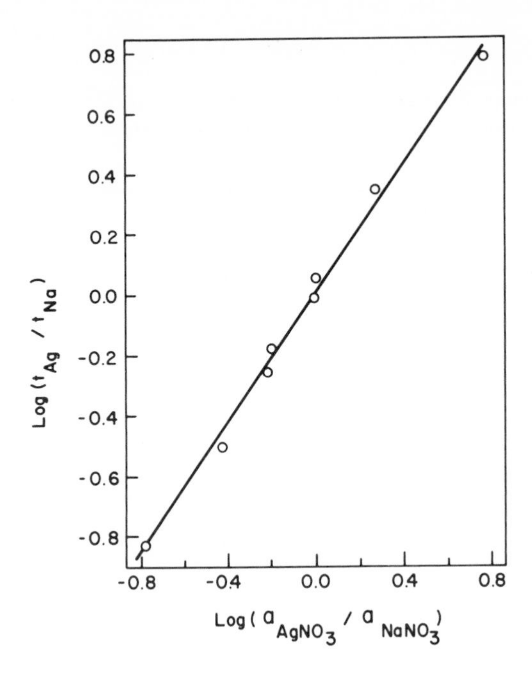

FIG. 24. Dependence of the transference numbers of silver and sodium on bath composition from the data in Ref. 68.

by Doremus (11) for the borosilicate glass. The transference number of sodium in the borosilicate glass is given by

$$t_{Na} = \frac{1}{1 + k_{NaAg}(a_{Na}/a_{Ag})} \tag{40}$$

Inspection of Eq. (40) shows that when k_{NaAg} is unity, the transference number of sodium is approximately equal to its mole fraction, and similarly for the transference number of silver. This is the result found by Keenan and Duewer. Thus, the ion-exchange model is consistent with both the emf and transference studies.

This analysis for Ag^+-Na^+ exchange neglects the two-phase nature of the borosilicate glass. Doremus (58) reported from the results of a study of diffusion of sodium and silver ions into the borosilicate glass at 335°C that although the tracer-diffusion coefficient of sodium is essentially independent of concentration in the glass, the tracer-diffusion coefficient of silver decreases when $\bar{N}_{Ag}$ in the glass is less than about 0.3. It was suggested that this behavior is related to the two-phase nature of the borosilicate glass. Although the high-silica phase rejects potassium, it has a very high affinity for silver. At low silver concentrations most of the silver is in the major phase. Tighter binding of the silver ions in this phase leads to lower diffusion coefficients of the silver. At high concentrations of silver most of the silver is in the borosilicate phase in which it has the same diffusion coefficient as sodium. Thus, it is evident from the different behavior of silver and potassium how the two-phase nature of the borisilicate glass affects the overall exchange process. In the next section some of the assumptions of the ion-exchange model for the membrane potentials observed with the borosilicate glass are examined further, in terms of the ion-exchange properties of the borosilicate membrane.

3. Ion-Exchange Properties

Change in membrane resistance, amount of potassium taken up (as described in Sec. IV, B), and the self-diffusion results shown in Fig. 5 were used to study ion-exchange kinetics in the borosilicate glass. The effect of temperature on the change in resistance of the exchanged borosilicate glass is shown in Fig. 25. The variation in the amount of potassium taken up by the borosilicate glass with temperature is shown in Fig. 26; the experimental points on the dotted line were taken above the strain point of the borosilicate glass. A similar discontinuity was reported by Wilson and Carter (71) for the temperature dependence of the tracer-diffusion coefficient in the borosilicate glass. The two-phase borosilicate glass appears to have ion-exchange properties that are averages for the two phases. A summary of the results for the borosilicate glass membrane is shown in Table 11. The functional form of the results does not differ from that obtained with homogeneous glass. Thus, from these data alone it is difficult to infer anything about phase separation in the borosilicate glass.

For the moment, let us focus our attention on the electrical properties of a homogeneous ion exchanger. More specifically, consider potassium-for-sodium ion exchange in which the glass is originally in the sodium form. If we assume that the total conductivity of the exchanger during or after ion exchange is the sum of the contributions from the sodium form of the exchanger and the potassium form of the exchanger, then at any point x

$$\sigma_T = \bar{N}_{Na}\sigma_{Na} + \bar{N}_K\sigma_K$$

where $\bar{N}$ is the mole fraction, and $\sigma_{Na} = \sigma_o$ is the specific conductivity of the sodium form of the glass, and σ_K is the specific conductivity of the potassium form of the glass. If it is assumed further that the self diffusion

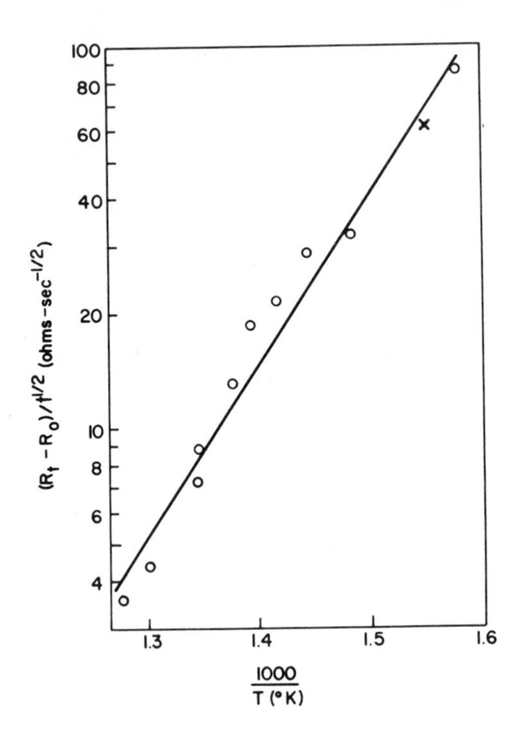

FIG. 25. Temperature dependence of the change in resistance of the exchanged borosilicate glass: x, during the course of exchange; o, exchanged glass disks.

of both sodium and potassium ions follows Eq. (18) with the same efficiency, then $\sigma_{Na}/\sigma_K = D_{Na}/D_K = \beta$. With these assumptions it can be shown from Eq. (15) that (36)

$$R_t - R_o = -\frac{(D_{Na}t)^{\frac{1}{2}} R_o}{L} \int_0^{L/(1)D_{Na}t^{\frac{1}{2}}} \frac{N_K(1-\beta)}{\beta + \bar{N}_K(1-\beta)} \, dy \tag{41}$$

If the fractional exchange of potassium at the surface is unity, and the glass is an ideal exchanger so that $\partial \ln \bar{a}/\partial \ln \bar{c} = 1$, then from Eq. (23) $\tilde{D}^o = D_{Na}$ and w = y in Eq. (13). The integral in Eq. (41), which we shall call $F(\beta)$, and the quantity $\frac{1}{C_{o,K}} (\partial \bar{c}_K/\partial w)_{w=o}$ were calculated for various values of the mobility ratio β and the diffusion-coefficient ratio $\tilde{D}^o/\tilde{D}^\infty$, respectively, with the aid of Fujita's solution (49) to the problem of diffusion in a

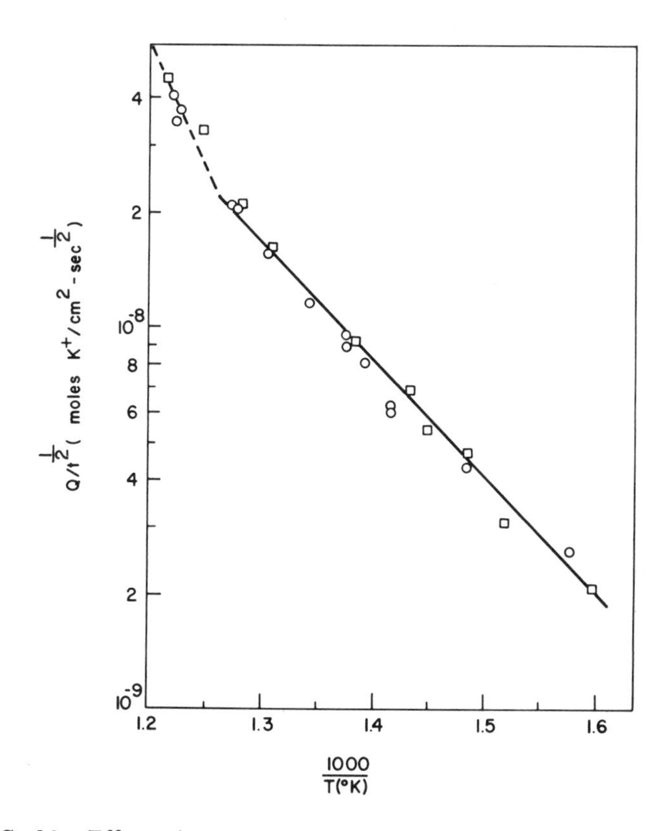

FIG. 26. Effect of temperature on the uptake of potassium in the borosilicate glass.

semi-infinite medium with variable diffusion coefficients. The quantity $\tilde{D}^{\circ}$ is, as before, the interdiffusion coefficient at the surface, and $\tilde{D}^{\infty}$ is the interdiffusion coefficient at $x = \infty$. These calculations were made with the IBM 360 computer, and some of the results are shown in Tables 12 and 13. The quantity $C_{0,2}$ in Table 12 is the equilibrium concentration of the foreign ion per unit volume of exchanger. In addition, the computer results showed that the integral $F(\beta)$ is a strong function of the fraction of surface exchange, especially as the mobility ratio β becomes very large.

Let us return now to an examination of the uptake and resistance results in terms of Eqs. (13) and (41) and the computer results in Tables 12 and 13. Doremus (36) reported good agreement for β determined from concentration profiles, weight change, and resistance change results for silver-for-sodium exchange in an ostensibly homogeneous soda-lime-silica

TABLE 11

Summary of Results for the Borosilicate Glass Membrane in Molten Sodium Nitrate-Potassium Nitrate Presented as:
$\log q = a + b/T$

Quantity, q	a	$-10^{-3}b$
■Electrode selectivity constant, k_{KNa}	6.063	3.053
■Mobility ratio, $\beta = u_{Na,II}/u_{K,II}$	7.06	3.67
▲ Potassium uptake, $Q/t^{\frac{1}{2}}$	-3.755	3.088
▲ Tracer-diffusion coefficient, D_{Na}	-2.638	4.750
▲ Dc resistivity	-1.075	-4.729
▲ Resistance change, $\Delta R/t^{\frac{1}{2}}$	-5.272	-4.600

■ Ref. 13; ▲ Ref. 46.

glass at 374°C, as already discussed in Sec. IV,C.2. Since in our case we do not have complete exchange of potassium at the surface, $(\partial\bar{c}_K/\partial y)_{y=o}$ does not equal $(\partial\bar{c}_K/\partial w)_{w=o}$. However, with the definition of w and y and Eq. (23), Eq. (13) can be modified to give

$$-\frac{1}{C_{o,K}}\left(\frac{\partial\bar{c}_K}{\partial w}\right)_{w=o} = \frac{Q_K[\beta + N_{o,K}(1-\beta)]^{\frac{1}{2}}}{2(D_{Na}t)^{\frac{1}{2}}C_{o,K}} \tag{42}$$

A value of $\beta \approx 43$ was calculated for the bulk glass at 380°C by iterative solution of Eq. (42) with interpolation of the results in Table 12 and the available data given in Tables 3 and 11. This is in fair agreement with the value of $\beta = 27$ at 380°C from Table 8. Although the agreement at higher temperatures was not as good, at least the sign of the temperatures coefficient of β was the same as that reported previously. Agreement between the values of the mobility ratio β determined from the resistance-change

TABLE 12

Parameter Calculated for Eq. (13) as a Function of the Ratio $\tilde{D}^{\circ}/\tilde{D}^{\infty}$ for a Homogeneous Exchanger[a]

$\tilde{D}^{\circ}/\tilde{D}^{\infty}$	$-\frac{1}{C_{o,2}}\left(\frac{\partial \bar{c}_2}{\partial w}\right)_{w=0}$	$\tilde{D}^{\circ}/\tilde{D}^{\infty}$	$-\frac{1}{C_{o,2}}\left(\frac{\partial \bar{c}_2}{\partial w}\right)_{w=0}$
0.010	0.836	17.6	0.269
0.050	0.788	20.0	0.261
0.100	0.753	28.7	0.229
0.250	0.690	59.9	0.178
0.500	0.632	97.5	0.147
0.667	0.605	170	0.118
1.00	0.565	655	0.0685
1.26	0.540	1,190	0.0533
1.50	0.524	4,880	0.0290
1.90	0.496	9,060	0.0221
3.00	0.448	38,790	0.0115
4.51	0.403	73,050	0.00866
10.0	0.324	321,500	0.00440

[a]Subscript refers to the species exchanging into the glass.

data and the results in Table 8 was even poorer. Values of F(β) calculated from the experimental results shown in Table 11 are given in Table 14 as a function of temperature. Comparison of these results with those shown in Table 13 indicates that the mobility ratio β decreases with increasing temperature. This, of course, is in contrast to the behavior reported previously (13).

TABLE 13

The Integral $F(\beta)$ Calculated as a Function of the Mobility Ratio β, for Complete Exchange at the Surface and Ideal Behavior

β	$-F(\beta)$	β	$-F(\beta)$
0.100	-3.77	28.7	3.99
0.250	-1.89	59.9	5.27
0.500	-0.848	97.5	6.60
1.00	0.00	170	8.30
1.50	0.442	655	14.5
1.90	0.691	1,190	18.7
3.00	1.16	4,880	34.6
4.51	1.59	9,060	45.5
10.0	2.46	38,790	87.5
17.6	3.23	73,050	118
20.0	3.36	321,500	271

TABLE 14

Values of $F(\beta)$ for the Borosilicate Glass Calculated from the Data in Table 11

Temperature, °C	$-F(\beta)$
360	92.9
373	79.6
400	57.6
420	46.0
432	40.4
445	35.6
453	32.9
471	27.7
491	23.1
511	19.5

The two-phase nature of the borosilicate glass introduces some definite uncertainties with respect to any of the above calculations. The quantity Q, for example, was determined on the assumption that the entire surface was taking part in the exchange; i.e., all the sites on the surface are equivalent and uniformly distributed. It is apparent that this is not the case for the phase-separated borosilicate glass. In fact it may have been the nonuniformity of sites that led to some difficulty in the reproducibility of the potassium-uptake results.

Assume that the glass consists of two co-continuous phases with no tortuosity so that the path length is the same. Then, in this simplest case, the ratio of the surface area of each phase to the total surface area is equal to the volume fraction of that phase. For the temperatures of interest here, all the ion exchange occurs in the borosilicate phase. The quantity $-\frac{1}{C_{o,K}}\left(\frac{\partial \bar{c}_K}{\partial w}\right)_{w=0}$, then, is independent of this assumption, because the surface area term, which appears in both Q_K and $C_{o,K}$, cancels in Eq. (42). However, the situation is not so straight forward for the resistance change. Consider that the two phases represent parallel electrical paths through the glass. Then the conductivity at any point x for the exchanged glass is given by

$$\sigma_T = V_{II}(\bar{N}_{Na}\sigma_{Na} + \bar{N}_K\sigma_K) + V_I\sigma'_{Na}$$

where $\bar{N}_i$ is the mole fraction and σ_i is the conductivity in the borosilicate phase, σ'_{Na} is the conductivity in the silica phase, and V_I and V_{II} are the volume fractions of the silica and borosilicate phases, respectively. The conductivity of the unexchanged glass is given by

$$\sigma_{o,T} = V_{II}\sigma_{Na} + V_I\sigma'_{Na}$$

Thus, the conductivity at any point x can be rewritten as

$$\sigma_T = V_{II}\bar{N}_K(\sigma_K - \sigma_{Na}) + \sigma_{o,T}$$

since it is assumed that exchange occurs only in the borosilicate phase. The resistance change, then, as a result of ion exchange is given by

$$\Delta R = -\frac{(D_{Na}t)^{\frac{1}{2}}}{L}\int_{o}^{L/(D_{Na}t)^{\frac{1}{2}}} \frac{N_{II}\bar{N}_K(1-\beta)\,dy}{N_{II}\bar{N}_K(1-\beta)+\beta(N_{II}+N_I B')} \tag{43}$$

where β' is the ratio of the sodium mobility in the silica phase to the sodium mobility in the borosilicate phase. Unfortunately, numerical values of the integral in Eq. (43) go in the wrong direction to explain the observed results shown in Table 14.

Perhaps this discrepancy is more fundamental than that discussed previously. Although constancy of the mobility ratio with changes in concentration may be established, the dependence of self mobility on concentration is not known. If the mobility of sodium, say, were dependent on concentration, then this dependence would have to be known explicitly in order to evaluate numerically the integral $F(\beta)$. On the other hand, the uptake results are not particularly sensitive to changes in self mobility. Such a mole-fraction dependence of mobilities has been reported (72) in a thin, hydrated glass membrane in which it appears likely that the mobility ratio is constant, since the product of mobility ratio and ion-exchange equilibrium constant has been observed to be constant. This effect can be viewed as a sort of "lower-order" mixed-alkali effect and may be related to the build up of elastic stresses in the exchanged phase (73). This suggests the need to investigate the resistance of thin, equilibrated borosilicate membranes with symmetrical solutions on both sides as a function of bath composition.

VI. SUMMARY

The network structure of glass produces cation-exchange properties characterized by relatively low cationic mobility in a rigid anion "lattice." Ion-exchange equilibria, at least in homogeneous exchangers, between glass and molten salt, can be adequately described by a relation similar to that proposed by Kielland for aqueous exchangers. Interdiffusion in glass, which is more like solid than liquid diffusion, determines the rate of exchange. The Nernst-Planck equations were found to be applicable to interdiffusion in glass for a variety of exchange reactions. Diffusion of a foreign ion into the exchanger, with mobility different from the original ion, can produce an electric potential, which depends upon ionic concentrations in solution, mobilities in the glass, and the ion-exchange equilibrium characteristic of the system. The results with some of the glass membranes suggest the need for considering possible effects of hydration in the glass in predicting membrane behavior in aqueous solution. Although the effect of hydration is eliminated in these exchange systems, the build up of stress, caused by the exchange of ions of different size, is substituted as a complicating feature.

REFERENCES

1. R. H. Doremus, in Modern Aspects of the Vitreous State (J. D. MacKenzie, ed.), Vol. II, Butterworths, London 1962, p. 1.
2. H. E. Rauscher, in Symposium sur la Surface du Verre et ses Traitements Modernes, Union Scientific Continentale du Verre, Charleroi, Belgium, 1967, p. 115.
3. S. S. Kistler, J. Am. Ceram. Soc., 45, 59 (1962).
4. M. E. Nordberg, E. L. Mochel, H. M. Garfinkel, and J. S. Olcott, J. Am. Ceram. Soc., 47, 215 (1964).
5. A. J. Burggraaf and J. Cornilissen, Phys. Chem. Glasses, 5, 123 (1964).
6. H. M. Garfinkel, Glass Ind., 50, 28; 74 (1969).
7. H. M. Garfinkel, Appl. Opt., 7, 789 (1968).
8. E. Thilo and F. Wodtcke, Z. Anorg. Allgem. Chem., 295, 247 (1957).
9. K. Notz and A. G. Keenan, J. Phys. Chem., 70, 662 (1966).
10. A. G. Keenan, K. Notz, and F. Wilcox, J. Phys. Chem., 72, 1085 (1968).
11. R. H. Doremus, J. Electrochem. Soc., 115, 924 (1968).
12. R. H. Doremus, J. Phys. Chem., 72, 2877 (1968).
13. H. M. Garfinkel, J. Phys. Chem., 73, 1766 (1969).
14. K. H. Stern, J. Phys. Chem., 72, 1963 (1968); 74, 1323 (1970).
15. K. H. Stern, J. Phys. Chem., 74, 1329 (1970).
16. R. H. Doremus, in Ion Exchange (J. A. Marinsky, ed.), Vol. 2, Dekker, New York, 1969, p. 1.
17. R. H. Doremus, in Glass Electrodes for Hydrogen and Other Cations (G. Eisenman, ed.), Dekker, New York, 1967, p. 101.
18. K. H. Stern, Chem. Rev., 66, 335 (1966).
19. E. C. Freiling and M. H. Rowell, in Ion Exchange (J. A. Marinsky, ed.), Vol. 2, Dekker, New York 1969, p. 43.
20. M. Blander, ed., Molten Salt Chemistry, Interscience, New York, 1964.
21. B. R. Sundheim, ed., Fused Salts, McGraw-Hill, New York, 1964.
22. J. Lumsden, Thermodynamics of Molten Salt Mixtures, Academic Press, London, 1966.
23. G. O. Jones, Glass, Methuen and Co., London (1956).
24. S. D. Stookey, Ind. Eng. Chem., 51, 805 (1959).
25. S. D. Stookey, Glastech. Ber., V. Sonderband, Internationaler Glaskongress, 32K Heft 5, 1 (1959).
26. S. D. Stookey and R. D. Maurer, Progress in Ceramic Science, Vol. 2, Pergamon Press, Oxford, 1961.
27. J. P. Williams, C. B. Carrier, H. J. Holland, and F. J. Farncomb, J. Mater. Sci., 2, 513 (1967).
28. A. R. Cooper and D. A. Krohn, J. Am. Ceram. Soc., 52, 665 (1969).
29. H. M. Garfinkel and C. B. King, J. Am. Ceram. Soc., 53, 686 (1970).

30. H. M. Garfinkel, J. Phys. Chem., 72, 4175 (1968).
31. V. Rothmund and G. Kornfeld, Z. Anorg. Allgem. Chem., 103, 129 (1918).
32. G. Karreman and G. Eisenman, Bull. Math. Biophys., 24, 413 (1962).
33. J. Kielland, J. Soc. Chem. Ind., 54, 232T (1935).
34. F. Helferrich, Ion Exchange, McGraw-Hill, New York, 1962.
35. R. M. Barrer and J. D. Falconer, Proc. Roy. Soc., A236, 227 (1956).
36. R. H. Doremus, J. Phys. Chem., 68, 2212 (1964).
37. C. B. Amphlett, Inorganic Ion Exchangers, Elsevier, Amsterdam (1964).
38. R. E. Tischer, Rev. Sci. Instr., 37, 431 (1966).
39. R. J. Charles, J. Am. Ceram. Soc., 47, 559 (1964).
40. R. H. Doremus and A. M. Turkalo, Science, 164, 418 (1969).
41. R. H. Doremus, J. Phys. Chem., 72, 2665 (1968).
42. W. Haller, J. Chem. Phys., 42, 686 (1965).
43. T. H. Elmer, M. E. Nordberg, G. B. Carrier, and E. J. Korda, J. Am. Ceram. Soc., 53, 171 (1970).
44. R. M. Garrels and C. L. Christ, Solution, Minerals, and Equilibria, Harper and Row, New York, 1965.
45. R. E. Tischer, to be published.
46. H. M. Garfinkel, Phys. Chem. Glasses, 11, 151 (1970).
47. G. Eisenman, in Advances in Analytical Chemistry and Instrumentation (C. N. Reilley, ed.), Vol. 4, Wiley, New York, 1965, p. 213.
48. H. M. Garfinkel and H. E. Rauscher, J. Appl. Phys., 37, 2169 (1966).
49. J. Crank, Mathematics of Diffusion, Oxford University Press, 1956.
50. C. Guillemet, J. M. Pierre-dit-Méry, and A. Bonnetin, in Symposium sur la Surface du Verre et ses Traitements Modernes, Union Scientifique Continentale du Verre, Charleroi, Belgium, 1967, p. 181.
51. J. R. Johnson, R. H. Bristow, and H. H. Blau, J. Am. Ceram. Soc., 34, 165 (1951); ONR Report, Contract No. N6onr-22522, NR032-304 (1950).
52. D. A. Duke, J. F. MacDowell, and B. R. Karstetter, J. Am. Ceram. Soc., 50, 67 (1967).
53. H. M. Garfinkel, unpublished results.
54. K. Compaan and Y. Haven, Trans. Faraday Soc., 52, 786 (1956).
55. Y. Haven and J. M. Stevels, Travaux IVe Congres Inter. du Verre, Imprimerie Chaix, 20, rue Bergene, Paris, 1957, p. 343.
56. Y. Haven and B Verkerk, Phys. Chem. Glasses, 6, 38 (1965).
57. R. H. Doremus, Phys. Chem. Glasses, 10, 28 (1969).
58. R. H. Doremus, Phys. Chem. Glasses, 9, 131 (1968).
59. R. W. Heckman, J. A. Ringlien, and E. L. Williams, Phys. Chem. Glasses, 8, 145 (1967).
60. R. Terai, Phys. Chem. Glasses, 10, 146 (1969).
61. S. Urnes, Phys. Chem. Glasses, 8, 125 (1967).
62. F. Helferrich and M. S. Plesset, J. Chem. Phys., 28, 418 (1958).
63. F. Helferrich, J. Phys. Chem., 66, 39 (1962).

64. G. Eisenman, Glass Electrodes for Hydrogen and Other Cations, Dekker, New York, 1967, p. 133.
65. G. Eisenman, Biophys. J., 2, 259 (1962).
66. H. M. Garfinkel and R. F. Bartholomew, unpublished results.
67. H. M. Garfinkel, J. Phys. Chem., 74, 1764 (1970).
68. A. G. Keenan and W. H. Duewer, J. Phys. Chem., 73, 212 (1969).
69. G. Eisenman, Advan. Anal. Chem. Instr., 4, 213 (1965).
70. H. de Waal, Phys. Chem. Glasses, 10, 108 (1969).
71. C. G. Wilson and A. C. Carter, Phys. Chem. Glasses, 5, 111 (1964).
72. G. Eisenman, J. P. Sandblom, and J. L. Walker, Science, 155, 965 (1967).
73. S. Urnes, Glastech. Ber., 42, 11 (1969).

Chapter 5

NEGATIVE CONDUCTANCE AND ELECTRODIFFUSION IN EXCITABLE MEMBRANE SYSTEMS

D. Agin

Department of Physiology
University of Chicago
Chicago, Illinois

I. INTRODUCTION

The physical theory of electrical excitability in nerve fibers is rather undeveloped. The most important advance, made by Hodgkin and Huxley in 1952 (1), has been the separation of the ionic currents and the formulation of empirical equations to describe their time and voltage dependence. From the beginning, the fundamental theoretical problem has been to understand these ionic currents in terms of first principles, to relate them to membrane structure, and to develop a molecular theory to account for the effects on nerve cells of pharmacological agents. Although a great deal of experimental information has accumulated in recent years, theoretical progress has been disappointing.

Excitable membranes, both biological and synthetic, show a distinct negative conductance. In nerve and muscle cells this negative conductance is responsible for much of the functionally important behavior of the cell membrane, and an understanding of its physical origin is of great biological importance. I want to discuss here some aspects of the problem from a general point of view and present an explanation, different from any published previously, for the negative conductance behavior of both natural membranes and synthetic black lipid bilayers. There is much that is unsatisfactory about the theory, but if the result is only a stimulation of new research I will consider the effort to have been worthwhile.

II. NEGATIVE CONDUCTANCE IN NONBIOLOGICAL ELECTRODIFFUSION REGIMES

Current-voltage curves are force-response diagrams for electrical systems and provide extremely useful phenomenological information. In general, there are three types of I-V curve:

1. Monotonic loci

2. Loci with a voltage maximum or minimum

3. Loci with a current maximum or minimum

Most systems exhibit curves of type (1). Type (2) systems show a differential negative resistance, and can be said to be current-controlled, since current is a multi-valued function of voltage and must therefore be an independent variable. Type (3) systems show a negative differential conductance, and are voltage-controlled. A good example of a type (2) system is the negative resistance thermistor (2). The Esaki tunnel diode (3) is a type (3) system.

We are here concerned with type (3) systems, but more specifically with the causes rather than consequences of negative conductance behavior. It may be useful to mention briefly certain type (3) systems.

A. Tunnel Effects

Tunneling phenomena are usually considered with respect to specific types of electronic semiconductor regimes (e.g., Esaki tunnel diode), but the quantum theoretical principle which is involved is not restricted to particular kinds of systems. The essential idea is quite simple: whether a particle is considered as a mathematical entity defined by the Schroedinger wave equation or mentally imaged as a wave-like disturbance of some sort, in both cases one concludes that a potential barrier of finite height becomes transparent as its thickness becomes small compared to the particle wavelength. Particle currents through the energy barrier (i.e., tunneling), as well as the ordinary ones over it, are therefore possible.

If the height and thickness of the energy barrier and the particle wavelength are known, it is possible to estimate the tunneling probability and from that the tunnel current as a function of applied emf. This is then a component of the total particle I-V curve, since the total particle current is the sum of the tunnel current and the ordinary current. In certain physical systems the tunnel current is not only large but has a well-defined maximum, so that a negative conductance appears.

Although the large masses of ions should preclude any significant probability for ion tunneling in biological membranes (4), there is really not enough information about the energy barrier to completely discount the possibility. Electron tunneling, however may easily exist in these membranes and may be of some importance for biochemical reactions associated with cell surfaces.

B. Oxide Films

An apparently unexplained electronic dc negative conductance has been observed in metal-oxide-metal sandwiches prepared from evaporated metal films 150-1000 Å thick (5). The amorphous oxide film is considered to be too thick for a tunnel effect. Several properties are of some interest:

1. The instantaneous I-V curve does not show a negative conductance.

2. The conductance appears to depend on the potential difference and not the electric field, since variation of film thickness produces no important change in behavior.

3. Changes in temperature over a wide range alter the absolute magnitude of current but not the shape of the I-V curve.

C. Passivated Iron

The negative conductance (and consequent electrical oscillations) shown by passivated iron has been of interest to biologists for many years (6), and also involves an oxide layer, in this case between a metal and an electrolyte solution. The system has been treated in great detail by others (7), and I only wish to point out the rather remarkable fact that although the conduction process in the metal-oxide-metal sandwich is believed to be electronic, conduction in the metal-oxide-electrolyte system is apparently now considered to be ionic. Ord and Bartlett (7) have recently suggested the following for passive iron:

1. An oxide layer exists on iron when it is passive.
2. An electric field drives ions through this layer.
3. The resistance to current flow is nonohmic.
4. In the steady-state the thickness of the oxide layer varies linearly with the passivating potential from about 10 to 70 Å.
5. When the system is in a non-steady-state there is a significant space charge within the oxide layer.

D. Teorell Oscillator

Another important and well-known system which shows a negative conductance is the Teorell hydraulic oscillator (8). As far as I am aware, this has been presented in its most concise form by Yamamoto (9), for a single glass capillary. The Teorell oscillator demonstrates that a rather simple coupling of electrodiffusion and mechanical hydrostatic pressure can produce a well-defined negative conductance. This system should serve as an archetype for the study of nonlinear electrokinetic cross-phenomena and merits much more general attention than it has received, particularly since linear cross-phenomena are probably of restricted relevance for biological processes.

E. Black Lipid Membranes and Organic Films

Under certain conditions, black lipid membranes separating two aqueous electrolyte solutions may exhibit a negative conductance (10). In all cases known to me, some organic material must be added to the system in order to produce the effect. For this reason I have been intrigued by the recent report of Stafeev et al. (11), that films of only cholesterol or cetyl alcohol between mercury electrodes show a pronounced negative conductance. The authors suggest that these films are bimolecular and that a voltage-dependent "melting" process may be involved.

III. NEGATIVE CONDUCTANCE IN THE SQUID AXON

Since there is much more experimental information concerning the squid axon membrane than any other biological preparation, this system is convenient as the basis for any theoretical discussion. Information appropriate to a physical analysis, however, does not always appear explicitly in the literature. Wherever possible, I have tried to compare theoretical predictions with experimental observations. In some cases, instead of experimental observations, generally accepted empirical equations have been used.

There appear to be two types of negative conductance exhibited by the squid axon. The first is transient and was discussed in detail by Hodgkin and Huxley in 1952 (1). The second, which appears under special conditions, is usually described as being steady-state, and was first reported by Moore in 1959 (12), although already implied by some experimental results of Segal (13) a year earlier.

Figure 1 shows the transient negative conductance of the squid axon as computed by me from the HH (Hodgkin-Huxley) equations, and Fig. 2 shows an idealized "steady-state" negative conductance observed experimentally with high external KCl.* It is not clear to me to what extent the latter situation, experimentally, is truly a steady state, but if one assumes that it is, and that at very long times for the given fluid compositions the system behaves as shown in Fig. 2, a great mathematical simplification results. We will probably learn in the future that there is in fact only one type of negative conductance in this system and that the situations differ only parametrically. At the moment, however, it seems reasonable to try to attack the simpler steady-state situation first.

The experimental results described by Fig. 2 have usually been interpreted as involving only a K^+ ion current, with the Cl^- permeability of the membrane being very small. Since this forces the analysis of a single-ion negative conductance, it is a critical interpretation. Other ions may be present in and about the membrane, but the assumption is that they are not responsible for any significant currents. Experimentally, there seems to be a requirement for trace amounts of Ca^{2+} in the external solution. I do not know to what extent Fig. 2 gives the current carried by only one ionic species, but I will assume that it does and later suggest how other ions may be involved.

---

*Without serious modification, the HH equations do not predict the results obtained by Moore (12). I will discuss this elsewhere.

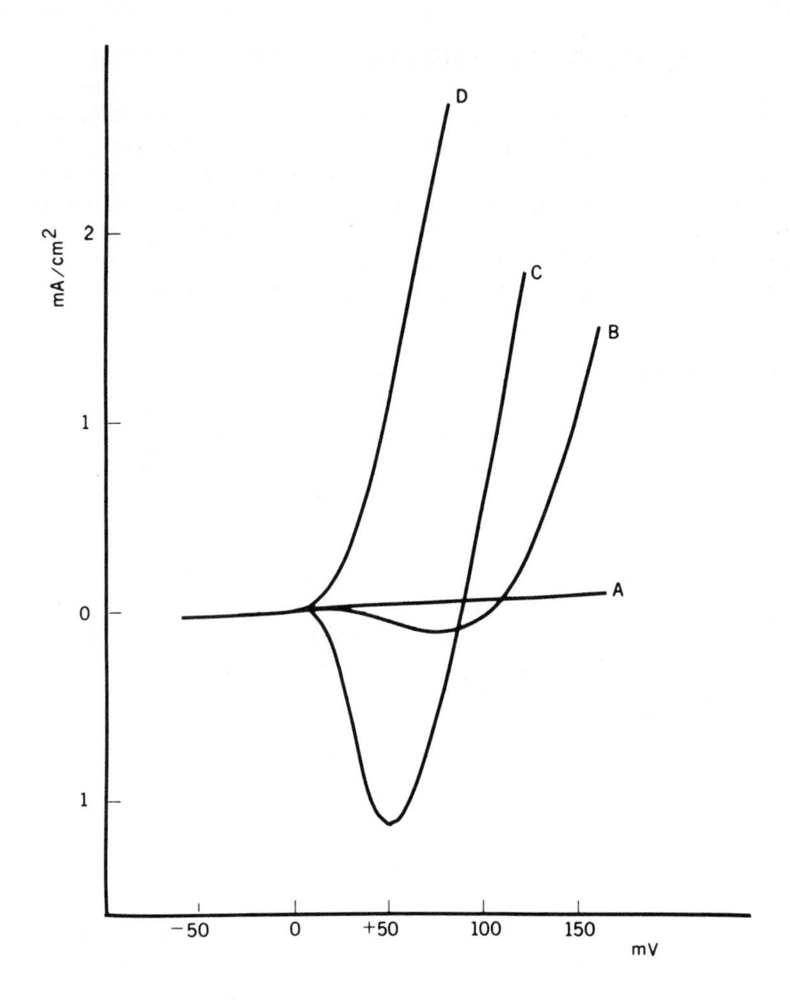

FIG. 1. Transient negative conductance of the HH axon under voltage clamp. the abscissa represents the displacement from the resting potential. A, .01 msec; B, 0.1 msec; C, 1 msec; D, 10 msec.

In black lipid membranes, as has been mentioned, a steady-state negative conductance is also observed (10), and the published I-V curves show a strong symmetry with respect to the first and third quadrants of the I-V plane (Fig. 3). Here also, the experimental information known to me is not complete enough to indicate that this is a single-ion current. For the squid axon there have been experimental suggestions of a partial symmetry (14). I will later suggest how the behavior of the squid axon and black lipid membranes may be related.

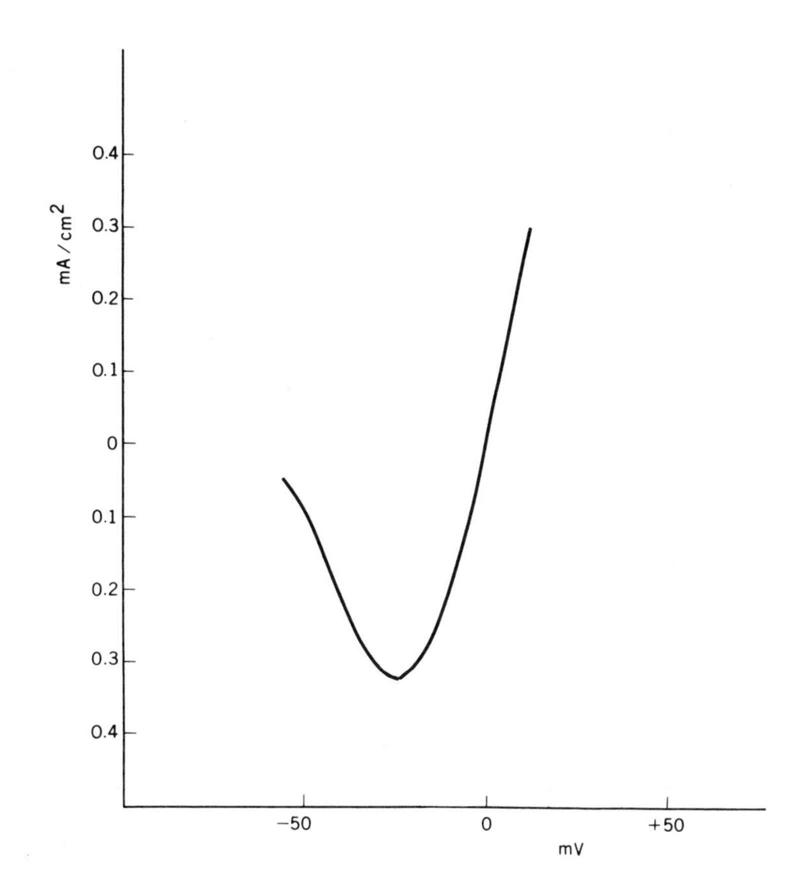

FIG. 2. Idealized steady-state squid axon I-V curve in high external K^+.

With these assumptions and interpretations, the problem, then, is to account for a single-ion steady-state negative conductance. All attempts to do this with the classical Nernst-Planck electrodiffusion equations have been unsatisfactory, and it appears necessary to look for some other approach, preferably one which can be tested by new experiments.

At the outset, it is necessary to recognize that definitive information concerning the structure of the membrane as an electrodiffusion regime is not available. There are several points, however, which I have considered to be important.

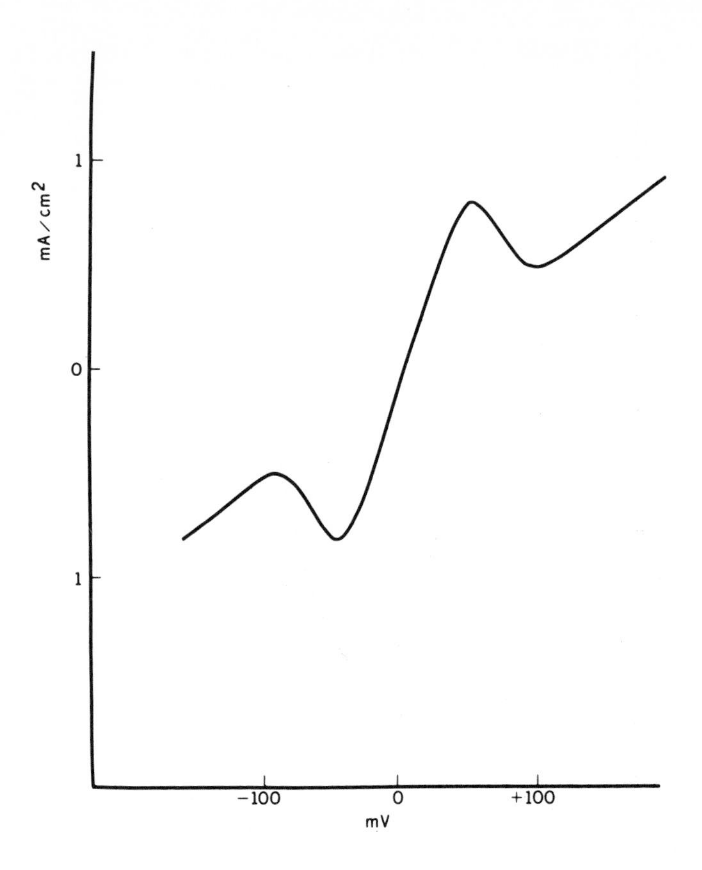

FIG. 3. Idealized I-V curve for an excitable black lipid membrane.

1. With the voltage clamp technique, the potential difference between one membrane boundary and the other is established within a few microseconds. At this time it is found that the instantaneous I-V curve of the squid axon is perfectly linear (15). Now, in general, there are two possible kinds of negative conductance phenomena: those associated with junctions (or membrane boundaries) and those related to bulk properties of the membrane. A junction nonlinearity should appear almost instantaneously, not long after the potential difference is established. The fact that the high frequency resistance of the squid axon membrane is linear suggests to me that the negative conductance in this system is not a junction phenomenon.

2. As can be seen in Fig. 2, in the steady-state a distinct and extensive negative conductance region exists in the squid axon in the absence of a K^+ concentration difference, with zero resting potential. Any theory must satisfy this requirement.

3. Also, in Fig. 2, it is seen that the current shuts off when the inside potential goes negative with respect to the outside and K^+ is supposedly moving inward. Thus the current is shutting off when the potential barrier is apparently being lowered for K^+. This suggests to me that if K^+ carries the current, its concentration profile is controlled by a negative charge distribution within the membrane.

4. The electrical behavior of Nitella is very similar to that of the squid axon (16), except that the current-time constants are increased by three orders of magnitude and the current densities decreased by three orders of magnitude. The plasma membrane supposedly has the same dimensions in both cases. I feel compelled to conclude from this that the electrodiffusion barrier involves more than just a lipid bilayer.

IV. THE GENERALIZED ELECTRODIFFUSION EQUATION

In the attempt to construct a useful theory, I have used a general expression for one-dimensional current flow:

$$I_i = Fc_i z_i v_i \tag{1}$$

$$= -Fc_i z_i \sum_k \tilde{u}_{ik} \frac{\partial U_{ik}}{\partial x} \tag{2}$$

$$\tilde{I} = \sum_i I_i = -F\sum_i c_i z_i \left[\sum_k \tilde{u}_{ik} \frac{\partial U_{ik}}{\partial x}\right] \tag{3}$$

where

I is the electric current density (C/cm^2 sec)
F is the Faraday constant (C/mole)
c is the concentration (mole/cm^3)
v is the velocity (cm/sec)
z is the valence (dimensionless)
$\tilde{u}$ is the joule mobility (cm^2 mole/sec J)
U is the molar energy (J/mole)

x is the position variable (cm)
i is the ionic species index
k is the molar energy index
$\tilde{I}$ is the total electric current density (C/cm^2 sec)

In Eq. (1) the current is defined, and in Eq. (2) it is assumed that the component velocities obey a superposition principle and that each can be expressed as the product of a mobility and a force, with the latter defined by a potential gradient. I will not discuss here the problem of the mean potential versus the potential of the mean force (17), but assume only that the boundary values of U define a measurable energy difference.

I have discussed elsewhere some possible assumptions concerning $\tilde{u}_k$ (18). Here I will assume that it is independent of position and that the Nernst-Einstein relation holds. For a single-ion steady-state current, involving a monovalent cation, we then have

$$I = -uc\frac{dU}{dx} \tag{4}$$

$$U = \Sigma U_k$$

$$u = F\tilde{u}$$

where u(cm^2/sec V) is an electric mobility, and this can be written as

$$I\int_{'}^{''}\frac{dx}{c} = -u(U'' - U') \tag{5}$$

Now, U must have at least one electrostatic component, $F\Psi$, where Ψ (V) is the electric potential. If we define

$$\Psi'' = V$$
$$\Psi' = O \tag{6}$$

we can write

$$I\int_{'}^{''}\frac{dx}{c} = -Fu(V-V^{\circ}) \tag{7}$$

where V° (V) is the potential difference at which the current vanishes. V° thus contains the boundary values of all terms due to concentration gradients, pressure gradients, temperature gradients, etc., whereas in a classical equation it would be related only to concentration gradients. The problem now is the determination of the concentration profile.

V. A NEW POSTULATE FOR THE SQUID AXON

I have assumed that negative charges, both fixed and mobile, are present in the system, and that the electric field is everywhere constant. I then ask: What relation between positive ions and negative ions produces a result which explains a variety of observations? The following postulate emerges:

$$\frac{[\text{mobile cations}]}{[\text{bound cations}]} = \frac{[\text{mobile anions}]}{[\text{fixed anions}]} \tag{8}$$

The brackets indicate concentrations. The bound cations C^* are assumed to be reversibly trapped on fixed negative sites ω so that in any region the concentration of free cations is given by

$$c = \omega - c^* \tag{9}$$

$$\frac{c}{c^*} = \frac{b}{\omega} \tag{10}$$

Since the mobile anions b, assumed to be of one species, carry little or no current in the steady state, they presumably are distributed according to

$$b = b^o e^{zFVx/RT\delta} \tag{11}$$

where b^o (mole/cm^3) is a boundary value; z (dimensionless) a valence: R (J/mole deg) the gas constant; T (deg) the Kelvin temperature; δ (cm) the thickness of the barrier. Combining Eqs. (9), (10), and (11)

$$\frac{1}{c} = \frac{1}{b(x)} + \frac{1}{\omega(x)} = \frac{e^{-zFVx/RT\delta}}{b^o} + \frac{1}{\omega(x)} \tag{12}$$

and from Eq. (7)

$$I = \frac{-Fu(V-V^o)}{-\frac{RT\delta}{zFVb^o}\left(e^{-zFV/RT}-1\right) + \int_{'}^{''} \frac{dx}{\omega(x)}} \tag{13}$$

The integral in the denominator depends on the fixed anion distribution and is not a function of V. When V is large and positive, a limiting resistance appears and is given by

$$R_{lim} = \int_{'}^{''} \frac{dx}{Fu\omega(x)} \tag{14}$$

For a step change from V^o, the instantaneous resistance should be

$$R_\infty = R_{lim} + \frac{RT\delta}{zF^2V^o ub^o}\left(1-e^{-zFV^o/RT}\right) \tag{15}$$

and when $V^o = 0$

$$R_\infty^o = R_{lim} + \frac{\delta}{Fub^o} \tag{16}$$

Equation (13) is then

$$I = \frac{-(V-V^o)}{\frac{-RT}{zFV}\left(R_\infty^o - R_{lim}\right)\left(e^{-zFV/RT} - 1\right) + R_{lim}} \tag{17}$$

VI. THEORETICAL RESULTS FOR THE SQUID AXON

Typical experimental values of R_∞^o and R_{lim} are 30 Ω cm² and 25 Ω cm², respectively. The computation of Eq. (17) for z=3 is shown in Fig. 4. As V^o is reduced, a marked negative conductance region appears, and for $V^o = 0$ the results are in good quantitative agreement with the experimental observations of Moore (12)*

The DC or chord resistance, R_o, is given by

$$R_o = \frac{-RT}{zFV}\left(R_\infty^o - R_{lim}\right)\left(e^{-zFV/RT} - 1\right) + R_{lim} \tag{18}$$

---

*Although electrophysiologists usually take inward current as negative, the sign of the cation current must always be opposite to the sign of the potential difference and the same as the sign of the electric field. Therefore inward current is here taken as positive.

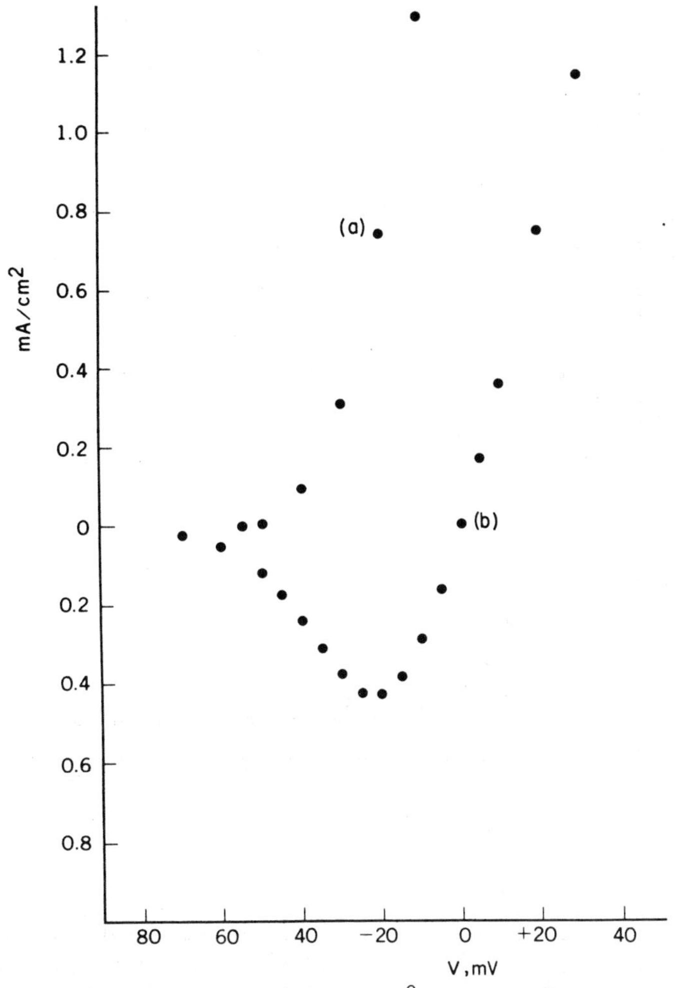

FIG. 4. Computation of Eq. (17). $R^{o}_{\infty} = 30\ \Omega\ cm^2$; $R_{lim} = 25\ \Omega\ cm^2$; $z = 3$; $T = 299.3K^{\circ}$, (a); $V^{o} = -55$ (b); $V^{o} = o$.

To examine Eq. (18), the most convenient data known to me are those given by Hodgkin and Huxley (1) for axon 17. For this axon

$V^{o} = -55$ mV

$I^{o} = -1.2$ mA/cm^2

$R^{o}_{\infty} = 45\ \Omega\ cm^2$

$R_{lim} = 41\ \Omega\ cm^2$

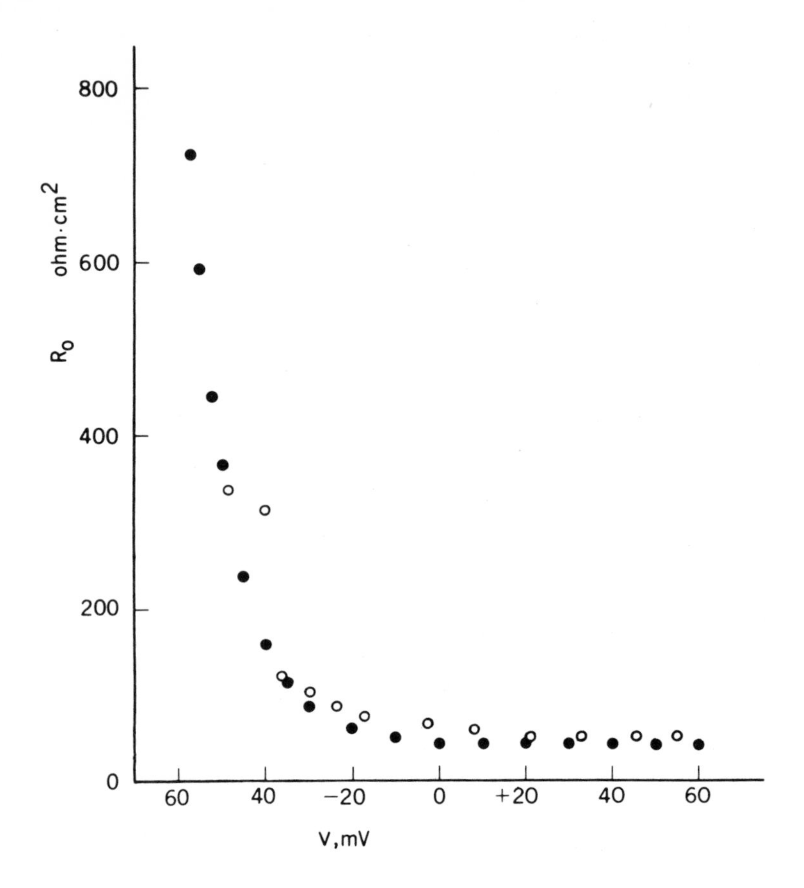

FIG. 5. Comparison of Eq. (18) with experimental results for Eq. (18), $R_{\infty}^{o} = 45\ \Omega\ cm^2$; $R_{lim} = 41\ \Omega\ cm^2$; T = 299.3 K°. The currents for axon 17 are recovered from Table 1, p. 509, of Hodgkin and Huxley (1) by multiplying the given conductances by $(V_A - 12)$, where V_A is the applied potential. The chord resistance is then obtained by dividing the applied potential by the measured current. Eq. 18, closed circles; axon 17, open circles.

Using these values, Eq. (18) is compared with axon 17 in Fig. 5. The agreement between theoretical predictions and experimental observations is apparently good.

VII. A NEW POSTULATE FOR BLACK LIPID MEMBRANES

In Eqs. (11) and (14) the mobile anions were assumed to obey a Boltzmann distribution without restrictive boundary conditions. Suppose, however, the <u>total</u> number of mobile anions remains fixed at some value B:

$$\int_{'}^{''} b\,dx = B\delta \tag{19}$$

This imposes a condition on b^{o}, so that, using Eq. (11)

$$b^{\circ} = \frac{zFVB}{RT(e^{zFV/RT}-1)} \tag{20}$$

$$b = \frac{zFVBe^{zFVx/RT\delta}}{RT(e^{zFV/RT}-1)} \tag{21}$$

and the current then becomes

$$I = \frac{-(V-V^{o})}{-(\frac{\delta}{FuB})(\frac{RT}{zFV})^{2}(e^{zFV/RT}-1)(e^{-zFV/RT}-1) + \int_{'}^{''}\frac{dx}{Fu\omega(x)}} \tag{22}$$

VIII. THEORETICAL RESULTS FOR BLACK LIPID MEMBRANES

In order to compute this equation, the simplifying assumption will be made that ω is constant and that $B = \omega$. We then have

$$A = \frac{I\delta}{Fu\omega} = \frac{-(V-V^{o})}{-(\frac{RT}{zFV})^{2}(e^{zFV/RT}-1)(e^{-zFV/RT}-1) + 1} \tag{23}$$

The result is shown in Fig. 6. A symmetrical negative conductance appears, not unlike that which has been observed experimentally.

IX. CONCLUSIONS

The critical assumptions for the theory of the squid axon are Eq. (8), and for black lipid membranes Eqs. (8) and (19). At present, these assumptions can be neither proved nor disproved, and the purpose of the theory presented here is to develop the consequences of these assumptions in the form of predictions which can be tested in the laboratory. A second purpose is to demonstrate areas where new experimental information is needed.

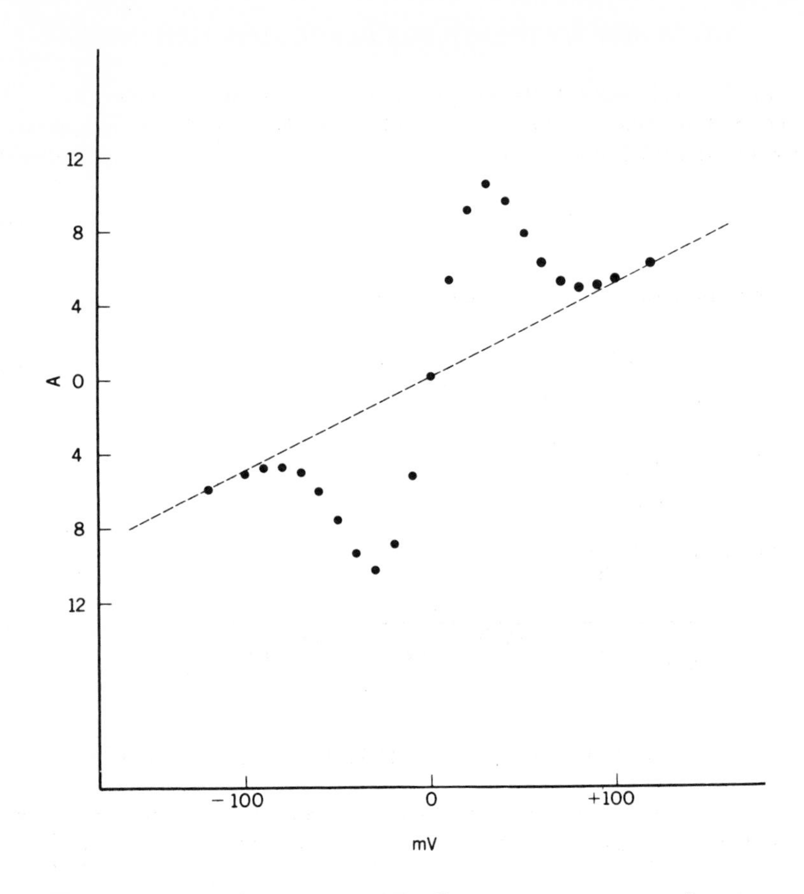

FIG. 6. Computation of Eq. (23). $V^{o} = o$; $z = 3$; $T = 330\ K^{o}$. A parallel leakage path, whose I-V curve is given by the dotted line, has been assumed.

Perhaps the most important theoretical prediction concerns the instantaneous resistance R_{∞}. According to Eq. (15), the instantaneous resistance is related in a specific manner to the limiting resistance, R_{lim}, and to V^{o}. Although there may be some technical difficulties, it should be possible to quantitatively test this relation experimentally. No matter what the ultimate fate of this theory, the relation between these three parameters should prove to be of great physical significance, and its quantitative description stands as a challenge for the laboratory.

REFERENCES

1. A. L. Hodgkin and A. F. Huxley, J. Physiol (London), 117, 500 (1952).
2. K. S. Cole, Membranes, Ions, and Impulses. Univ. of California Press, Berkeley, 1968, p. 167.
3. L. Esaki Phys. Rev., 109, 603 (1958).
4. D. Agin, Proc. Nat. Acad. Sci. U.S., 57, 1232 (1967).
5. T. W. Hickmott, J. Appl. Phys. 33, 2669 (1962).
6. R. S. Lillie, Biol. Rev. 11, 181 (1936).
7. J. L. Ord and J. H. Bartlett, J. Electrochem. Soc. 112, 160 (1965).
8. T. Teorell, J. Gen. Physiol. 42, 831, 847 (1959).
9. K. Yamamoto, J. Phys. Soc. (Japan) 20, 1727 (1965).
10. P. Mueller and D. O. Rudin, J. Theor. Biol. 18, 222 (1968).
11. V. I. Stafeev et al., Sov. Physics-Semiconductors 2, 642 (1968).
12. J. Moore, Nature 183, 265 (1959).
13. J. Segal, Nature 182, 1370 (1958).
14. K. S. Cole, Membranes, Ions, and Impulses, Univ. of California Press, Berkeley, 1968, p. 458.
15. K. S. Cole, Membranes, Ions, and Impulses, Univ. of California Press, Berkeley, 1968, p. 288.
16. U. Kishimoto, J. Cellular Comp. Physiol. 66 (Suppl. 2), 43 (1965).
17. R. H. Fowler and E. Guggenheim, Statistical Thermodynamics, University Press, Cambridge, 1956, p. 405.
18. D. Agin, Biophys. J., 9: 209 (1969).

Chapter 6

SOME OBSERVATIONS ON THE ELECTRICAL PROPERTIES OF BIOLOGICAL MEMBRANES*

C. S. Spyropoulos

University of Chicago, Department of Physiology
and Democritus Nuclear Research Center
Athens, Greece

PART I

EXPERIMENTS RELATED TO THE IONIC DEPENDENCE OF THE MEMBRANE POTENTIAL OF THE LOLIGO AXON AND CHARA INTERNODE

*Supported in part by PHY NS-06719, AF AFOSR--1128-66, Research Grants and C. S. Spyropoulos Funds.

PART II

INSIDE-OUT GIANT AXONS OF LOLIGO VULGARIS

PART I

EXPERIMENTS RELATED TO THE IONIC DEPENDENCE OF THE MEMBRANE POTENTIAL OF THE LOLIGO AXON AND CHARA INTERNODE

I. INTRODUCTION

At the turn of the century, Bernstein (1) proposed that the potential difference existing between the interior of cells and the extracellular fluid is due to the intraextracellular concentration gradient of potassium. He assumed that the membrane is selectively permeable to potassium and not to chloride and sodium ions. Since potassium is more concentrated in the interior of the cell than in the extracellular fluid, the cell is continuously polarized, the intracellular compartment being negative with respect to the extracellular fluid. Later evidence for this relation was obtained in large part by changing the external potassium concentration $[K]_o$ [for review, cf. Ref. (2)] and the internal potassium concentration $[K]_i$ (3,4). The relation at higher $[K]_o$ fits the Nernst equation where the membrane potential $Vm = -58 \log [K]_i/[K]_o$ mV. An observed deviation ("saturation" effect) at low $[K]_o$ or high $[K]_i$ from the theoretical straight-line relation was explained at various times by invoking permeabilities or leaks of the membrane to Na or Cl ions, proposing that the K permeability falls as the resting potential rises, etc. [e.g., (2, 3-5)]. A variety of equations usually based on the constant field theory of Goldman (6) have been used to fit the experimental values of the saturation effect by choosing appropriate permeability coefficients for K, Na, and Cl.

Ion transport via potassium-specific permeability "channels" is almost generally regarded as being responsible for the resting potential. In the present paper, some experiments will be presented that are related to this general problem. These in part involve some experiments on the effects of ionic strength (electrolyte concentration EL) on the membrane potential and a comparison of these effects with the effects on the membrane potential of varying cation mixtures.

II. METHODS

A. Loligo Axon

The giant axon of the stellate nerve of Loligo vulgaris was used. With some exceptions to be noted, the methods for recording membrane potentials and internally perfusing the Loligo axon have been described previously (7). The double cannulation procedure [Fig. 1(b), Ref. (7)] for internal perfusion was employed in most experiments. One exception involved the electrodes used internally in internal perfusion experiments. The electrodes were filled with 1M KCl. After long exposures of the internal pipet to low ionic strength sucrose solutions inside the perfused axon, the tip potentials were observed to be changed. This may have been due to a dilution of 1M KCl in the tip with isotonic sucrose solutions of low ionic strength. Baker, et al. (4) offer an alternative interpretation of a change in tip potential; but similar tip potential changes were seen in vitro on prolonged exposures of the tip to isotonic sucrose solutions of very low ionic strength. This time-dependent tip potential difficulty was reduced partly by plugging the tip with agar - 1M KCl and partly by withdrawing the pipet from the axon, measuring the tip potential in vitro, and reinserting the electrode in the axon.

Two experimental difficulties, both involving Ca, should be stressed in the experiments on Loligo axons. One was the slow reduction in the membrane potential (Vm) at low ionic strengths. This problem has been discussed in a previous communication (7). The second difficulty involved the existence of spontaneous action potentials at low external calcium in isotonic Na

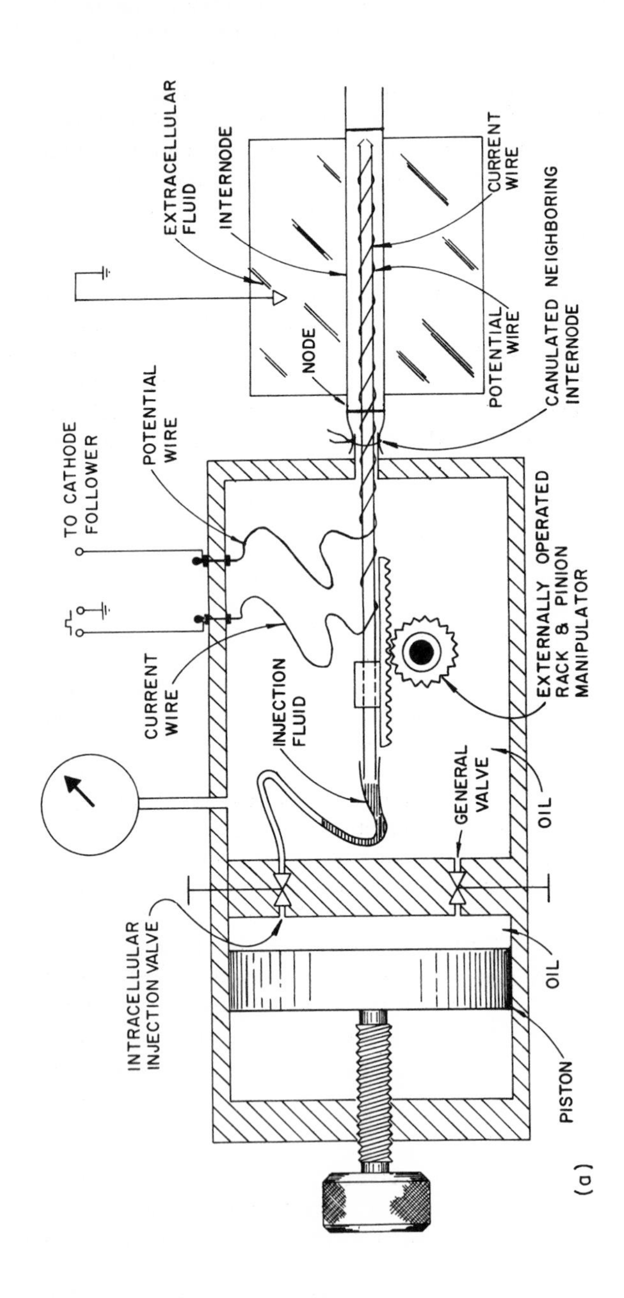

FIG. 1. (a) Diagrammatic illustration of device used for intrainternodal injection and measurement of electrical properties of the internode of Chara or Nitella. See text for further details. (b) Diagrammatic illustration of tapering in pipets used for internal recording of membrane potential in Chara. Diagram is not to scale. (cf. text for further details.)

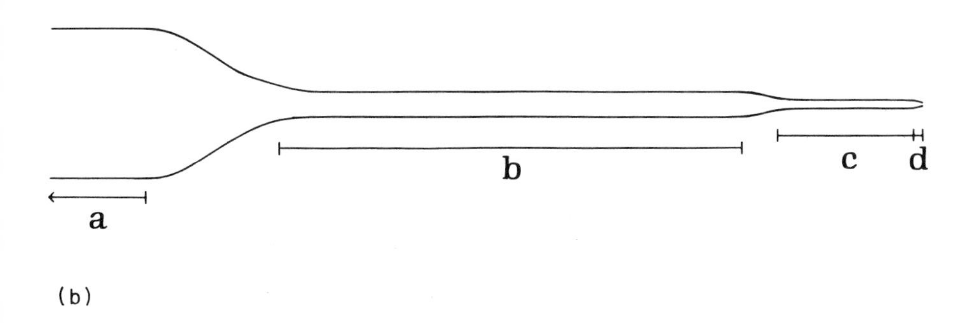

(b)

FIG. 1 (continued)

or Li. When spontaneous firing was observed at 550 mM NaCl or LiCl, the Vm was estimated in the following way: (a) Firing was prevented with tetrodotoxin or a narcotic. (b) During the firing the lowest (inside most negative) potential was used. (c) Occasionally firing stopped or did not occur and Vm was compared to the Vm ascertained with other methods. (d) Extracellular NaCl or LiCl was reduced to a concentration where spontaneous firing was not observed and that concentration was used in estimating the membrane potential dependence upon ionic strength. (e) The extracellular calcium was increased to a concentration where spontaneous firing ceased. The Vm at this point was compared to the Vm ascertained with the other methods mentioned above. As mentioned in an earlier communication (7), in the experiments involving changes in the ionic strength of the intracellular or extracellular medium, stirring of the fluid compartments appeared to be necessary. This was particularly important with solutions of very low ionic strength. Stirring was accomplished by attaching one end of a probe to the vibrator of a high frequency speaker (tweeter) and the other end was immersed in the medium bathing the axon. Finally the author experienced certain difficulties in axons perfused with chlorides. This was particularly disturbing since some of the changes in membrane potential were time dependent. The survival of the axons was such that the experiments often could not be completed. This difficulty was partly overcome by periodic perfusion with isotonic KF. Still the results were subject to error. One hundred and forty-eight axons were used in this study. These were in addition to those previously used in a relevant study with Teorell (7). Experiments were carried out at room temperature (19-22°C).

B. Chara Internode

Chara foetida grown in the University of Chicago Greenhouse was used. In the early experiments, the Osterhout air-gap technique and the microelectrode technique were used (the microelectrode being inserted through internodal or nodal surfaces). Upon varying the ionic strength, pH, etc., the membrane potentials were not stable or reproducible. Only qualitative or semiquantitive results could be obtained. In the case of the microelectrode technique, this appeared, in large part, to be due to variations in the sealing of the tip of the micropipet. Both the Osterhout and the microelectrode technique were, nevertheless, occasionally used throughout the study, by way of controlling the results obtained with a new alternative technique. With this new technique, stable reproducible measurements of the electrical properties of Chara (and Nitella flexilis) could be made. In this method an internode with implanted electrodes gave membrane potentials that were stable for weeks and new internodes even sprang from the nodal cells of the internode under study. Naturally, in these preparations no severe experimental perturbations had been applied. The internode could be space clamped, intrainternodal injections made, and membrane potentials measured with pipet electrodes. In the present study, data are presented only on measurements of membrane potentials. The device used in this technique is illustrated diagrammatically in Fig. 1(a) and an actual photo is presented in Fig. 2. The collapsed internode immediately adjacent to the internode under study was cannulated with a blunted syringe needle whose outside diameter was slightly smaller than that of the internode. Its length varied with the size of the internodes involved. The syringe needle was first electrically insulated inside and outside with a very thin coat of G. E. Enamel 9825. Its other surface was then coated with a thin layer of rubber and finally, over this, with a thin layer of a mixture of "sticky wax," bee's wax, and Vaseline. The threads used for the ligatures were threads from a nylon stocking. These threads were coated with rubber cement and mixtures of wax and Vaseline. These precautions appeared necessary in order to avoid "cutting" of the thin cell wall upon applying the ligatures during cannulation.* Three successive ligatures were applied around the internode at the point of cannulation. They were tighter progressing toward the pressure device. The ligatures were made taut after the contents of the collapsed internode (about to be cannulated) were replaced with a Vaseline-heavy mineral oil mixture by injection through the syringe needle. At this

---

*In a few instances cannulation was accomplished by using Alpha Wire heat-shrinkable tubing. A short piece of this tubing (whose inside wall was coated with wax-Vaseline mixtures) was slipped into the syringe needle all the way to the base. The syringe needle was inserted into the collapsed internode and the thermoplastic tubing slipped over the internode and gently heated.

FIG. 2. Photograph of device illustrated in diagram of Fig. 1(a) and described in text. (1) Syringe needle used for cannulation of neighboring collapsed internode. (2) Carriage of rack and pinion device holding electrode or injection pipet. (3) Rack of the rack and pinion device. (4) Knob for turning pinion device. (5) One of pair of electrical terminals. (6) Screw cap for access to the interior of the chamber. (7) Knob for micrometer driven piston. (8) Valve between piston chamber and injection pipet. (9) Valve between piston chamber and general chamber. (10) Piston. (11) Pressure gage.

stage the collapsed internode was somewhat inflated. The pressure in the cannulated neighboring internode was increased by advancing the piston, with the injection valve closed, until it was approximately the same as that of the adjacent (under study) intra-internodal pressure. At this point, wire electrodes (for space clamp) or pipet (for intra-internodal recording or injection) were inserted into the internode through the intervening nodal cells with an externally operated rack and pinion device located in the pressure chamber. Clogging of the tip of the electrode with Vaseline was circumvented by filling it with agar or by blowing out the Vaseline in the tip by closing the general valve and opening the injection valve. Minor adjustments were made upon puncturing the nodal wall (increasing or decreasing the pressure of the chamber) depending upon whether the internodal contents tended to enter the cannulated adjacent internode, or whether oil tended to enter the internode under study. When oil or light Vaseline-oil mixtures were used, some pressure had to be applied. Occasionally the pressure chamber filled with Vaseline or oil merely acted as a closed system and little or no pressure was applied. During the course of the experiments, minor pressure adjustments were needed when large variations in ionic strength were studied. When the external tonicity was held at approximately the tonicity of the contents of the internode by using sucrose, the readjustments were less. The nodal region penetrated was covered on the outside with Vaseline or a mixture of Vaseline and oil. The same piston was used both for building up the pressure in the chamber and for intra-internodal injection. This was accomplished by using two needle valves at the outlet of the piston chamber. When the desired pressure in the main chamber was obtained, the injection valve was closed and the general pressure chamber valve opened. After injection the injection valve was closed, the general valve was opened, and the pressure readjusted (by withdrawing the piston).

Since tip diameters were larger than the thickness of the cytoplasmic layer, the membrane potential measured was the difference in potential, presumably between the sap and the outside medium. The pipets were filled with 100 mM or 200 mM KCl or K gluconate. The pipet electrode used is illustrated diagrammatically in Fig. 1(b) (the figure is not to scale). Shank portion (a) is held on the carriage of the rack and pinion device in the pressure chamber. The dimensions of pulled portions (b) and (c) varied with the size of the collapsed internode and the size of the intact internode under study. The diameter of (b) portion was slightly smaller than the internal diameter of the syringe needle. Most of portion (c) was inside the internode under study. Its diameter was 10-50 μ. The tapered portion (d) was sometimes not included. With some "sturdy" nodal walls it appeared to facilitate puncturing and penetration. An action potential obtained under current clamp in an internode of Nitella flexilis is presented in Fig. 3, using the technique outlined above. The internode was in tap water. The pulse was 2 msec in duration and 0.5 $\mu A/cm^2$. At least 600 internodes were used. Some were used with this new technique and some were used with the Osterhout and microelectrode techniques. Experiments were carried out at room temperature (19-24° C).

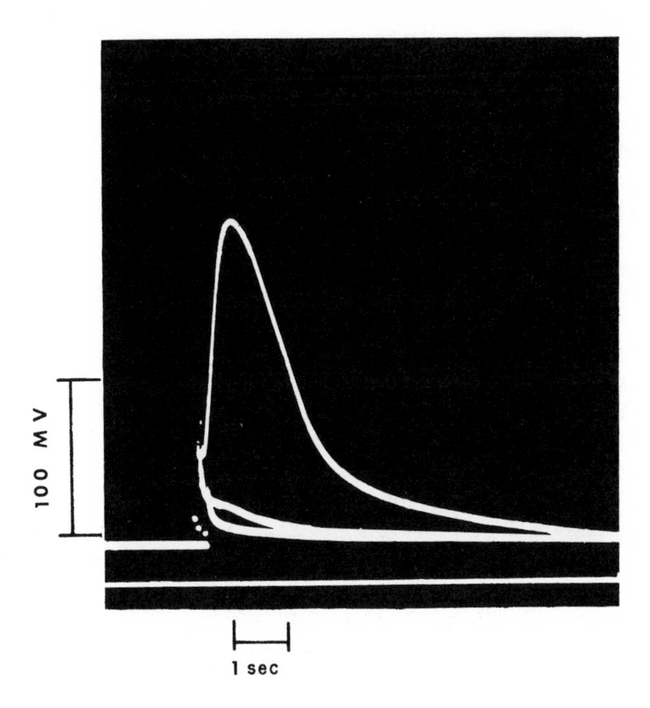

FIG. 3. Record of action potential under current clamp of Nitella flexilis using device described in text and Figs. 1 and 2. Nitella internode diameter was 0.65 mm and its length 2.3 cm. The length of the current wire was 2.1 cm. Exposed part of potential wire was in the middle of the current wire. Calibrations: 1 sec, 100 mV.

III. RESULTS

A. Effects of Ionic Strength (electrolyte concentration) on Membrane Potential

The effects of ionic strength on the membrane potential were studied both intracellularly and extracellularly, both in Loligo axons and Chara internodes. The cations employed include H, K, Rb, Cs, Na, Li, choline, and tetraethylammonium. This section demonstrates that the relation between the membrane potential and the logarithm of the concentration of a cation in or out of the cell in Loligo axon or the Chara internode is essentially a straight, saturation-free, line over a wide concentration range.

1. Loligo Axons

a. Extracellular Ionic Strength. The effects of extracellular ionic strength on the membrane potential were studied on (1) axons with normal axoplasmic constituents and (2) on axons with altered internal media. The tonicity was always held constant with sucrose at a tonicity equivalent to 540 mM KCl.

(1) Axons with Normal Axoplasmic Constituents. These effects were first studied in collaboration with T. Teorell about a decade ago and reported elsewhere (7). The earlier article should be consulted for further details of some of the work presented here. The membrane potential of the giant axon of Loligo pealii or Loligo vulgaris was increased by a reduction in the concentration of extracellular electrolytes (Fig. 4). Over a wide concentration range the relation between the membrane potential Vm and the logarithm of the extracellular electrolyte concentration $[EL]_0$ was a straight line (i.e., no appreciable "saturation" effect). The Vm-$[K]_0$ relation depicted in Fig. 4 shows some deviation from the theoretical Nernst relation. For example, a tenfold decrease in $[KCl]_0$ resulted in a hyperpolarization ranging in different fibers from 42-56 mV while the theoretical slope is 58 mV at this temperature. In individual fibers the slope is defined as the average slope of one or more cycles over the concentration range of 550 to 0.55 mM. At $[K]_0$ =550 mM, Vm varied from 3-32 mV inside positive. The reversed Vm was usually between 12 and 15 mV (inside positive). For the other group, 1A cations, the deviation from the ideal Nernst slope is greater. A tenfold reduction in $[RbCl]_0$ resulted in a hyperpolarization of 41-52 mV. A tenfold reduction in $[TEA\ Cl]_0$ resulted in a hyperpolarization of 28-38 mV. A tenfold reduction in $[CsCl]_0$ concentration resulted in a hyperpolarization of 27-41 mV. A tenfold reduction of $[NaCl]_0$ resulted in a hyperpolarization of 17-39 mV. A tenfold reduction in $[LiCl]_0$ resulted in a hyperpolarization of 14-27 mV. A tenfold reduction in $[Choline\ Cl]_0$ resulted in hyperpolarization of 6-21 mV. The hyperpolarization obtained with a tenfold reduction in the concentration of extracellular divalent electrolytes was 15-28 mV. At pH below 4.5-5.0, the range of slopes obtained with hydrogen ion (i.e., acetic acid) varied from 24 to 55 mV. Occasionally, upon decreasing pH, the change in membrane potential was discontinuous: namely, at a particular $[H]_0$ the Vm changed monotonically within 1-10 min to a lower (inside less negative) potential level. An example of the effects of $[H]_0$ is presented in Fig. 6*. (See page 288). The Vm obtained at pH 5-7 was not stable. The internal

---

*In order to facilitate a comparison of single-cation effects to mixed-cation effects later on in the text, a sequential reference to figures is not followed in this presentation.

negativity slowly (1-15 min) declined by 5-12 mV. The change in solutions took 30 sec to 1 min and during this period the Vm may have been higher. The data on $[H]_0$ in fibers other than those presented in Fig. 6 (15 axons) are presented in Table 1 (axons 1-15).

[A few hints are needed in the use of this table, and Tables 2 and 3. The labels in the column under the heading "Axon Code" define the history of the axon used. The sequential label is an arbitrary identification. If alone, it means the axon had not been used previously. If the label is compound, the second symbol indicates the experiment number and the table number where the previous treatment may be found is shown in parentheses. The sequence of individual experiments presented in Table 1 is: either measurements at neutral pH, abruptly to extreme acid, abruptly to extreme alkaline, and stepwise to extreme acid (axons coded 11-15, 24-30 and 40-42); or measurements at neutral pH, abruptly at the lowest pH, stepwise to extreme alkaline, abruptly to neutral. In Table 2 the sequence is: measurements at concentrated solutions, step-wise to dilute, abruptly at concentrated solutions. In Table 3 the sequence is: measurements at high [K]/[Na], stepwise to low [K]/[Na], and abruptly to high [K]/[Na]. Only experiments reversible within about 10% are included in the table. The arrows in Tables 2 and 3 are interposed between an initial reading of Vm and a later semi-steady-state reading. The letter "d" above the arrow indicates that a discontinuity in the Vm - log [EL] relation was observed. In Table 1 a discontinuity is indicated by repeating the same pH and the corresponding Vm following the discontinuity.]

Table 1 presents some axons with more points in the Vm - log $[H]_0$ relation than those shown in the experiment of Fig. 6. The fiber presented in Fig. 6 was also used to demonstrate the effects on the Vm of variations in H-cation mixtures. The author was concerned about its survival and limited the number of points. This procedure was followed in other types of experiments, where contrasting effects were demonstrated. The cationic selectivity sequence observed by measurement of Vm at 550 mM concentrations of 1A Alkali cations and choline was K > Rb > Cs > Na > Li > choline. Vm was lowest in K. The slopes of the Vm - $[EL]_0$ relation showed the same sequence as the aforementioned selectivity, namely the highest slopes were obtained with KCl and the lowest with choline Cl (cf. Fig. 4). With the exception of those involving H, the data are not tabulated since over 200 axons were used.

(2) Axons with Altered Internal Media. Two types of experiments were undertaken: experiments on axons internally perfused with (a) K_2SO_4 or KCl, and (b) NaCl.

(a) Axons Perfused with Isotonic K_2SO_4 or KCl. When the axoplasm was replaced with isotonic K_2SO_4, the range of slopes of the Vm - log $[NaCl]_0$ or Vm - log $[KCl]_0$ were similar to the slopes obtained with normal cytoplasmic constituents. They were straight-line relations. For details of these experiments, Ref. (7) should be consulted.

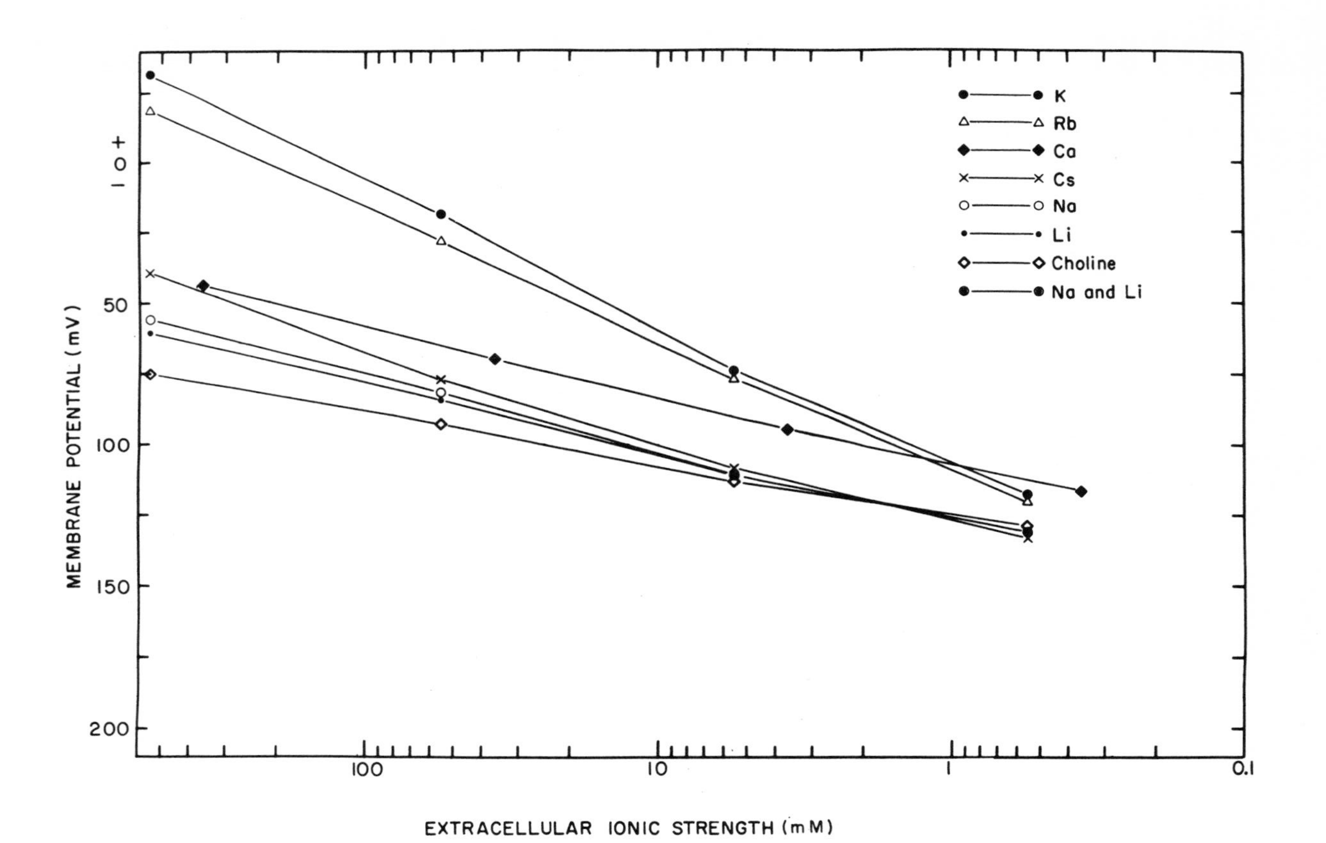
K
Rb
Ca
Cs
Na
Li
Choline
Na and Li
MEMBRANE POTENTIAL (mV)
+
0
−
50
100
150
200
EXTRACELLULAR IONIC STRENGTH (mM)
100
10
1
0.1

(b) Axons Perfused with Isotonic NaCl. Two types of experiments were carried out: $[K]_o$ was varied or $[Na]_o$ was varied. An example of the effects of varying $[KCl]_o$ is presented in Fig. 7, quadrant 4, solid line curve (see page 292). The dependence appears to be a straight line. At 550 mM $[K]_o$ (with 3 mM $CaCl_2$) the internal positivity declined within 5-8 min by 3-12 mV. At low $[K]_o$ no Ca was included and measurements were made rapidly (1-2 min), since the Vm declined irreversibly [cf. Ref. (7)]. Five other axons were used. In all axons, the Vm-log $[KCl]_o$ relation was approximately a straight line. The data for these additional axons are presented in Table 2 (axons 1-5). The Vm - log $[NaCl]_o$ relation with isotonic NaCl inside was approximately a straight line. With 550 mM inside and 550 mM NaCl (containing also 3 mM $CaCl_2$) externally, Vm was 0 ± 3 mV. Measurements at low $[Na]_o$ were made in the absence of Ca. The measurements were those at 1-2 min after changing solutions. A total of five fibers were employed and the data for the additional axons presented in Table 2 (axons 6-10).

b. <u>Intracellular Ionic Strength</u>. Two types of experiments were undertaken: (1) The axon was immersed in isotonic Na Cl (with 3 mM Ca Cl_2) and $[K\ Cl]_i$ or $[Na\ Cl]_i$ varied; or, (2) the axon was immersed in isotonic K Cl (with 3 mM Ca Cl_2) and $[K]_i$ was varied. The experiments of varying $[Na]_i$ with the axon immersed in isotonic K Cl was not undertaken.

(1) Axons Immersed in Isotonic NaCl. An example of these experiments is presented with solid lines in quadrant 1, Fig. 7. In this axon and the remaining five used in these experiments, the Vm at 550 mM $[K\ Cl]_i$ declined with time. The Vm - log $[KCl]_i$ relation was not quite a straight line; it had higher slopes at lower $[KCl]_i$. In either case there was no saturation at low $[KCl]_i$. The data on the remaining fibers were similar and are presented in Table 2 (axons 11-15). The effects of varying $[Na\ Cl]_i$ in axons bathed in isotonic Na Cl were not as consistent as one would wish. A total of five axons were used. Only two points in the Vm - log $[Na\ Cl]_i$ were obtained, one at 550 mM and one at 2 mM. Naturally, one

---

FIG. 4. The dependence of the membrane potential of the giant axon of Loligo vulgaris upon extracellular ionic strength. With the exception of the curve involving choline, two curves were obtained on the same axon. K and Rb, Ca and Cs, and Na and Li. The curves were reproducible in two cycles within 2 mV. The experiments in this respect were atypical and selected. Data were obtained commencing at the most concentrated region proceding stepwise to the most dilute and abruptly returning to the high concentration. Two min are allowed at each step. The points are single-run measurements. Tonicity was kept constant with sucrose at a tonicity of 550 mM NaCl.

TABLE 1

The Effects of pH on the Membrane Potential of the Giant Axon of Loligo Vulgaris

Axon Code	pH and Corresponding Membrane Potential																Cations	
	pH	mV	pH	mV	pH	mV	pH	mV	pH	mV	pH	mV	pH	mV	pH	mV	Varied	Fixed
Data in "Pure" H-Ion Solution																		
1	5.0	-115	4.1	-91	3.2	-69											$[H]_o$	NONE
2	5.2	-132	4.5	-118	4.2	-116	3.5	-87	3.0	-71	2.9	-69					"	"
3	5.1	-124	4.1	-105	3.6	-94	3.2	-82	2.9	-76							"	"
4	5.4	-131	4.6	-121	4.1	-105	3.4	-68	3.8	-92							"	"
5	5.4	-125	4.5	-110	3.5	-89	2.8	-74									"	"
6.27	5.1	-130	4.7	-119	4.0	-102	3.0	-71									"	"
7.28	5.1	-126	4.2	-99	3.4	-72	2.9	-56									"	"
8	5.4	-128	4.4	-113	3.7	-93	3.1	-69									"	"
9.25	5.1	-140	4.4	-122	3.8	-94	3.2	-66									"	"
10	5.2	-136	4.8	-128	4.5	-120	3.5	-88	2.9	-64							"	"
11	5.0	-141	4.5	-129	3.6	-96	3.0	-80	2.8	-72							"	"
12	5.0	-116	4.5	-106	4.0	-96	3.2	-80	2.9	-68							"	"
13	5.3	-144	4.5	-132	4.0	-120	3.8	-117	3.8	-89	3.0	-64	2.8	-54			"	"
14	5.1	-136	4.4	-123	4.0	-113	4.0	-86	2.9	-53							"	"
15	5.1	-122	4.0	-99	3.4	-73	3.2	-65	3.0	-52	2.8	-37					"	"

TABLE 1 (continued)

The Effects of pH on the Membrane Potential of the Giant Axon of Loligo Vulgaris

Axon	pH and Corresponding Membrane Potential																Cations	
Code	pH	mV	pH	mV	pH	mV	pH	mV	pH	mV	pH	mV	pH	mV	pH	mV	Varied	Fixed
Data in H-Cation Mixtures																		
16	10.2	+9	7.3	+12	5.2	+11	4.4	+12	3.1	+17							$[H]_o$	$.55M[K]_o$
17	11.5	+10	7.3	+10	5.2	+10	4.4	+12	3.1	+14							"	"
18.30	10.2	+11	7.3	+13	5.2	+16	4.4	+17	3.1	+19							"	"
19.38	11.5	+14	9.0	+14	7.5	+18	5.2	+18	4.4	+16	3.5	+21					"	"
20.41	10.2	+12	9.0	+13	7.3	+13	5.2	+15	4.4	+15	3.1	+15					"	"
21	10.2	+6	7.3	+8	5.2	+9	4.4	+9	3.1	+11							"	"
22.1	10.2	-65	7.0	-54	5.0	-53	4.2	-49	4.0	-47	3.5	-40	2.8	-13			$[H]_o$	$.55M[Na]_o$
23	10.5	-60	8.0	-55	6.2	-53	5.0	-51	4.0	-32	3.5	-24	3.0	-22	2.8	-18	"	"
24.2	10.2	-63	8.0	-56	7.5	-55	6.0	-54	5.0	-52	4.0	-47	3.2	-29	2.8	-11	"	"
25.3	11.1	-60	8.0	-53	5.2	-53	4.5	-47	3.5	-42	3.5	-20	2.8	-6			"	"
26.4	11.1	-58	7.0	-51	6.2	-50	4.2	-48	3.6	-44	2.8	-8					"	"
27	11.1	-61	10.2	-59	8.0	-55	5.1	-53	4.0	-51	3.5	-41	3.2	-30	2.8	-15	"	"
28	10.2	-61	8.0	-57	6.0	-56	4.5	-52	2.8	-19							"	"
29	11.1	-62	6.0	-56	5.0	-54	4.5	-53	4.0	-45	2.8	-16					"	"
30	11.1	-64	9.2	-59	8.2	-56	5.5	-55	4.1	-54	4.1	-30	2.8	-11			"	"
31	11.1	-66	9.2	-60	8.6	-54	5.0	-52	4.0	-47	3.5	-40	2.8	-10			"	"
32.14	8.0	-55	6.0	-53	4.2	-50	3.8	-45	2.8	-20							"	"
33.8	11.1	-61	7.0	-56	4.5	-54	3.8	-51	2.8	-13							"	"

TABLE 1 (continued)

The Effects of pH on the Membrane Potential of the Giant Axon of Loligo Vulgaris

Axon Code	pH and Corresponding Membrane Potential																Cations	
	pH	mV	pH	mV	pH	mV	pH	mV	pH	mV	pH	mV	pH	mV	pH	mV	Varied	Fixed
Data in H-Cation Mixtures (continued)																		
34	11.1	-64	9.2	-60	8.6	-56	7.5	-57	6.0	-56	5.5	-56	4.7	-38	2.8	-15	$[H]_o$	$.55M[Na]_o$
35	9.2	-60	7.0	-57	5.5	-57	4.1	-32	2.8	-15							"	"
36.13	7.0	-59	6.0	-58	5.0	-55	4.6	-54	3.2	-33	2.8	-22					"	$.55M[Chol]_o$
37	10.5	-71	6.0	-58	4.0	-52	3.3	-39	2.8	-14							"	"
38	10.1	-79	6.5	-69	5.5	-69	5.0	-58	4.0	-53	3.7	-50	3.1	-36	2.8	-22	"	"
39.10	7.5	-62	5.0	-59	3.8	-51	2.8	-26									"	"
40	7.5	-59	5.5	-57	3.8	-49	3.5	-41	3.5	-21	2.5	-11					"	"
41.15	11.1	-71	7.0	-57	5.5	-55	5.0	-57	4.5	-56	3.6	-44	3.1	-31	2.8	-16	"	"
42	11.1	-79	6.4	-68	5.0	-57	4.0	-50	2.8	-27							"	"
43	11.1	-70	5.5	-58	4.5	-54	3.5	-41	2.8	-14							"	"
44.26	6.0	-56	4.5	-52	2.8	-29											"	"

TABLE 2

Effects of Ionic Strength on the Membrane Potential of the Giant Axon of Loligo Vulgaris

Axon Code	Ionic Strength (mM)												
	550	333	100	55	33.3	10	5.5	3.3	2	1	.55	Fixed	Varied
1	71→59		38			-4					-35	.55M$[Na]_i$	$[K]_o$
2	66→56		35			-7					-32	"	"
3	73→67		45			-8					-43	"	"
4	64→61		37			-5					-28	"	"
5	70→65		40	24							-38	"	"
6.3(3)	1			-16			-36			-39		.55M$[Na]_i$	$[Na]_o$
7	3			-15			-31			-42		"	"
8	2			-12			-27			-31		"	"
9.5(3)	-3			-14			-28			-36		"	"
10	0			-10			-35			-39		"	"
11	-72→68		-61		-45			-8			15	.55M$[Na]_o$	$[K]_i$
12	-69→64	-53	-30		-18						33	"	"
13	-75→64	-60	-57			-8	-3				29	"	"
14	-70→67		-53	-47		-22	-18				12	"	"
15	-74→72	-61	-48		-37	-19		-5			19	"	"
16	1								32			.55M$[Na]_o$	$[Na]_i$
17	2								20			"	"
18	0								42			"	"
19	-1								17			"	"

TABLE 2 (continued)

Effects of Ionic Strength on the Membrane Potential of the Giant Axon of Loligo Vulgaris

Axon Code	Ionic Strength (mM)											Fixed	Varied
	550	333	100	55	33.3	10	5.5	3.3	2	1	.55		
20	3								28			.55M$[Na]_o$	$[Na]_i$
21.16	32								1			2mM$[Na]_i$	$[Na]_o$
22.18	42								-1			"	"
23.19	17								3			"	"
24	-3		20		38	58		75			99→96	.55M$[K]_o$	$[K]_i$
25	-2		21		40	57					94→91	"	"
26	-1		18		42	55		73			101→89	"	"
27	0				37	60					97	"	"
28	-5		16		35	55		69			91→87	"	"

cannot say anything about the configuration of the Vm - log $[Na\,Cl]_i$ relation. One can only say that there may be an effect of dilution (ionic strength). The data on the fibers used are presented in Table 2 (axons 16-20). Initially, the perfusate contained 550 mM NaCl and the Vm was -1 to +3. When the internal perfusate was switched to 2 mM Na Cl within 3-30 min, the internal potential became positive by 14-43 mV. This positivity was abolished (within 4 mV) in 1-2 min by immersing the axon in 2 mM NaCl (with zero calcium). Three axons were used in the latter experiments and the data presented in Table 1 (axons 21-23). As mentioned in the Methods section, stirring was important for Vm measurements at low ionic strength; this was especially true for these particular experiments. The maximal internal positivity at 2 mM [Na] was decreased by one-third to one-quarter in the absence of stirring.

(2) Axons Immersed in Isotonic KCl. The external medium also contained 3 mM $CaCl_2$. An example of these experiments is presented in quadrant 3, Fig. 7 (solid line). The Vm - log $[K\,Cl]_i$ was approximately a straight line. As $[K\,Cl]_i$ decreased, internal positivity was increased. At 550 mM $[K\,Cl]_i$ Vm was 0-8 mV (inside negative). Five other fibers were used. The data are presented in Table 2 (axons 24-28).

2. Chara Internode (Extracellular Ionic Strength)

In the Chara internode most of the experiments to be presented here were carried out with gluconate as the counterion instead of chloride. The measurements made with gluconate were more reproducible and stable and the survival time of the preparation was much longer than when chloride was employed. An increase in electrophysiological and morphological survival time has been observed in the squid axon using gluconate as the anion. The extracellular ionic strength was varied from 0.1 to 100 mM. In some experiments the tonicity was not held constant. In some it was held constant with sucrose. The maximal ionic strength used in individual experiments varied from 1 mM salt to 100 mM. The other steps were 0.33 mM, 0.5 mM, 3.3 mM, 5 mM, 10 mM, 33.3 mM, and 50 mM. In various experiments tonicity was held constant with sucrose at all these salt concentrations. No striking differences were observed in the results on the effects of ionic strength when sucrose was used as an osmotic substitute, and the results when tonicity was not held constant. This was not the case in the experiments on Loligo axons, which deteriorate rapidly in solutions of low tonicity. The extracellular ions varied were K, Na, Ca, and H. Hydrogen was introduced to the extracellular medium in the form of gluconic acid. Fresh solutions were used since something grows in gluconic acid within a day or so. The membrane potential at any given concentration of Na, Ca gluconate, and gluconic acid varied in different preparations (all excitable) by as much as 70-80 mV. Preparations with very high membrane potentials were not stable, and experiments of the type desired could not be carried out. On this account, the data presented here may be subject to some error. The experiments within the same internode involving, for example, a dilution of a salt from 1 to 100 mM with three intervening steps, were reproducible within 2 to 4 mV. If the variation was more than 5 mV the internodes were discarded. Absolute Vm measurements could not be compared in different fibers; for this reason all the experiments on Chara are illustrated on one internode. Figure 5 and part of Fig. 6 show the results obtained on one internode. The internode was the best of hundreds that were used. The membrane potential was not the highest one observed, but the results were the most reproducible and the measurements the most stable. Tonicity was held constant (equivalent to 100 mM monovalent electrolytes) with sucrose. The sequence of ascertaining the Vm - $[EL]_0$ relation was the following: After implanting the pipet electrode the internode was "equilibrated" in 10 mM Na gluconate for 5 hr. The membrane potential was then measured at different $[EL]_0$. At each concentration, time was allowed for attainment of a steady state. This was arbitrarily chosen to be 5-15 min for the concentrated solutions. The time required for changes in Vm upon changing the extracellular Na, Ca, or H was usually much shorter. Only instantaneous measurements (within 1-2 min) were made in dilute solutions of Na or H due to the slow decline in Vm. For Na these concentrations were below 1 mM and for H above pH 5. If longer periods of time were allowed, the Vm--upon returning to a higher concentration--usually was not the same. The Vm during the time required to change the solution (approximately 30 sec) may have been higher. The concentrations at which Vm is not stable varied from fiber to fiber. Occasionally, at 0.04 mM Nagluconate, stable Vms were obtained. It can be seen from Fig. 5 and part of Fig. 6 that the Vm - log $[EL]_0$ relation for K, Na, Ca, and H is approximately a Nernst straight-line relation, with no appreciable saturation. Since hundreds of internodes were used, the data are not tabulated for the present paper. This was the case also for the effects of extracellular ionic strength on the normal Loligo axon, where only the ranges of the slopes were given. The variation in the slopes of the Vm - log $[EL]_0$ relation using gluconate as the counterion in Chara internodes was the following: for K, 46-58 mV; for Na, 42-56 mV; for Ca, 20-29 mV; for H below pH 4.0, 33-58 mV. In order to attain a pH of 2.8, approximately 40 mM gluconic acid was required; however, since the slopes obtained with acetic acid and HCl as the source of H were approximately similar, the effect of undissociated gluconic acid per se may have been obviated. Frequently, discontinuities were observed in the Vm - $[EL]_0$ relation in many internodes of Chara; these discontinuities were either monotonic and time dependent (1-10 min) or reflected the arrest of an action potential at the level of the plateau. These discontinuities, however, will be discussed in detail elsewhere. In the Chara internode that had been stored in tap water, the slopes of the Vm - log $[EL]_0$ using chloride as the counterion were usually found to be much lower than those where gluconate

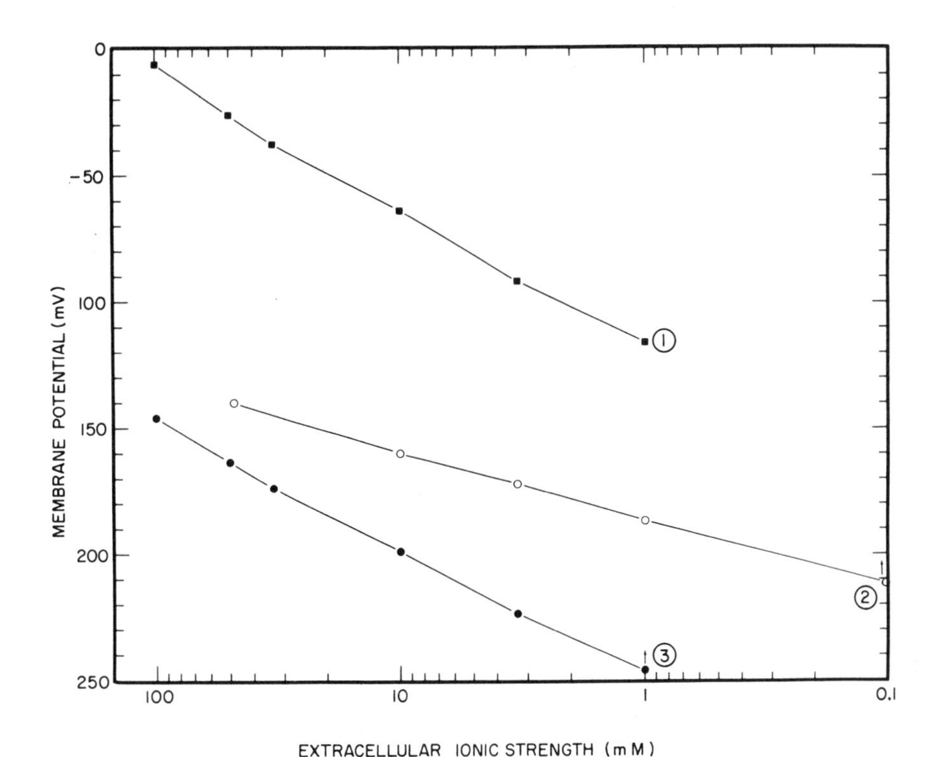

FIG. 5. The dependence of the membrane potential of the internode of Chara foetida on extracellular ionic strength. All the curves in this figure and curves 7 and 8 in Fig. 6 were obtained on the same internode (1.2 mm in diameter and 2.5 cm in length). Membrane potentials were measured with the new technique described in Figs. 1 and 2 and the text. The sequence of the experiments represented by the curves was the following: (I) Ca gluconate curve (No. 2), (II) Na gluconate curve (No. 1), (III) curve 7 of Fig. 6, (IV) curve 8 of Fig. 6, and finally, (V) K gluconate curve (No. 3) of the present Figure. Vm measurements commenced at lowest concentrations proceeded stepwise to the highest and abruptly again at the lowest. Tonicity was kept constant with sucrose at that equivalent to 100 mM K gluconate or Na gluconate. This is not a typical internode since no other internode was observed in which all those experiments could be completed; but any particular Vm-log [EL] relation displayed in Figs. 5 and 6 could be obtained in other internodes and the ranges of the slopes of the Vm - log [EL] relations were within the ranges expressed in the text. (Cf. text for further details.)

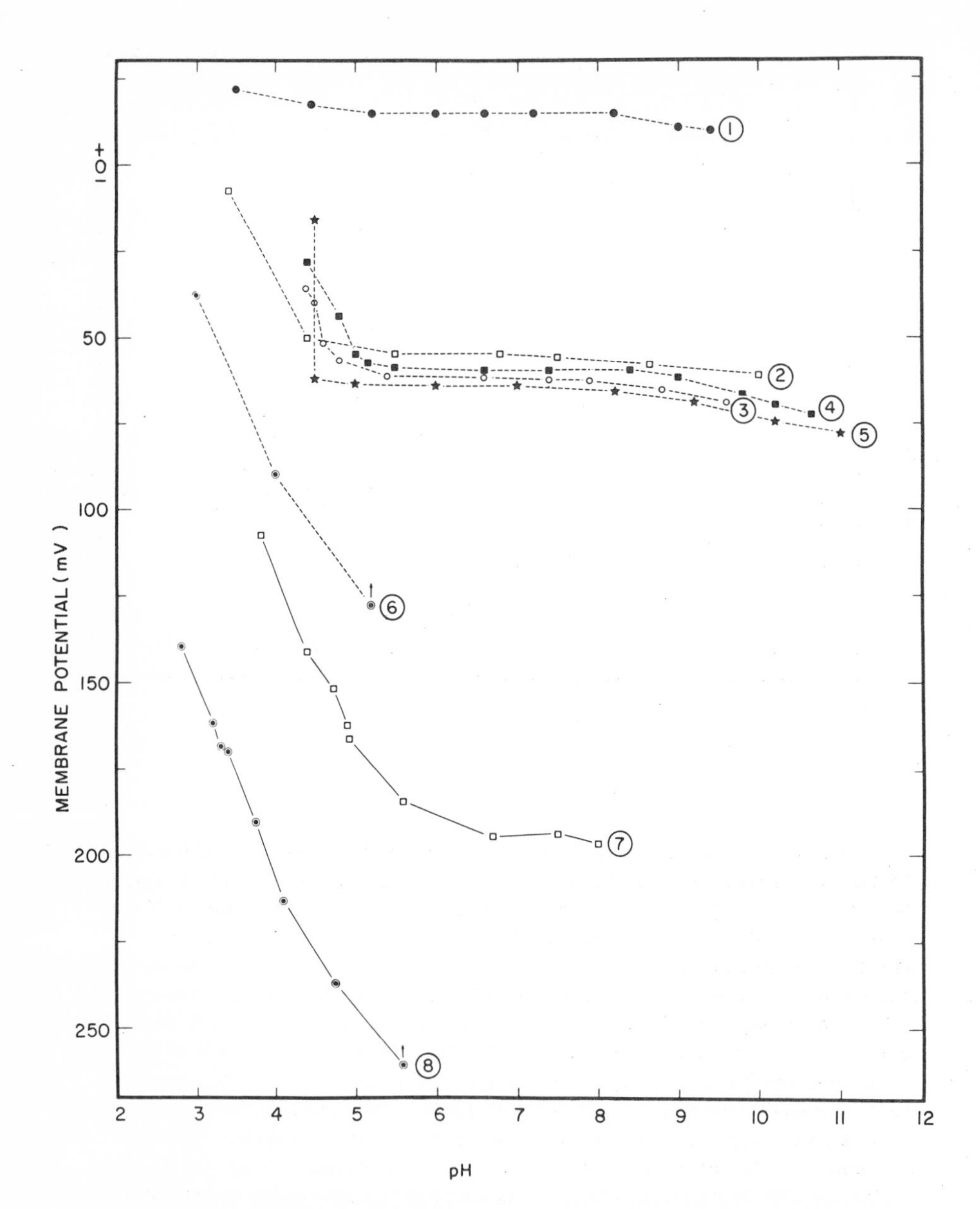

FIG. 6. The effects of external pH on membrane potential of the giant axon of Loligo vulgaris (curves 1-6) and the internode of Chara foetida (curves 7, 8). (a) Loligo vulgaris--curves 1-6. Curves 1-6 were carried out on the same axon (820μ in diameter). The axon was thoroughly cleaned

was used as a counterion; however, until current experiments are complete the author does not wish to commit himself on this problem of anionic contribution or imply that gluconate acts merely as a more impermeable anion. At 100 mM $[EL]_o$ the Vm of the internode depended upon the cation. Vm was highest in choline and lowest in K. The selectivity sequence was choline > Li > Na > Cs > Rb > K.

B. Effects of Varying Cation Mixtures on Membrane Potential

In Sec. III. A the author presented experiments indicating that the Vm - log [EL] relation of a cation in single-ion "pure" solutions was a straight-line (saturation-free) relation even at low [EL]. This appeared to be true for intracellular and extracellular variations in [EL] and for both the Chara internode and the Loligo axon. The author also presented experiments indicating that the slope of the Vm - log [EL] relation occasionally was near the theoretical Nernst slope, not only for K ions, but also for Na, Ca, and H. In the present section, experiments are presented indicating that when [K] or [H] are varied (in the Loligo axon or in the Chara internode) in the presence of other cations, the Vm - log [K] or the Vm - log [H] is not a straight line (it shows saturation). The experiments involving variations in [K] were carried out in the presence of Na. The total ionic

---

of adhering tissue. In curve 1 are studied the effects of pH in the presence of 550 mM $[KCl]_o$, in curve 2 the effects in the presence of 550 mM $[NaCl]_o$, in curves 3-5 in 550 mM $[\text{choline Cl}]_o$, and in curve 6 in pure H-ion (acetic acid) solution. Tonicity was held constant with sucrose. The sequence of the experiments was the following: (I) seawater, (II) lowest pH-sucrose solution, (III) stepwise to highest pH - sucrose solution (curve 6), and (IV) sea water. In experiments 1 through 5 initial readings were at neutral pH; subsequently stepwise measurements were made at low pH, abruptly at neutral pH, stepwise through the alkaline range, and again abruptly in sea water. High pH was obtained with tris, phosphate, and NaOH, depending upon the pH. (b) Chara Internode--curves 7-8. The internode was the same as was used in the experiments of Fig. 5. In the experiment of curve 7 is studied the effect of pH (gluconic acid) on Vm in the presence of 33.3 mM Na gluconate. In the experiment of curve 8 is studied the effect on the Vm of pure H-ion solutions (gluconic acid) with sucrose to maintain tonicity constant. The sequence of the measurements of Vm was the following: (I) at pH 6.7 (in 33 mM Na gluconate), (II) abruptly at lowest pH (in 33 mM Na gluconate), (III) stepwise to the highest pH (in 33 mM Na gluconate), (IV) abruptly at pH 6.7 (in 33 mM Na gluconate), (V) at highest pH in pure H-ion solution, (VI) stepwise to the lowest pH in pure solution, and finally, (VII) abruptly to the highest pH in pure H-ion solution.

strength was kept constant. [H] was varied in the presence of constant [Na] or [K]. The ionic strength also was held approximately constant, since in the Loligo axon K and Na were held constant at 550 mM and in Chara, Na was held constant at 33.3 mM and the lowest pH was around 3.

1. Loligo Axon

a. Variation of Extracellular Cation Mixtures. Two types of experiments were carried out: Experiments on (1) axons having the normal axoplasmic constituents and (2) axons internally perfused with isotonic NaCl.

(1) Normal Axoplasmic Constituents. The experiment in which $[K]_o$ was varied, with total ionic strength kept constant by varying $[Na]_o$, is the classic experiment by Curtis and Cole (8), in which saturation was described. It is presented in Fig. 7, quadrant 2, dashed line. These data are included for the purpose of comparing these previous results to some related experiments presented in this paper. In the present study hydrogen ion was varied in the presence of 550 mM KCl, choline chloride*, and NaCl. All contained 3 mM Ca Cl_2. Solutions of low pH were made with appropriate concentrations of acetic acid. Examples of these experiments are presented in Fig. 6 (dashed curves). Three curves are shown with choline to illustrate the types of variability frequently encountered. Often discontinuities in the Vm - [H] relation are seen. As yet the variability in the results was too great and the points too few in any given fiber to make many quantitive stipulations. However, a saturation of a type is evident. It is not very clear from Fig. 4, but it is more evident in Table 1. The fiber in Fig. 6 was chosen primarily to show the pH dependence of Vm in "pure" H-ion solutions. The same fiber was used in five other runs displayed in the figure (cf. Fig. legend). The increased sensitivity to H-ion usually starts at about pH 4-5. At pH about 2.8-3.3 it becomes "close" to a Nernst dependence in some of the fibers. In others, where it did not approach a Nernst dependence, it may have done so, had it been possible at the time to do experiments at even lower pH. But a further appreciable reduction in pH was not feasible by increasing acetic acid (it was about 10 mM at that time). In addition the fibers do not recover very well from a lower pH. The problem is being pursued. Thirty axons were used. The data for the fibers not included in Fig. 6 are presented in Table 1 (axons 16-44). With some fibers at 550 mM $[KCl]_o$ no appreciable decrease was observed at low pH. At high pH with NaCl and choline chloride, an increase in Vm was observed. High pH was obtained by adding appropriate amounts of tris or phosphate buffer or NaOH.

---

*In Ref. (9), p. 1257, there is a typographical error: pH 5.7 should read pH 4.7.

TABLE 3

Effects of Varying Cationic Mixtures on the Membrane Potential of the Giant Axon of Loligo Vulgaris

$\left(\frac{[K]}{[Na]}\text{ as mM, Vm as mV}\right)$

Axon Code	$\frac{[K]}{[Na]}$ $\frac{550}{0}$	$\frac{333}{217}$	$\frac{100}{450}$	$\frac{55}{495}$	$\frac{33.3}{517}$	$\frac{10}{540}$	$\frac{5.5}{545}$	$\frac{3.3}{547}$	$\frac{1}{549}$	$\frac{.55}{550}$	Cation Varied	Cation Fixed
1	69→65		36	25	20	9	5		4		$[K]_o/[Na]_o$	$.55M[Na]_i$
2.3(2)	67→61	50	30		15	7	6		2		"	"
3	63→59		29		19	4	2		2		"	"
4.4(2)	61→53	48	26		13	3	4			0	"	"
5	65→57		27	20	13	5		5		-1	"	"
6.5(2)	64→55		23		18	4		3		-2	"	"
7	-68→63	-60	-58 $\xrightarrow{d}$ 40	-32	-16	-7	-6			-3	$[K]_i/[Na]_i$	$.55M[Na]_o$
8	-62→59	-57	-55	-52 $\xrightarrow{d}$ 33	-20	-7		-7		-4	"	"
9.14(2)	-65→60	-54	-45	-30	-15	-9		-6		0	"	"
10	-63→56		-40	-28	-12	-8	-7			1	"	"
11.15(2)	-59→54	-51	-36	-24	-7	-5		0		3	"	"
12	-61→57		-39	-20	-9	-6			1		"	"
13	-54→53	-47	-29		-8		-5			0	"	"
14	1	21	40		54	71	72			77→74	$[K]_i/[Na]_i$	$.55M[K]_o$
15.25(2)	-4		37	41	46	51	53			61→53	"	"
16	-5	14	34		49	59	58		65→61		"	"
17	0		38		56	67		70		73→69	"	"
18.27(2)	-7	8		34	47	55	55			66→57	"	"
19	-8	12	23		40	49		51		59→51	"	"

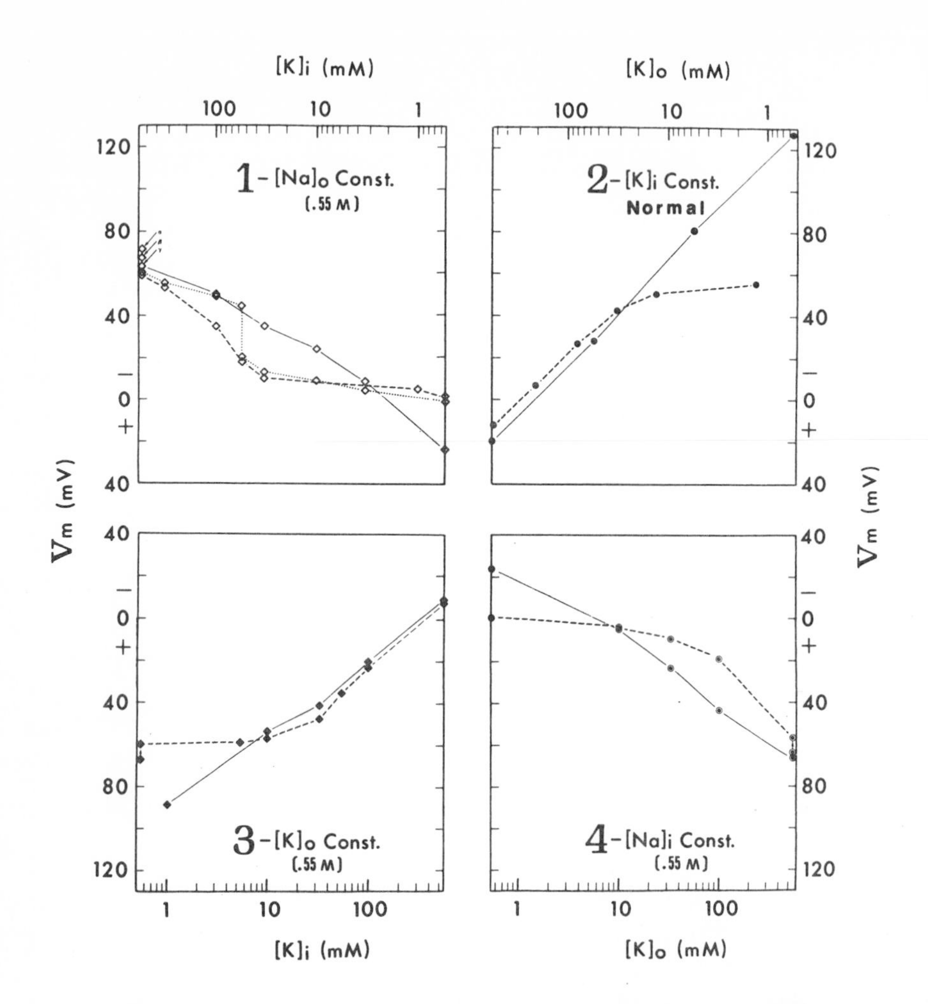

FIG. 7. The effects on the giant axon of Loligo vulgaris of [K]/[Na] mixtures (dashed and dotted lines) and of [K] in pure solution using sucrose to maintain tonicity (solid line). In each of the 4 quadrants the experiments were carried out on the same axon. In all quadrants experiments commence at 550 mM [K], proceed stepwise to the lowest [K], and abruptly end at 550 mM again. The experiment where [K] was varied in pure solution was carried out first. In quadrant 1, three experiments were carried out on the same fiber. The experiment represented by the dotted line was carried out first. In quadrants 3 and 4 the arrows at 550 mM $[Na]_i$ or $[K]_o$ indicate the difference between the initial measurement of Vm and the final semi-steady-state measurement. In quadrant 1 the initial Vm is indicated with α, β, γ, respectively, for the Vm of the solid, dotted, and dashed lines. In

(2) Axons Perfused with Isotonic Na Cl. In these experiments only $[K]_o$ was varied in the presence of NaCl. An example of these experiments is presented in quadrant 4, Fig. 7. Saturation is evident. At high (550 mM) $[KCl]_o$ (containing also 3 mM $CaCl_2$), the internal positivity was not stable and the reversed Vm tended to decrease. Six additional fibers were used and saturation was seen in all. The data, with the exception of those of the fiber presented in Fig. 7, are presented in Table 3 (axons coded 1-6). At 550 mM $[K]_o$ the internal positivity decreased within 3-7 min by up to 9 mV.

b. Variations of Intracellular Cation Mixtures. Only K was varied in these experiments in the presence of Na. Two types of experiments were undertaken: (1) Intracellular cation mixtures were varied in axons immersed in isotonic NaCl containing 3 mM $CaCl_2$; or, (2) in isotonic KCl with 3 mM $CaCl_2$.

(1) Axons Immersed in Isotonic NaCl. In six out of seven axons, the external medium contained 3 mM $CaCl_2$. In five of these, spontaneous firing was seen. The problems of the control of spontaneous firing are mentioned in Sec. II (Methods). An example of these experiments is presented in quadrant 1, Fig. 7. One run showed discontinuities in the Vm - $[K]_i$ relation; the other did not. A saturation is evident at low $[K]_i$ concentration, though in some fibers saturation at high concentration was also noted. Of the axons investigated, in addition to those presented in the Fig. 7, discontinuities were observed in two. Data for the additional axons are presented in Table 3 (axons coded 7-13). In axons where discontinuities were observed, the slopes of the Vm - log $[K]_i$ relation at higher $[K]_i$ were less. The initial Vm at 550 mM $[K]_i$ was transient. Within 2-5 min after changing the solutions the Vm (internal negativity) declined by up to 7 mV.

(2) Axons Immersed in Isotonic KCl. The external medium also contained 3 mM $CaCl_2$. An example of this type of experiment is presented in quadrant 3, Fig. 7, dashed line. A saturation is evident at low $[K]_i$. Saturation appeared in the other fibers used in this type of experiment.

---

quadrant 1 $[NaCl]_o$ is kept constant at 550 mM and the internal composition varied as indicated. In quadrant 2 the axons contain the normal axoplasmic constituents and the extracellular medium is varied as indicated. The experiment indicated by the dashed line in quadrant 2 is taken from Curtis and Cole (8). In quadrant 3 $[KCl]_o$ is kept constant at 550 mM and the internal composition varied as indicated. In quadrant 4 $[NaCl]_i$ is kept constant at 550 mM and the extracellular composition varied as indicated (cf. text for further details.)

Six other fibers were used. Data for these additional axons are presented in Table 3 (axons coded 14-19). The initial Vm at 0.55 mM $[K]_i$ was higher (by up to 9 mV) than the later (at 2-4 min) semisteady-state value.

2. Chara Internode (Variations of Extracellular Cation Mixtures)

In these experiments only H-Na mixtures were investigated. Low pH was usually controlled with fresh gluconic acid. To attain a pH of 3.7, approximately 40 mM of gluconic acid were required in the presence of 33.3 mM Na gluconate. The effect of this high concentration of gluconic acid was controlled by attaining low pH with HCl and acetic acid. The results with the three acids were grossly similar, at least within the variability seen in the same internode, and from internode to internode. Sixty-two internodes were employed. As in all Chara experiments, it was difficult to compare internodes, since the Vm of internodes varied considerably. In Fig. 6, solid line, square points, an atypical experiment is presented. Actually, this experiment was carried out on the same internode, where data were collected for Fig. 5 and for the experiment represented by the solid line and concentric circle points in Fig. 6. It is noted that some saturation may be evident. With the exception of this one fiber the results on Chara will be presented elsewhere. They are being held back until the large variabilities in Vm are understood. Below about pH 4.0-5.0 an increased sensitivity to H was usually observed. Around pH 3.0 the Vm-log $[H]_o$ relation was close to a Nernst relation. Again, in this region discontinuities were occasionally observed in the Vm - log $[H]_o$ relation.

IV. DISCUSSION

A. Observations by Other Investigators

There are some almost hidden experiments in the literature that may be interpreted as indicating that ionic strength affects the membrane potential. Lorente de Nó has observed an increase in the demarcation potential of nerve immersed in isotonic sucrose (10). Curtis and Cole (8) and Hodgkin and Katz (11) recorded a small hyperpolarization when NaCl in sea water was substituted with dextrose. They seemed to have dismissed it. In their experiments, divalent ions in the external medium were retained and from Fig. 4 it can be seen that the hyperpolarization under these circumstances may be small. In experiments using the sucrose-gap technique [e.g., (12-18)], a number of investigators have observed very large hyperpolarizations. This hyperpolarization was dismissed as being an artifact inherent in the technique and due to the sucrose-axon-sea water liquid junction contribution [e.g., (18)]. The hyperpolarization observed in experiments using the sucrose-gap technique is probably due to hyperpolarization of the membrane under the sucrose solution (7). Stämpfli (19), in the node of Ranvier, substituted 80 mM of the NaCl in Ringers with sucrose, in an attempt to ascertain the contribution of chloride to the resting potential. He found instead a hyperpolarization of 19 mV. This is probably an ionic-strength effect since this author, in unpublished experiments in the node of Ranvier (three preparations only), found that the Vm - log $[KCl]_o$ in pure solutions shows little saturation at 0.1 mM $[K]_o$, and that the Vm - log $[NaCl]_o$ showed slopes of 20-29 mV. The effects of dilution of extracellular divalent ions (cf. Fig. 4) on the Loligo axon have recently been repeated by Tasaki (20). Finally, some of the experiments presented in Fig. 7 have been reported previously by Baker et al. (3, 4). These included some of the measurements in the experiments in quadrant 1, and some of the experiments represented by the dashed curves in quadrants 3 and 4. In experiments similar to those represented by the solid line in quadrant 1, Fig. 7, they occasionally observed an increase in internal negativity as $[K]_i$ was decreased. This they found to be limited to the higher $[K]_i$ ranges. This was not observed in the present experiments; maybe the number of our fibers was limited. In the present study (cf. Table 1) a "saturation" at higher $[K]_i$ was usually more pronounced in those fibers showing discontinuities (see Tables 1 and 2).

B. Possible Artifacts or Misinterpretations

The possibility that the hyperpolarization upon dilution of $[EL]_o$ is an artifact, due to some liquid junction potential, as has been suggested [e.g., (18)], was probably excluded in control experiments reported previously (7). To these control experiments may be added the experiments on the Chara internode which was immersed in its entirety in solutions of varying ionic strength. The possibility that in Chara the effects of ionic strength are through an effect on the "sealing" of the pipet (even when introduced with the new technique through the nodal cells and having the external nodal region immersed in oil or Vaseline) is obviated by the observation that the effects of ionic strength were similar using the Osterhout air-gap technique. The possibility that the dilution effects are an effect of sucrose per se is unlikely, since the effect is observed with glycerol (as the osmotic substitute) and transiently in the Loligo axon (and irreversibly) when no nonelectrolyte was employed (when tonicity was not held constant). Similarly, in Chara the dilution effect is "steady" and is observed whether or not sucrose or glycerol are employed to maintain tonicity constant. The possibility that the ionic strength observations are artifacts is almost excluded by the following experiments reported in a previous communication (7): treatment of the axonal surface with heavy metals [$AgNO_3$ or AuCl, Fig. 2, Ref. (7)], perforation of the membrane with a sharp instrument, or high intracellular pressures resulting in a reduction or reversal of the membrane potential. The observed potentials were of the order expected for diffusion potentials (7). The results with sharp instrument perforation or heavy metal treatment were irreversible. The effects with high internal pressures were frequently reversible. At high internal pressures or following heavy metal treatment, the effluxes of anionic tracers were equal to those of cationic tracers (7). That high internal pressures "perforate" the "membrane" was evidenced by the following: (a) Tritiated water effluxes were markedly increased. (b) A sudden efflux of large molecules (i.e., radioactive starch and dextran) was observed at high pressures. This abrupt efflux was "shut off" abruptly upon returning to low pressures. This is reminiscent of the state of the erythrocyte membrane during and following hemolysis. (c) The electrical resistance of the membrane was markedly reduced. The effects of heavy metals are ambiguous. They probably do not merely neutralize charges, since the membrane electrical resistance using the pulse technique is decreased [see Ref. (7)]. At low $[EL]_o$ the Vm (internal negativity) was increased by up to 30 mV upon violently stirring the fluid compartment containing the preparation. (Usually the increase is smaller.) This may be due to film diffusion [e.g., (21)].

C. General Interpretations

The pertinent results are the following: (a) The Vm - log $[EL]_o$ and the Vm - log $[EL]_i$ relations in pure single-ion solutions are an almost straight, saturation-free line over a wide concentration range. This holds true for other cations than K as well. (b) The Vm - log $[EL]_o$ and Vm - log $[EL]_i$ relations for K and H in the presence of other cations show saturation at low [K] or [H]. Results (a) and (b) are contrasted in Fig. 7 (for K). Occasionally (e.g., Figs. 5 and 6), the slope of the Vm - log $[EL]_o$ relation is near theoretical Nernst for ions other than K as well (e.g., H, Na, and Ca). These results are not implicit in the independent channel idea [for review cf. (22)]; but, they are not excluded from this approach either. Alternatively, if the less than ideal Nernst slopes in single-ion solutions were proved to be due to something like an anionic permeability contribution, then these results would be expected from the ion-exchange cation selective approach (23-26), though, even then it would not prove that an ion-exchange system is involved. Either the Goldman (6), or something like the Eisenman (26), equation could be appropriately modified and used to describe such results as presented in this paper. The presence of discontinuities in the Vm - log [EL] relations is superficially inconsistent with a simple ion-exchange approach.

The ion-exchange properties may reside in the membrane and/or the axoplasm. If one assumes that at rest, part of the axon membrane proper is impermeable to Na and anions and part permeable to K and chloride, the resting potential may be regarded in part as a gel potential (something like a Donnan potential) of the axoplasm (in perfused axons of the residual axoplasm); however, this is highly speculative.

D. Other Comments

1. In Fig. 7, quadrants 1 and 4, the straight solid lines at dilute [K] cross over the dashed curved lines. This was a consistent finding (axons coded 1-13, Table 3 and axons coded 1-5, 11-15, Table 2). This may be explained as follows: The system in quadrants 1 and 4, when $[K]_i$ or $[K]_o$ is 0.55 mM (which is close to zero) may be behaving as a Na concentration cell. This is consistent with two other sets of experiments mentioned in Sec. III. A. 1.a. (2)(b) (and axons 6-10, Table 2) and Sec. III. A. 1. b. (1)(and axons coded 16-23, Table 2). These other experiments were undertaken to demonstrate in a pure sodium system, that the Vm may be dependent upon the "transmembranal" Na concentration gradient. Ignoring the 0.55 mM $[K]_i$ or $[K]_o$, and considering the inside or the outside of the axon as having zero Na (or 2 mM), the range of values given in Table 3 for the experiments in quadrants 1 and 4 correspond roughly to the range of values in Table 2 for the experiments presented in Secs. III. A. 1. b(1) and A. 1. a(2)(b). The range of values for the Vm given in Table 2 for the experiments in quadrant 1, Fig. 7, and in Sec. III. A. 1. b(1)--with practically zero $[NaCl]_i$ and $[KCl]_i$ and isotonic $[NaCl]_o$--is somewhat different than the range of values given by Baker et al. (4) for the Vm in similar experiments. The internal positivity in our experiments was appreciably larger (Table 2). One partial explanation may be that, at these dilute solutions, violent stirring is important and Baker et al. did not stir their preparation aside from the stirring incident to the flow of the internal perfusate. Another possibility is that the author waited for 3-30 min for this effect. In this connection, a parenthetical notation of Narahashi (32) may be relevant. He noticed in one axon that, if he waited long enough, the Vm was 36 mV (inside positive) with $[K]_i$, $[K]_o$, and $[Na]_i = 0$, and $[Na]_o$ = that of sea water. This is within our range of values.

A simple visual manipulation of Fig. 7 may be helpful in making some of the points of this paper clearer and may reveal, in the process, a kind of "transmembranal" symmetry. The quadrants of Fig. 7 have been disjointed for the sake of clarity in the presentation of the data. Now, let us join them and idealize the data somewhat. They should be brought together so that the zero Vm point is shared by all quadrants. One notes that the dashed lines (the lines showing saturation) meet at zero Vm. The dashed lines of all quadrants now constitute a figure resembling a swastika. The solid lines of quadrants 2 and 3 should be (or probably are) continuous and should intersect the swastika at zero Vm, when $[K]_i = [K]_o$ (see other part of this Discussion section on the experimental conditions of quadrants 2 and 3). The aforementioned essentially states that the membrane-residual axoplasm complex is (with respect to these measurements) symmetrical. The solid lines of quadrants 1 and 4, however, are not continuous, do not go through zero Vm, and if extended diagonally, are parallel. This may be only a visual lack of diagonal symmetry. In the opinion of the author, (see other part of the discussion), this parallelism or this difference in Vm between the two points in the V-axis intersected by the solid lines of quadrants 1 and 4 (at zero $[K]_i$ or $[K]_o$, respectively) is due to the potential difference of the Na concentration cell existing at this point across the cell. Some experimental evidence for this is mentioned in the discussion.

One last point in reference to Fig. 7--there is some asymmetry between the experiments presented in quadrants 2 and 3. Several explanations are possible: (a) The axons in the experiments of quadrant 2 had "normal" axoplasmic constituents and lower $[K]_i$ than in quadrant 3, 550 mM KCl. (b) It may in part be due to an asymmetry in the membrane-residual axoplasm complex. (c). It may in part be due to unknown experimental differences. (d) It may in part be related to the fact that the dashed curve in quadrant 2 is taken from Curtis and Cole's experiment (8) on Loligo pealii and the solid line from this author's experiment on Loligo vulgaris. (e) It may be related to the fact that in internal perfusion experiments (with KCl) the axon is poor.

2. The selectivity sequence K > Rb > Cs > Na > Li in Chara and Loligo is similar to Eisenman's selectivity sequence IV for glass electrodes (26). Frequently, in the older literature and more recently [e.g., (27)], the sequence K > Rb > Cs > Na > Li was reported as the sequence for decreasing ability for depolarization of biological membranes by extracellular alkali cations. Baker et al. (3) have also observed it in internal perfusion experiments. They compared Vm obtained with isotonic alkali metal chloride solutions in the perfusate. For a recent review on biological specificity sequences, one may refer to Diamond and Wright's article in the Annual Review of Physiology (28).

3. In the experiments on the effects of pH on Vm--both in pure solution and in the presence of other cations--the region of the pH below which an increased sensitivity to H was observed (or the region below which the axon began to behave like a poor pH electrode) was around pH 4.0-5.0. This coincides with the pK of such groups as carboxylic and phosphoric [e.g., 29, 30)]. The experiments of the author are probably related to the experiments of Rojas (31), who reports a decrease in conductance for hyperpolarizing currents of the membrane of the squid axon at pH below 4.5-5.0, with isotonic K as the only internal and external cation.

4. One obvious interpretation of the high membrane potential of Chara and Nitella internodes is the low ionic strength of the normal bathing medium (fresh water). The membrane potential of the internode in isotonic sodium chloride is within the range seen in Loligo axons or vertebrate cells.

5. There are a number of shortcomings in the experiments or generalizations presented in this paper. Currently, some of these are in the process of being rectified.

a. The pH effect may be considered by some as a "rotting" of the membrane. The observation that the slopes of the Vm - log $[H]_o$ relations sometimes become increasingly closer to the Nernst limits at lower pH, may be regarded as being circumstantial. This may be the case. In favor of this idea is the fact that the measurements of Vm at low pH are scattered and not very reproducible. The pH effect however is reversible. More experiments are needed before one proposes that the axon behaves as a pH electrode. For example, I hope to study the Vm - log $[H]_o$ and Vm - log $[H]_i$ relations in pure solution and in H-cation mixtures simultaneously (i.e., by changing both $[H]_i$ and $[H]_o$ in the same cell). It is important also to extend the pH range to lower pH regions, especially in the Chara internode, and to measure internal pH while varying external pH. The internal pH may change upon changing $[H]_o$.

b. Baker et al. (4) occasionally found, at higher $[K]_i$ ranges, an increase in Vm as $[K]_i$ was reduced. Although this author did not see this, it is of some concern.

c. The experiments on internal perfusion, involving sucrose as an osmotic substitute, are very disturbing since the time involved (as long as 30-45 min) for steady-state measurements is much greater than one would casually expect from diffusion processes. Evaluation of some results should probably be held in abeyance until this point is made clear. In artificial ion-exchange membranes the effect of stirring, on the in-series potential drops in the stagnant layers adjacent to the membrane, may take time. This time constant may be accentuated in highly viscous sucrose solutions. In dilute [EL], stirring affected the Vm of the axon of Loligo.

d. Practically all the experiments on the squid axon were undertaken with chloride as the counterion. The survival time was shorter and the results more variable, less stable, and less reproducible than when other, more "favorable", anions were used. When this was realized, the author was too deep in the problem to start from the beginning again and he compromised or improvised by increasing survival time by periodic internal treatment with KF. Still the experiments are not completed. Had the experiments been done with the more "favorable" anions, anionic contribution might have been evaluated as well. Preliminary experiments show that the survival time in gluconate is much longer than in chloride. Presently an attempt is being made to amend this mistake. Hopefully, the resulting data will be less variable and more quantitive than presented in this paper.

e. In Chara, the internal composition is unknown. This is especially important in experiments like the one presented in Fig. 5 and part of Fig. 6. Hours were allowed for equilibration in gluconate and the experiment lasted for 27 hr. At these times, internal anions may have exchanged with gluconate and the concentration of the cations is unknown. This may in part account for the extremely large variations in Vm in different preparations and in the same preparations as a function of time. These large variations made impossible any comparison of data between fibers or any quantitative analysis of the data. It is contemplated, instead of measuring the internal composition, to actually control it by using the internal perfusion technique in Chara (which apparently is feasible).

V. SUMMARY

A. The relation between the membrane potential and the logarithm of the concentration of electrolytes (Vm - log [EL]) was studied in Loligo giant axons and Chara internodes. Hydrogen, alkali metal alkaline earth, choline, and tetraethylammonium cations were used. The counterions were chloride and gluconate.

B. In Chara internodes, the Vm - Log [EL] relation was studied only in the extracellular medium. Vm in Chara internodes was measured using a new technique which is described above. In Loligo axons the Vm - log [EL] relation was studied both in the extracellular medium and the intracellular medium. The latter was accomplished with the "internal perfusion" technique.

C. The Vm - log [EL] was studied in "pure" single-ion solutions and in solutions containing mixtures of cations.

D. In "pure" single-ion solutions the Vm - log [EL] relation was a straight-line relation, a saturation-free relation. Under certain conditions the slope of this relation is near Nernst for all cations examined. Under other conditions it is near the theoretical Nernst only for K and sometimes, at low pH ranges, for H ions.

E. The Vm - log [K] or the Vm - log [H] relation in solutions of mixed cations did not appear to be a straight line; it showed saturation at the lower concentration range. Only at the higher concentration ranges did the slope of this relation often tend to approach the theoretical Nernst slope.

F. At isotonic $[EL]_0$ the Vm of the internode of Chara and the axon of Loligo varied with the cation employed. The selectivity sequence was choline > Li > Na > Cs > Rb > K. Vm was lowest in K.

PART 2

INSIDE-OUT GIANT AXONS OF LOLIGO VULGARIS

I. INTRODUCTION

Recently, the intracellular composition of some cells (e.g., squid giant axon, barnacle muscle fiber) was altered with the injection and perfusion techniques. The effects of intracellular addition of calcium (20, 33), TEA (34), TTX (35), enzymes (36), redox agents (37), reduction of ionic strength (38), changes in pH (39), substitution of internal KCl with other alkali cations (3), and a comparison of all these effects to those obtained with similar extracellular perturbations prompted most investigators to regard the membrane per se as being asymmetric.

Certain models of biological membranes also show asymmetries (e.g., glass electrodes, bilayers, fixed or mobile ion-exchange artificial membranes). In the glass electrode the asymmetry is somewhat obscure [see Ref. (40) for a recent review]. In bimolecular lipid membranes the asymmetry is revealed by addition of certain substances that presumably adsorb to one side of the membrane (41). In artificial ion-exchange systems certain asymmetries are revealed when the charged system shows gradients in charge density or when a positive and negative membrane are juxtaposed [e.g., rectification, Ref. (42)].

In this paper a method is presented whereby the giant axon of Loligo vulgaris is everted, turned inside out (much like a stocking). The results obtained are consistent with the idea that the membrane-residual axoplasm complex is asymmetric.

II. METHODS

The lateral (tunneled) or stellate giant axon of Loligo vulgaris was employed. Tapered axons were selected. Such axons were found more frequently in "fat" squid. The maximal diameter (excluding bulges) was from 900-1900 μ. Bulges 2-2.75 mm in diameter and 4-8 mm in length were found in three squid near the stellate ganglion in the stellate nerve and the lateral axons. The larger diameters are often found in Loligo vulgaris, whose mantel length is about 1 $\frac{1}{2}$ ft or over (in one instance, 2 ft long). The investigator reverted to this problem upon obtaining large squid. This was somewhat infrequent, so the pursuit of the problem was intermittent. The initial length of the axon was 4-6 cm. The tapering varied from 10 to 40% (of the diameter) per cm axonal length. Chances of obtaining tapered fibers were greater in lateral tunneled axons (radiating from the stellate ganglion) than in the stellar nerve. The tapering usually was not smooth.

The axon was first cleaned circumventing a small amount of connective tissue every 1 $\frac{1}{2}$-2 $\frac{1}{2}$ mm or so on opposite sides of the axon. Subsequently, the axon was perfused by the techniques shown in Ref. (7), Figs. 1B or 1E. A photo of the cannulae employed is shown in Fig. 8(a). The amount of axoplasm was reduced by erosion and sometimes also by addition of chymotrypsin to the perfusate. Fine (20-100 μ) nylon threads (from a nylon stocking) were employed to apply ligatures around the periodic remnants of connective tissue, as illustrated in Fig. 9. The same types of threads were used for cannulation. The threads were previously coated with bee's wax or rubber cement or soaked in melted mixtures of dental sticky wax, parafin, and mineral oil. In conjunction with the rubber coat on the cannulae, this also prevented cutting or damage to the axonal surface upon cannulation. Added insurance during cannulation was achieved by having two constrictions in the cannula [see Fig. 8(a)]instead of one. Figure 11 illustrates another method employed in applying periodic ligatures. The threads were glued to one end of a tungsten wire (25 μ diameter and about 5 mm long). The other end of the tungsten wire was sharpened to a point electrolytically,

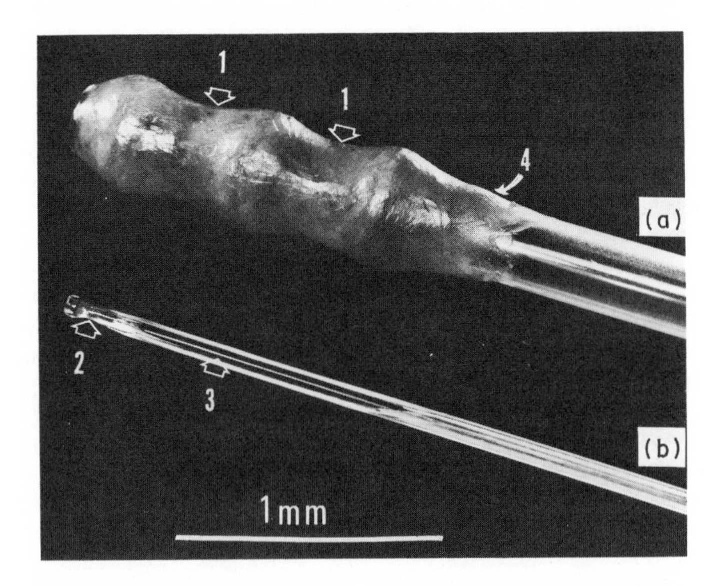

FIG. 8. (a) Cannula. (b) Probe. 1 and 2, constrictions for applying ligatures. 3, slit in probe for perfusion. 4, Rubber cement. Diameters varied to suit diameter of the axon.

a technique similar to that used in making metal microelectrodes. In later experiments Ethicon Inc. (Somerville, N. J.) ophthalmic micro-needles--electrolytically thinned and sharpened, with thread attached--were used. The needle, during "sewing" of the tissue surrounding the axon, was manipulated with a "fine" pair of forceps whose sharpened tips were coated with rubber. This was done on three axons. The ligatures were applied first and subsequently the connective tissues surrounding the ligatures were removed. The axon was subsequently mounted [as indicated in Fig. 9(a)] on a chamber composed of two compartments. The nylon threads from the ligatures were fixed to plastic screws (mounted on the chamber platform) and made taut by advancing or turning the screws.

The partition between compartments was made of thin rubber sheets (20-50 μ in thickness) manufactured by Young's Drug Products Corp. (Trenton, N. J.). The rubber was frozen solid with Dry Ice or liquid nitrogen and a hole was drilled, whose diameter was close to the diameter of the axon. In some instances thin partitions were prepared by drying Plexiglass or Formvar previously thinned with ethylene dichloride.

The hole in the partition was threaded with the axon. Since the axon was tapered it could be pulled through until the rim of the pore made a slight indentation in the surface of the axon. In this manner the partition was made almost leak proof. In this region the axon was not completely cleaned. KF (0.54M, pH adjusted to 7.35 with 4mM phosphate) was added to compartment A and sea water to compartment B. A probe of the type shown in Fig. 8(b) was used, made either of glass or Plexiglass [see Ref. (7) for pulling Plexiglass]. The proximal (near the partition) cannula was removed, the distal retained. At the onset of insertion of the probe into the axon (through the proximal large end), the axon was perfused from the distal cannula with KF. After partial insertion of the probe, perfusion was commenced also through the lateral slit of the probe with KF. The outlet was the open proximal end. Perfusion resulted in slight inflation of the axon. This was necessary in order to avoid shearing the axonal wall with the probe during insertion. When the probe reached the distal cannula [stage (b), Fig. 9], a ligature was applied around the axon in the region of the constriction of the probe [stage (c), Fig. 9]. Previously, the constricted portion of the probe had been painted with rubber. This, in conjunction with the wax or rubber cement on the threads, again prevented cutting of the axonal surface when the latter was secured to the probe. The distal cannula was removed by cutting the axon between the cannula and the probe. At this stage, the axon was perfused (with KF) only through the lateral slit [stage (c), Fig. 9]. The outlet for the perfusate was the proximal large opening of the axon. The degree of inflation was controlled by a makeshift "valve" near the opening and the pressure head. This was accomplished in part with a semi-loose thread knot around the axon in conjunction with the two

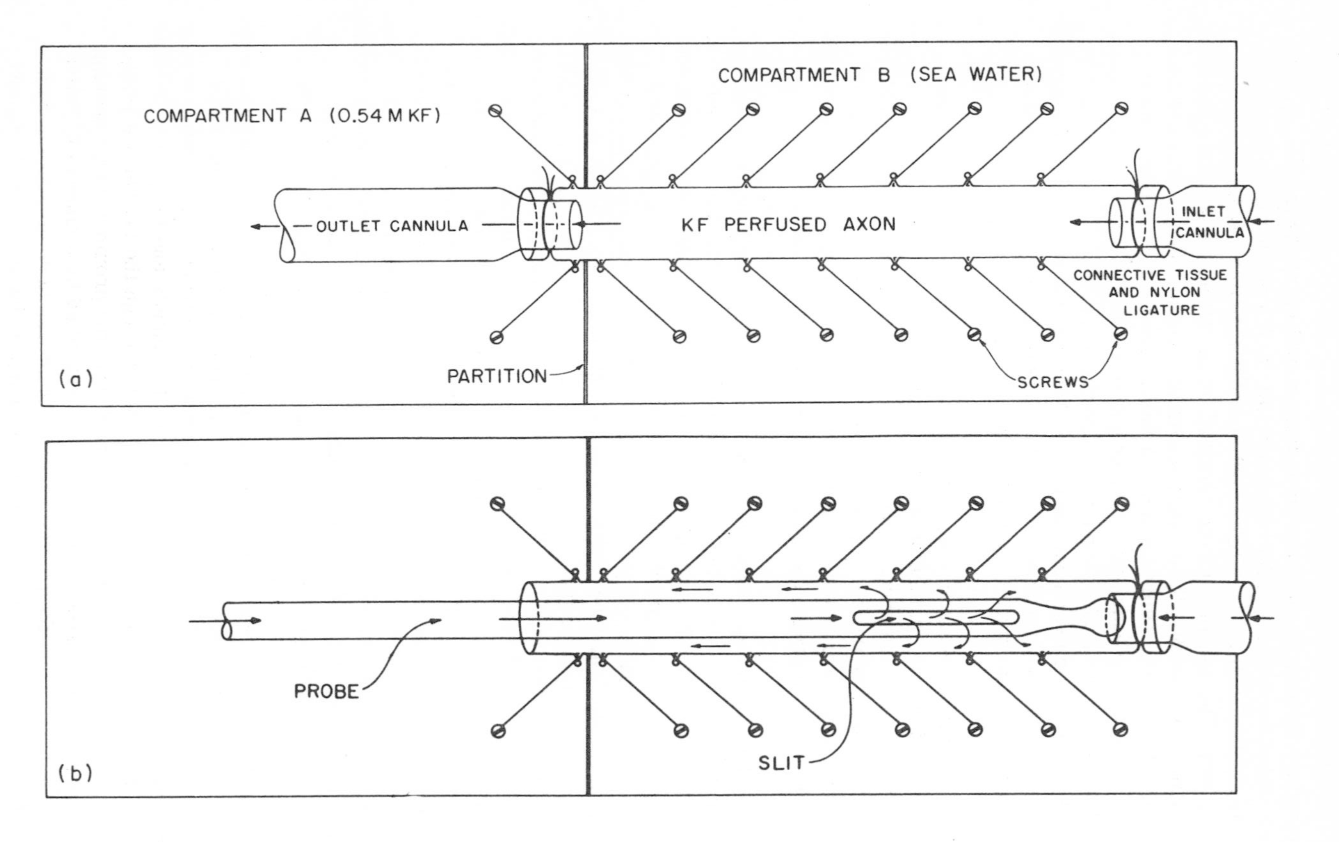

FIG. 9. Diagrammatic illustration of method of eversion of axon.

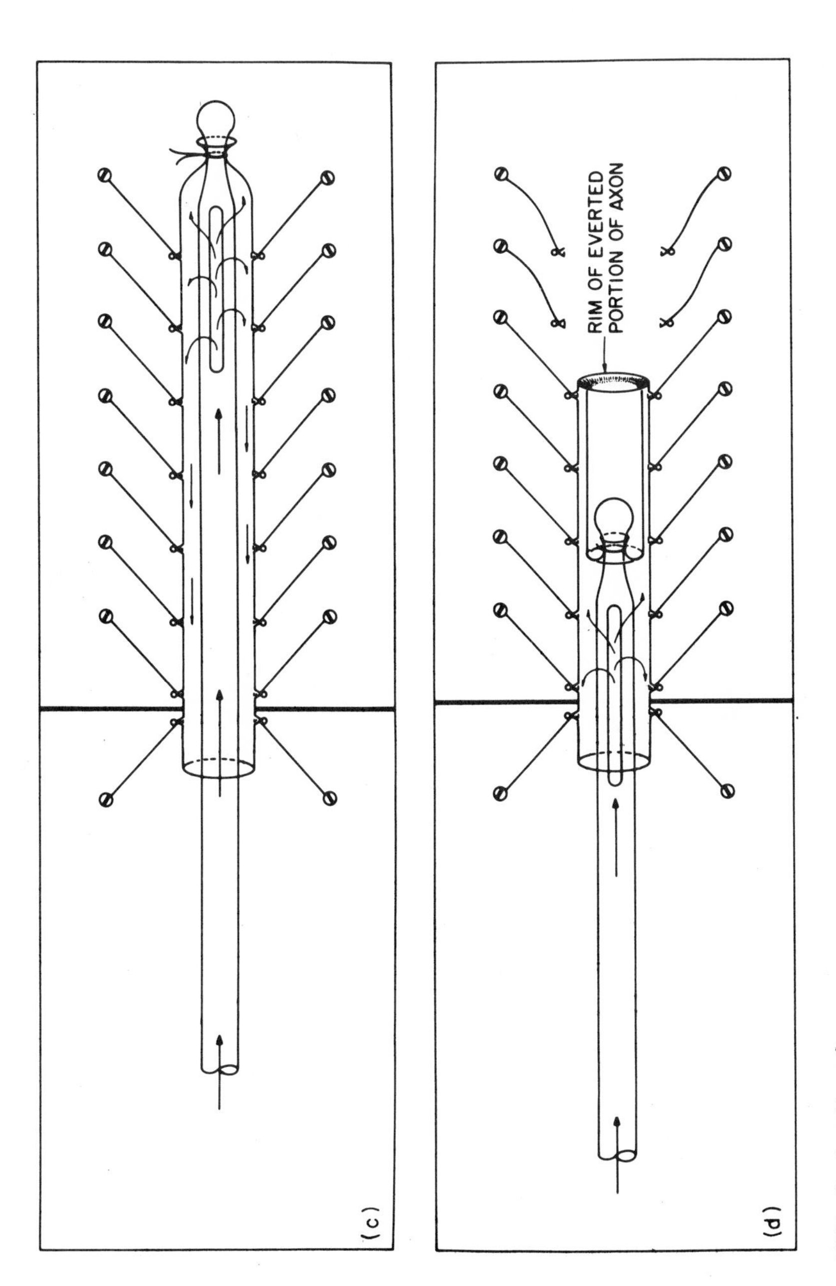

FIG. 9. (continued)

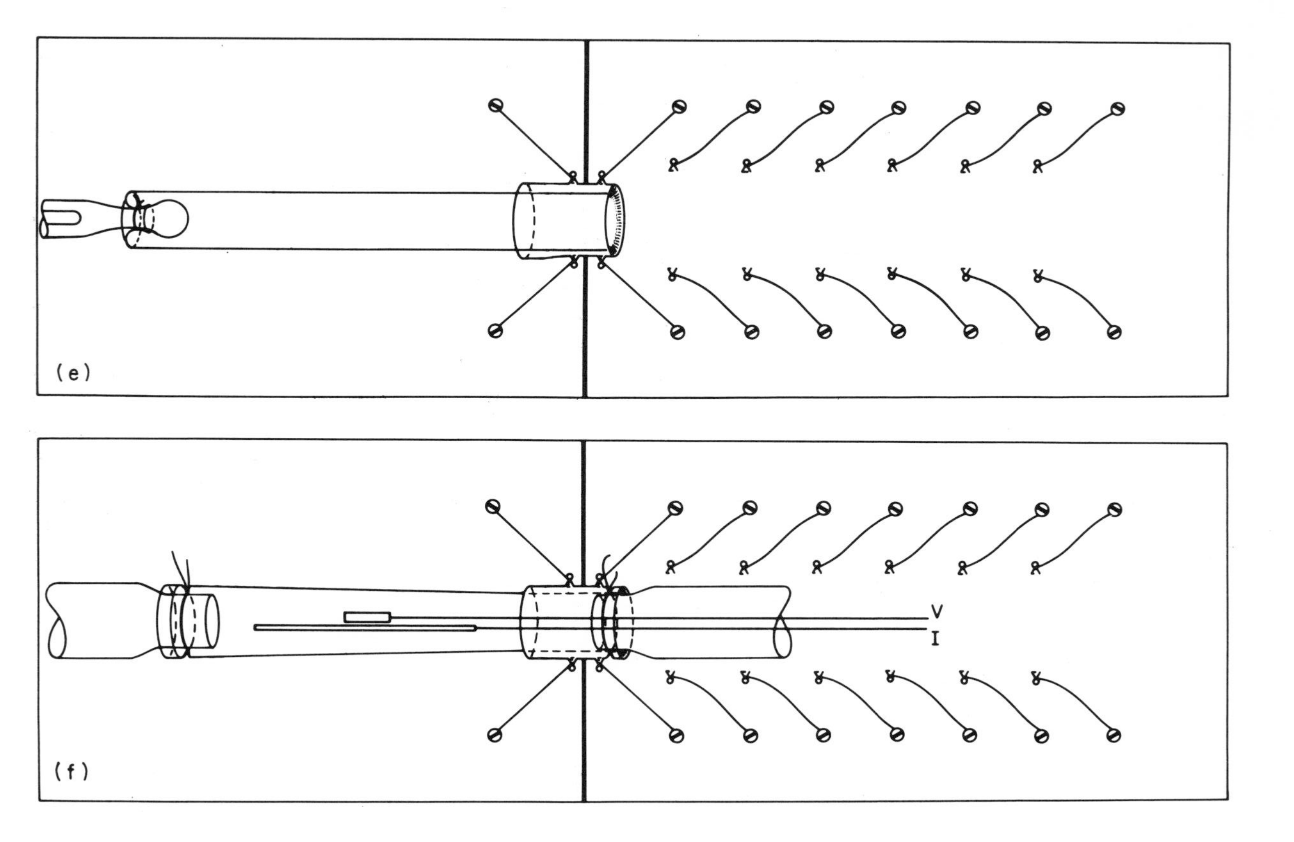

FIG. 9. (continued)

screws in the vicinity of the outlet and the partition. Slowly the probe was withdrawn, resulting in eversion of the terminal portion of the axon [stage (d), Fig. 9]. The withdrawal was momentarily stalled upon reaching the first pair of ligatures on the axon. The ligature together with the connective tissue and fiber remnants were cut with a needle or fine iridectomy scissors. Thus, the axon was more thoroughly cleaned than before. The withdrawal of the probe was then resumed with the concomitant eversion of the axon. The inflation of the axon was continuously adjusted (with the "valve" and pressure head) so that the everted portion of the axon was not inordinately squashed or wrinkled. The periodic ligatures served to prevent the entire axon from assuming an accordion or snake-like shape upon withdrawal of the probe. It appeared that a small amount of axoplasm and a very small amount of connective tissue on the Schwann layer side were beneficial in that they prevented complete collapse of the axon while adjusting the internal pressure.

The sequence of cutting a pair of ligatures and then withdrawing the probe was repeated until the entire axon, with the exception of the portion near the partition, was turned inside out. In the interim, it was occasionally necessary to adjust the lateral tension on the axon with the screws to which the threads of the ligatures were attached. In addition, occasionally it was necessary to elevate the axon. This was accomplished by applying wedges below the axon or by lifting the axon by pulling the threads that held the superficial connective tissue. The above precautions were taken in order that the everted and normal portions of the axon would be more nearly concentric.

When the eversion of the axon was completed [stage (e), Fig. 9], the inside-out portion was located in compartment A and the axoplasmic face (now outer) was bathed in 0.54 M KF (pH adjusted to 7.35 with 4 mM phosphate). The probe was disconnected and the axon double cannulated [stage (f), Fig. 9]. Perfusion was initiated with sea water from the previously distal, now proximal, terminal of the axon. A double platinum wire electrode for current clamp was inserted through the outlet cannula [stage (f), Fig. 9]; one wire (insulated except for 1.5 cm length) served as a current electrode, the other (insulated except for 0.5mm) as a potential electrode.

A variation of the above method of eversion was also employed (Fig. 10). Following the preliminary perfusion of the axon, the distal (to the partition) cannula was removed while the proximal was retained. Perfusion through the proximal cannula (with the proper pressure head) kept the axon inflated. The probe described previously [Fig. 8(b)] was inserted through the distal opening until the constricted portion was in the interior of the axon. A ligature was applied around the axon in the region of the

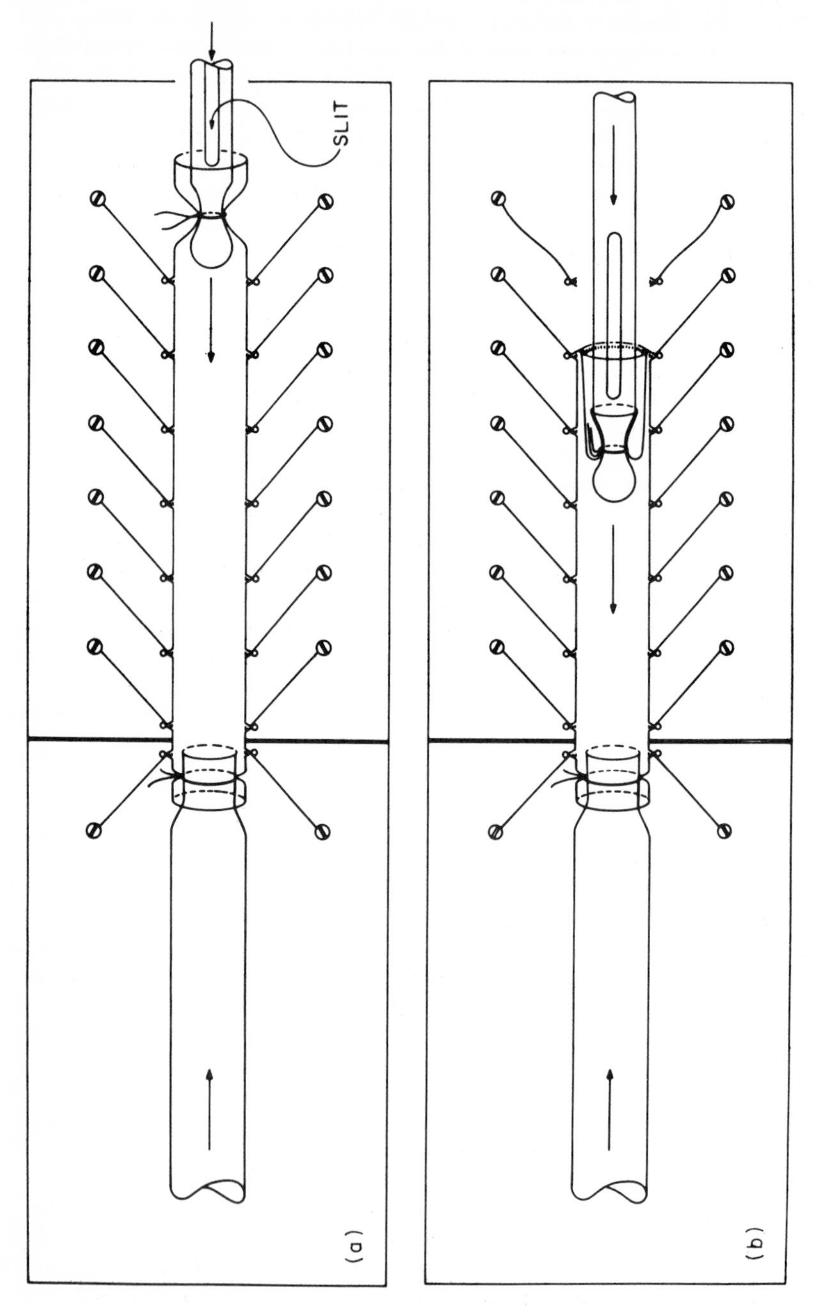

FIG. 10. Diagrammatic illustration of second method of eversion.

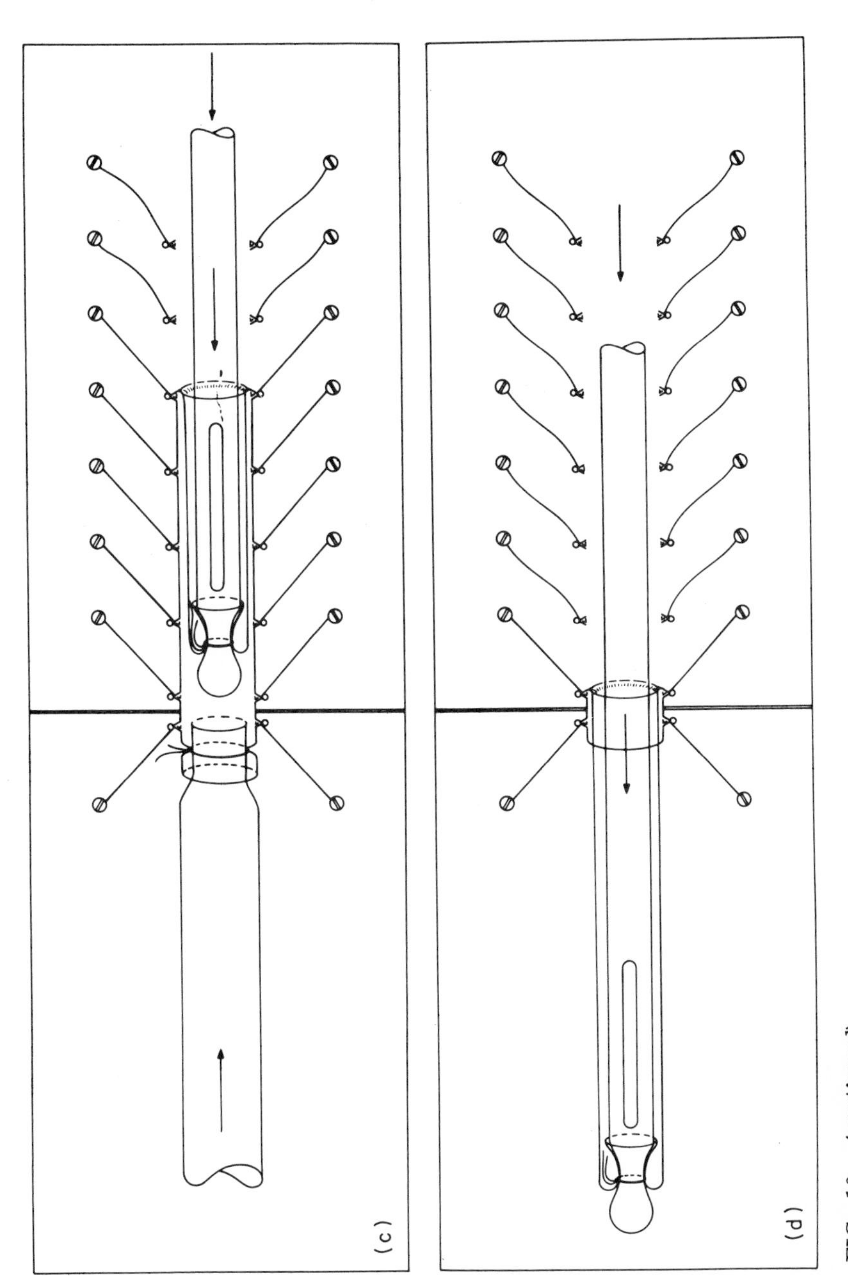

FIG. 10. (continued)

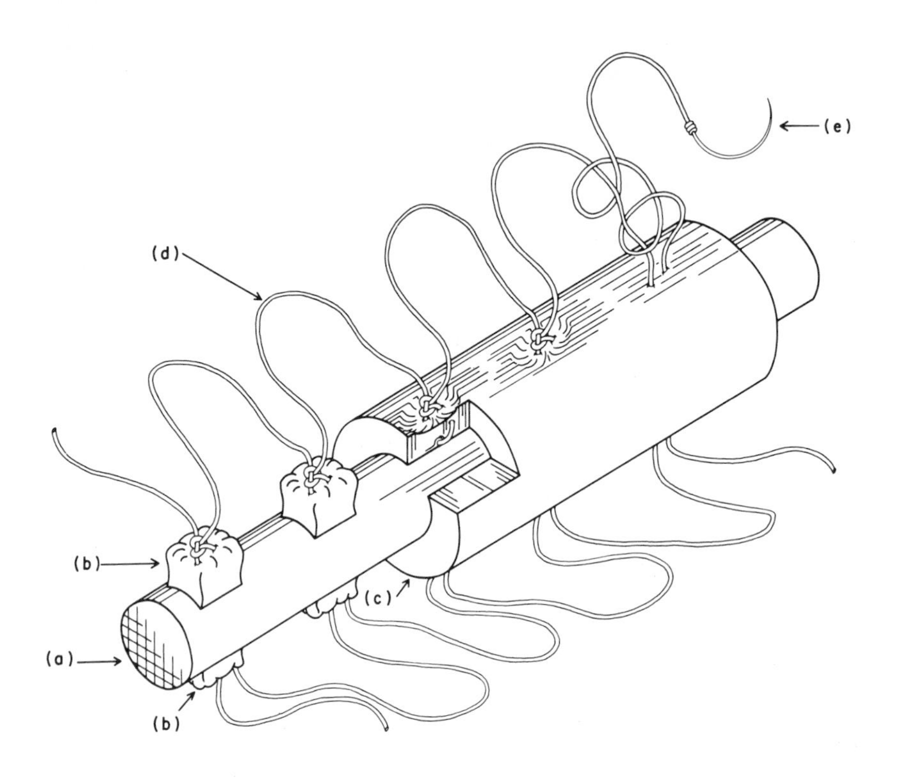

FIG. 11. Diagrammatic illustration of method for applying periodic ligatures to axon. (a) axon. (b) ligature applied to adhering connective tissue. (c) connective tissue and nerve fibers. (d) thread. (e) microneedle.

constricted portion. In this manner the axon was secured to the probe [stage (a), Fig. 10]. Eversion was initiated by advancing (instead of withdrawing) the probe into the axon [stage (b), Fig. 10]. After a certain length of axon was everted, sea water was introduced through the lateral slit of the probe. The outlet was the rim of the axon where actual eversion occurred. Proper inflation of the concentric portions of the axon was controlled by adjusting the pressure head applied to the interior of the axon through the proximal cannula, together with the pressure head for perfusion with sea water through the lateral slit of the probe. When the probe reached the proximal inlet cannula [stage (c), Fig. 10], the latter was removed. The remaining eversion was somewhat difficult since inflation of the outer of the concentric cylinders was not controlled and there was a tendency for it to collapse. When the axon was almost completely everted [stage (d), Fig. 10], and perfusion through the slit was discontinued, the probe was detached from the axon and withdrawn slightly. The proximal (previously distal) portion was cannulated and perfusion with sea water was initiated through the cannula and reinstated through the slit of the probe. Thus at this time, the "interior" (Schwann layer side) was perfused both through the lateral slit of the probe and the now proximal cannula. At this time the axon was properly inflated so that the probe could be removed safely. Upon removal of the probe, a cannula was installed in the now distal, large end of the axon. In two axons the procedure was stopped at stage (d) of the figure. Perfusion was continued. One axon lasted 25 min; the other $1\frac{1}{2}$ hr.

Other techniques (e.g., current clamp, etc.) employed have been described previously [e.g. (7)]. In the everted axon the external solution was electrically ground. At the onset of perfusion of the inside-out axon, the Schwann layer side (now facing the interior of the axon) was exposed to sea water. The axoplasmic side (now facing the outside of the axis cylinder) was bathed in 0.54 M KF.

III. RESULTS

The following experiments were performed on everted axons:

A. Resting and Action Potentials with Sea Water "Inside" and Isotonic KF "Outside" the Everted Axon

Initially in all successful experiments, the axoplasmic side (now facing the outside of the axon) was bathed in KF and the Schwann layer side (now facing the interior of the cell) was bathed in sea water. Under these conditions 13 out of 31 attempts were successful. I suspect that either with more practice or some improvement in the procedure [probably stopping at stage (d), Fig. 10], these somewhat poor yields may be improved.

Action potentials were observed that lasted, if at all, for 1-12 hr. The amplitude of the action potentials was 100-130 mV, the resting potentials 53-70 mV. The undershoot was somewhat lower and the threshold membrane potential under current clamp somewhat higher. The polarity was the reverse of the normal one (Fig. 12). The "inside" was positive at rest. These action potentials were not compared to the action potentials obtained in the normal axon before eversion. This would have necessitated insertion of a current clamp electrode, a step that would have complicated an already tedious procedure and, considering the tediousness of subsequent manipulations, would have been meaningless anyhow. In one axon heart-like action potentials were observed similar to those of Adelman (43).

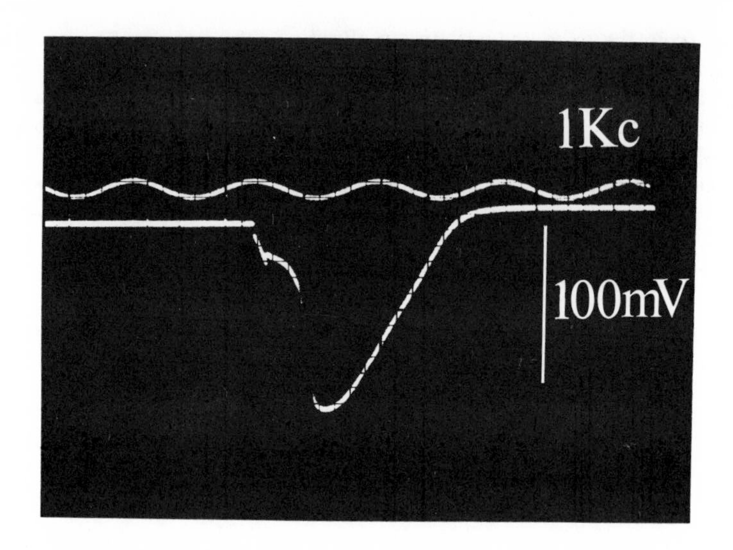

FIG. 12. Action potential from everted axon. Downward deflection "inside" negative. 15° C. 1 Kc. Maximal axonal diameter 1560 μ.

B. The Effects of Sea Water on the "Outside" (Axoplasmic Face) and Isotonic KCl on the "Inside" (Schwann Face) of the Everted Axon

These experiments were performed on two axons. 0.54 M KCl with 4 mM phosphate buffer was used at pH 7.3. Resting potential decreased by 30-40 mV within 15 min. No action potential could be elicited. The results were irreversible.

C. The Dependence of the Amplitude of the Action Potential on the Concentration of Sodium on the "Internal" or Schwann Layer Side of the Everted Axon

These experiments were performed on three axons. One axon had been used in a previous experiment to be described later (Sec. III.D). This procedure of employing used axons in repeating an experiment was necessitated in this project by the fact that the availability of successful preparations was limited. The results of all three experiments were similar, and similar to those of Hodgkin and Katz (11) on the dependence of the amplitude of the action potential of normal axons on extracellular sodium. The results obtained from one of the axons is presented in Fig. 13. Details are

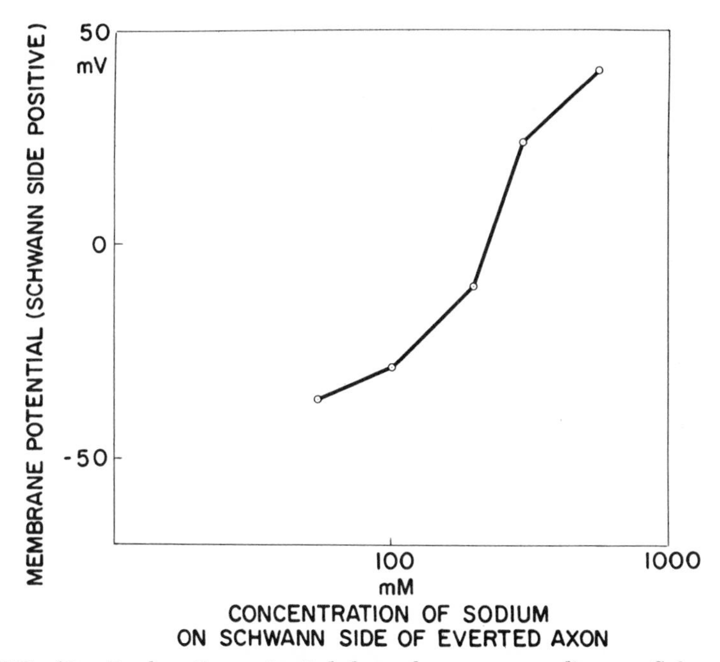

FIG. 13. Peak action potential dependence upon sodium on Schwann side. NaCl in sea water was substituted with choline chloride. Maximal diameter of axon, 1520μ 17° C.

given in the Fig. legend. The resting potential was not measured always; an internal pipet electrode would have complicated an already complicated procedure.

D. The Dependence of the Resting and Action Potential of the Everted Axon on "Internal" Potassium

Three axons were employed. One had been used previously (Sec. III. C). The effects of potassium on the "interior" or Schwann layer side were similar to previous experiments [e.g. (2)] on normal axons. The everted axon was inexcitable at high potassium concentration to "inward" current. With "outward" current, hyperpolarizing responses were observed. The results on the membrane potential are presented in Fig. 14.

E. The Effects on the Resting and Action Potential of the Everted Axon of "External" Calcium

Three axons were employed. One had been used previously (Sec. III. D). The "interior" (the Schwann layer side) was perfused with sea water and the

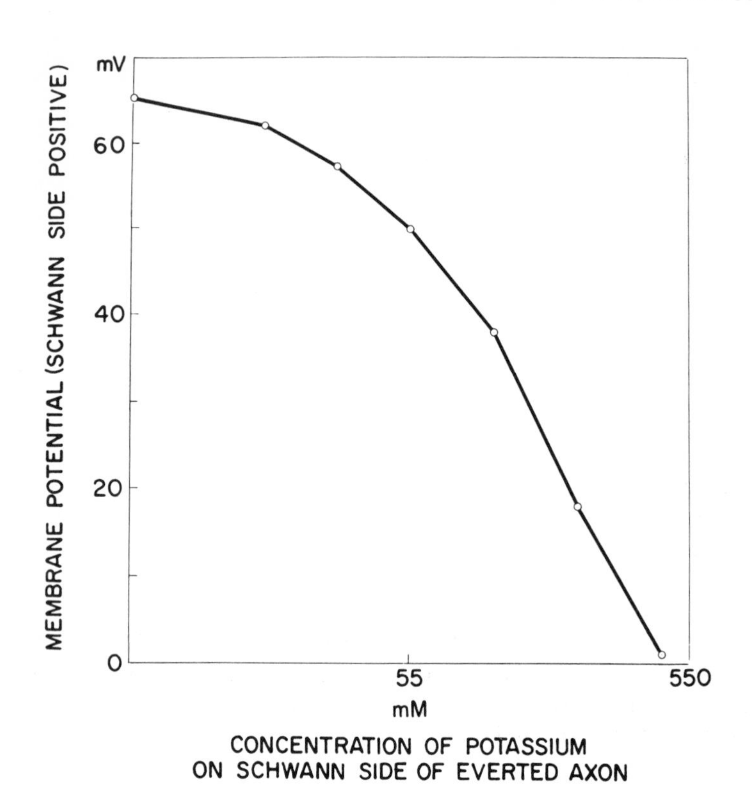

FIG. 14. Resting potential dependence upon potassium on Schwann side. NaCl in sea water was substituted with KCl. Maximal diameter of axon, 1280μ 18° C.

"exterior" (axoplasmic layer) with isotonic KF, during preparation and subsequently with KCl, with 1 mM phosphate to maintain pH at 7.3. Addition of 7 mM $CaCl_2$ to the "external" fluid blocked excitability within 2-3 min in all three experiments. Before calcium was added, the axon was kept for 5-12 min in KCl with little or no change in resting potential. These results are similar to those observed previously upon intracellular injection of calcium in normal axons [e.g., (20)].

F. The Effects in the Everted Axon of Lowering "Internal" Calcium

Three axons were used. One had been used previously (Sec. III.C). These particular experiments were chosen because the author has been interested in the effects of low extracellular calcium on the electrical properties of normal axon (9). Among other effects, it was found that at low calcium the resting potential was reduced by approximately 15-40 mV. The everted axons were bathed in KF solutions (see Methods, Sec. II). Initially, the perfusate bathing the Schwann layer side was 0.5 M choline chloride and 40 mM $CaCl_2$ buffered at pH 8.0 with 1 mM tris. When the perfusate was calcium-free, the resting potential was reduced by 20-41 mV within 3-18 min, reversibly.

G. The Effects of "External" TEA Chloride on Everted Axons

Only one axon was employed. 75 mM TEA was employed and pH was maintained at 7.3 with 4 mM phosphate. Tonicity was kept at 0.54 M with KF. The spike was prolonged to 250 msec. No prolongation was observed with 85 mM TEA chloride applied "internally" (Schwann layer side).

IV. DISCUSSION

I presently cannot think of a problem where the eversion technique would be preferable to the normal internal perfusion, but certain features of the eversion technique may be useful for particular types of experiments. For example, the axoplasmic side is more accessible to experimental attack. Much higher internal (Schwann layer) perfusion rates may be achieved in the everted axon than in the perfused normal axon without apparent mechanical injury; however, this can be attained also in the Schwann face in normally perfused axons. Aside from the mechanical perturbation incurred during the setting up of the preparation, the axoplasmic face (presumably more susceptible to mechanical injury) is not inordinately disturbed by internal perfusion.

There are also obvious disadvantages with the eversion procedure. The eversion technique includes, in the initial stages, a perfusion of the normal axon; even disregarding this the remaining stages are more tedious than the perfusion technique itself. As yet the possibility of obtaining successful preparations with this technique is less than that with perfusion of the normal axon. As yet the results are not appreciably different than those obtained with the perfusion technique.

The conclusion from these experiments, that the membrane-residual axoplasm complex is asymmetric seems reasonable; however, this conclusion is somewhat predicated on previous experiments using the perfusion technique on normal axons.

Whether the asymmetry investigated by observing the effects of the electrical properties is related to a macroscopic structural asymmetry remains to be seen. The axonal surface does show a macroscopic structural "asymmetry" (sequence of Schwann layer-"railroad track"-residual axoplasm). Schwann layers, however, do not exist in other cells (e.g., barnacle) which also show analogous asymmetries. This would imply that the Schwann layer per se is not implicated. Some may argue that residual axoplasm is appreciable and may still be an integral part of the "membrane." A 1 μ-thick layer of residual axoplasm may comprise an extremely small fraction of the volume of a 1000-μ axon; but it is more than enough to fill a 2-μ fiber.

V. SUMMARY

The giant axon of Loligo vulgaris was turned inside out and the Schwann layer side (now facing the interior of the axon) was perfused. With the exception that the polarity (with respect to what is now labeled the "interior" of a cylinder) was reversed, the results obtained were similar to those obtained on the normal axon internally perfused. Each face of the axonal surface retains its identity upon eversion. The results are consistent with the concept that the membrane-residual axoplasm complex is assymmetric.

REFERENCES

1. J. Bernstein, Pflügers Arch. Ges. Physiol., 92, 521 (1902).
2. A. L. Hodgkin, Biol. Rev., 26, 339-409 (1951).
3. P. F. Baker, A. L. Hodgkin, and T. I. Shaw, J. Physiol., 164, 355 (1962).
4. P. F. Baker, A. L. Hodgkin, and H. Meves, J. Physiol., 170, 541 (1964).
5. A. L. Hodgkin and B. Katz, J. Physiol., 108, 37 (1949).
6. D. E. Goldman; J. Gen. Physiol., 27, 37 (1943).
7. C. S. Spyropoulos and T. Teorell, Proc. Natl. Acad. Sci. U.S., 60, 118 (1968).
8. H. J. Curtis and K. S. Cole, J. Cellular Comp. Physiol., 19, 135 (1942).
9. C. S. Spyropoulos, Federation Proc. 27, 1252 (1968).
10. R. Lorente de Nó, A Study of Nerve Physiology, I and II, Rockefeller Inst. Med. Research, New York, 1947.
11. A. L. Hodgkin and B. Katz, J. Physiol., 108, 37 (1949).
12. F. J. Julian, J. W. Moore, and D. E. Goldman, J. Gen. Physiol., 45, 1195 (1962).
13. F. J. Julian, J. W. Moore, and D. E. Goldman, J. Gen. Physiol., 45, 1217 (1962).

14. J. W. Moore, T. Narahashi, and W. Ulbricht, J. Physiol., 172, 163 (1964).
15. T. Narahashi, J. W. Moore, and W. R. Scott, J. Gen. Physiol., 47, 965 (1964).
16. R. Stämpfli, Helv. Physiol. Pharmacol. Acta, 21, 189 (1963).
17. R. Stämpfli, Experientia, 10, 508 (1954).
18. M. P. Blaustein, and D. E. Goldman, Biophys. J., 6, 453 (1966).
19. R. Stämpfli, Ann. N.Y. Acad. Sci., 81, 265-284 (1959).
20. I. Tasaki, A. Watanabe, and L. Lerman, Am. J. Physiol., 213, 1465 (1967).
21. F. Helfferich, Ion Exchange, McGraw-Hill, New York, 1962.
22. B. Hille, Progr. Bioph. Molec. Biol., 21, 1 (1970).
23. T. Teorell, Progr. Biophys. Biophys. Chem., 3, 305 (1953).
24. T. Teorell, Proc. Soc. Expt. Biol. Med., 33, 282 (1935).
25. K. H. Meyer and J. R. Sievers, Helv. Chim. Acta, 19, 649 (1936).
26. G. Eisenman (Ed.), Glass Electrodes for Hydrogen and other Cations, Dekker, New York, 1967.
27. W. R. Adelman, and J. P. Senft, J. Gen. Physiol., 51, 102 (1968).
28. J. M. Diamond and E. M. Wright, Ann. Rev. Physiol., 31, 581 (1969).
29. H. A. Nash, and J. M. Tobias, Proc. Natl. Acad. Sci. U.S., 51, 476 (1964).
30. E. Rojas and J. M. Tobias, Biochim. Biophys. Acta, 94, 394 (1965).
31. E. Rojas and I. Atwater, J. Gen. Physiol., 51, 131 (1968).
32. T. Narahashi, J. Physiol., 169, 91 (1963).
33. S. Hagiwara, J. Gen. Physiol., 48, 55-57 (1965).
34. R. O. Brady, C. S. Spyropoulos, and I. Tasaki, Amer. J. Physiol., 194, 207-213 (1958).
35. T. Narahashi, N. C. Anderson, and J. W. Moore, Science, 153, 765-767 (1966).
36. E. Rojas and M. Luxoro, Nature, 199, 78-79 (1963).
37. F. Huneeus-Cox, H. L. Fernandez, and B. H. Smith, Biophys. J., 6, 675-689 (1966).
38. W. K. Chandler, A. L. Hodgkin, and H. Meves, J. Physiol., 180, 821-836 (1965).
39. I. Tasaki, A. Watanabe, and T. Takenaka, Proc. Natl. Acad. Sci., 48, 1177-1184 (1962).
40. R. H. Doremus, in Glass Electrodes for Hydrogen and other Cations (G. Eisenman, Ed), Dekker, New York, 1967, Chap. 4.
41. P. Mueller and D. O. Rudin, J. Theoret. Biol., 18, 222-258 (1968).
42. A. Mauro, Biophys J., 2, 179-198 (1962).
43. W. J. Adelman Jr., F. M. Dyro, and J. Senft, J. Gen. Physiol., 48, 1-9 (1965).

AUTHOR INDEX

Underlined numbers give the page on which the complete reference is listed.

E

F

G

H

SUBJECT INDEX

T

U

V

W

Z